3	$(16)\ P(A \text{ or } B) = P(A) + P(B)$ $- P(A \text{ and } B)$	Compound probability rules
3	$(17)\ P(A \text{ and } B) = P(A)P(B|A)$	
3	$(18)\ \mu = \Sigma x \cdot P(x)$	To find the mean of a probability distribution
3	$(19)\ \sigma = \sqrt{\Sigma x^2 \cdot P(x) - [\Sigma x \cdot P(x)]^2}$	To find the standard deviation of a probability distribution
3	$(20)\ P(x) = w \cdot p^x \cdot q^{n-x}$, where w is found from Pascal's triangle or by using the formula $$w = \frac{n!}{x!(n-x)!}$$	To find the probability of x successes in n trials of a binomial distribution
3	$(21)\ \mu = np$	To find the mean of a binomial distribution
3	$(22)\ \sigma = \sqrt{npq}$	To find the standard deviation of a binomial distribution
4	$(23)\ SD_{\bar{x}} = \dfrac{\sigma}{\sqrt{n}}$	To find the standard deviation of a frequency distribution of $\bar{x}$ for samples of size n drawn from a very large, or infinite, population
4	$(24)\ SD_{\bar{x}} = \dfrac{\sigma}{\sqrt{n}}\sqrt{\dfrac{N-n}{N-1}}$	Same as formula (23), but used when the population size N is not very large in comparison to n, the size of each of the samples
5	$(25)\ \bar{x} \pm zSD_{\bar{x}}$	A confidence interval estimate for μ
5	$(26)\ SE_{\bar{x}} = \dfrac{s}{\sqrt{n}}$	Formula for the standard error of the sample mean
5	$(27)\ n = \left(\dfrac{z\sigma}{E}\right)^2$	Formula for determining the sample size when estimating a sample mean; generally rounded up to the next whole number
5	$(28)\ SD_p = \sqrt{\dfrac{PQ}{n}}$	To find the standard deviation of the frequency distribution of p
5	$(29)\ p \pm zSE_p$, where $SE_p = \sqrt{\dfrac{pq}{n}}$	A CIE for P, a population proportion
5	$(30)\ n = \dfrac{z^2}{4E^2}$	To determine a sample size when estimating a proportion and you have no idea of the rough size of P
5	$(31)\ SE_s = \dfrac{s}{\sqrt{2n}}$	To find the standard error when e standard deviat follow

LEON F. MARZILLIER

Elementary Statistics

LEON F. MARZILLIER
Los Angeles Valley College

Elementary Statistics

Wm. C. Brown Publishers

Book Team

Editor *Earl McPeek*
Developmental Editor *Theresa Grutz*
Art Editor *Donna Slade*
Photo Research Editor *Carol M. Smith*
Permissions Editor *Mavis M. Oeth*
Visuals Processor *Vickie Werner*

 Wm. C. Brown Publishers

President *G. Franklin Lewis*
Vice President, Editor-in-Chief *George Wm. Bergquist*
Vice President, Director of Production *Beverly Kolz*
Vice President, National Sales Manager *Bob McLaughlin*
Director of Marketing *Thomas E. Doran*
Marketing Communications Manager *Edward Bartell*
Marketing Manager David F. Horwitz
Executive Editor Edward G. Jaffe
Production Editorial Manager *Colleen A. Yonda*
Production Editorial Manager *Julie A. Kennedy*
Publishing Services Manager *Karen J. Slaght*
Manager of Visuals and Design *Faye M. Schilling*

Cover design by Mark Elliot Christianson

Library of Congress Catalog Card Number: 89–61330

ISBN 0-697-05931-3

Printed in the United States of America by Wm. C. Brown Publishers, 2460 Kerper Boulevard, Dubuque, IA 52001

10 9 8 7 6 5 4 3 2 1

I dedicate this book to my father, Frederick Marzillier,
who, while he was still alive, encouraged me to embark
on this project in the first place and whose generosity
helped me accomplish it.

CONTENTS

Preface

"Why another statistics text?" That is the question I anticipate many who hold this book in their hands will ask themselves. It is true, there are many already on the market; but, after seventeen years of teaching the subject at the community college level, I have yet to find a text that does the job satisfactorily. Those texts that have been written for this audience, I have found, fit into one of two categories—those written by mathematicians and those written by behavioral scientists. The former produce texts that are overly mathematical and structured in the way algebra texts are. The latter give nice overviews of the subject but avoid computational material, often to the detriment of understanding.

The background that I possess may be uniquely suited to writing a text that cuts across these two classes. The combination of a bachelor's degree in mathematics and a master's degree in statistics from a social science institution, the London School of Economics, together with extensive experience in teaching students, many of whom dislike the subject, has given me rare insights into methods of presenting the material found in no other text on the market today. I hope the final product, which you now hold in your hands, will meet the objective I had when I started this project—to bring real understanding of statistics to the college student.

The text is designed for use in the standard lower division course in statistics, required of so many disciplines these days. Community college students are, of course, the users I had in mind when I wrote this text. However, it is most certainly appropriate for all lower division university students as well. No particular major field of study that uses statistics is either targeted or excluded by this book. One can find examples of applications of statistics to many diverse fields of study in the text and it would certainly be appropriate for majors in psychology, sociology, biology, economics, business, education, physics, engineering, and health, to name but a few. I have assumed that the student has had at least two years of high school algebra, or the equivalent in college. Without this background, the reader is too often sidetracked by struggles with manipulations, resulting in frustration, rather than having what I feel should be a pleasurable experience! I have long thought that a logic course would also be a very useful weapon in the student's arsenal, especially when the material on hypothesis testing is reached.

I have included those topics in the text that are found in most introductory statistics texts on the market today. There is, however, too much material to cover in a semester's or quarter's course of standard length, unless the class is above average and is able to move with rapidity. Some of the chapters in Part 3 can, therefore, be omitted without loss of continuity. These chapters are intentionally short compared to the longer, earlier chapters and are designed to give a sample of different inferential techniques to the student, without going into great depth. After all, this is a *first* course in statistics! Some instructors may feel, in this age of computers, that the coverage pertaining to frequency distributions (sections 1–2, 2–2, 2–3, and 2–6) is unnecessarily extensive. I disagree, since I feel it is important for the user of statistics to understand what it is that high-powered calculators and computers do before learning how to push the right buttons. However, I do give alternative plans for covering the first two chapters, in the Instructor's Manual that accompanies the text. I urge you to cover the chapters as written, though, and I think you'll find that they are "user-friendly"!

Another dislike that I have for most statistics texts is that the authors split the treatment of bivariate data, covering some of it with descriptive statistics in Part 1 of the book and the rest at the end, after covering the ideas of inferential statistics. I prefer the more logical approach, which you will find in this text, covering it all together in Part 4, "Analysis of Bivariate Data."

Each statistical technique described in the book is followed by an example or examples with complete solutions, an exercise set follows each section, and a further set of exercises comes at the end of each chapter. These examples and exercises are designed to illustrate for the user how statistical techniques can be applied to many diverse fields and help put the student into the driver's seat, where real learning takes place. Also at the end of each chapter is a summary of terms and formulas, with which a student should be familiar before going on to the next chapter. Each chapter then ends with a set of exercises on all the ideas included in the chapter and an optional "Hands-on Class/Student Project," which is a set of exercises that enables the student to obtain a firmer grasp of the ideas of the chapter by using data that he/she has collected or generated. Answers to the odd-numbered problems appear at the end of the book so that students can self-check their work. To aid the instructor, an Instructor's Manual and TestPak accompany the text. Included in the former are the answers to the even-numbered problems, so that the instructor can use these as homework problems, if desired, as well as my ideas on the structure and teaching of the course. The latter is a resource for testing purposes. A study guide is also available for students but is not a necessary purchase for a complete understanding of the subject.

In conclusion, I would like to take this opportunity to acknowledge the professional way in which this book was brought to the marketplace by the staff of Wm. C. Brown. In particular, I would like to thank Ed Jaffee, Earl McPeek, and Theresa Grutz. Without their help, the project could never have been completed. Finally, I would like to express my sincere gratitude to my colleague Phil Clarke for working *all* the problems in the text, thereby producing an invaluable error check. I know that nothing irritates students and instructors more than finding numerous errors in the back of the book. As a result of Phil's work, errors should be very few and far between!

Leon F. Marzillier

Reviewers

Michael Bonanno
Suffolk County Community College

Vern Crandall
Novell, Inc.

Mario Fiondella
Manchester Community College

Jim Foster
Weber State College

Mark Goldstein
West Virginia Northern Community College

Carroll Hall
New Mexico State University

Michael Karelius
American River College

Eric Lubot
Bergen Community College

Joseph L. Mazanec
Delta College

Alfred Milligan
Western New Mexico University

Lawrence Ringer
Texas A&M University

Jean Burr Smith
Middlesex Community College

Harold Thomas
Pittsburg State University

Froylan Tiscareno
Mt. San Antonio College

Bruce Vermeulen
Colby College

To the Student—What Is Statistics?

Some of you may have picked up this book simply because you wanted to know more about statistics. If so, I applaud you and am flattered that you chose my book with which to learn it. However, I suspect that the real reason you have this book in your hands is that you have been told you need to know more about this mysterious field to have a complete and thorough knowledge of your chosen course of study.

"But what does statistics have to do with psychology or sociology?" you might ask. "Isn't statistics part of mathematics?"

Statistics is a branch of mathematics, but it is also a branch of psychology and of sociology. Research is the lifeblood of these fields and it would be very difficult, to say the least, to interpret any research without statistical theory. There are, however, many other diverse fields that use statistics: biology, medicine, business, politics, economics, computer technology, weather forecasting, sports, education, and even such unlikely fields as English and history. Any field in which experiments are performed or data is collected needs statistical theory to analyze and interpret the results. It would be useless to perform a psychological experiment and have pages and pages of results if the researcher can't draw any conclusions from them.

Enter statistical theory.

Some questions that might be answered by statistical theory are:

What is the portrait of a city, ethnically and by socio-economic background?

How are the graphs printed in newspapers and magazines read and interpreted?

What is the average score on a psychological test for a group of children?

What is the chance of rain tomorrow?

What percentage of the voters in Chicago think the president is doing a good job? And how can this question be answered if a statistician has time and money to interview only a few of these voters?

How can you check if an advertisement is telling the truth?

Is the number of deaths from cancer related to the distance between victims' homes and a nuclear power plant?

These are the types of questions that can be answered using statistical theory. As you read and study this book, you will learn about methods that will allow you to answer such questions.

It would be impossible, in one course or one book, to discuss all the topics of statistical theory. The material you are about to study is the usual material required for a first course in statistics presented in a clear, comprehensible manner. It is not addressed to any particular group of people, such as psychologists or sociologists. Rather, it presents statistical procedures and techniques anybody can use. Examples and exercises are taken from many fields to illustrate the numerous and diverse applications of statistics.

The book is divided into four parts. The first part presents descriptive statistics—that is, the techniques required for describing and presenting data. This is followed by Part 2, a transition to inferential statistics—methods by which facts about a large population can be inferred or deduced from a small sample of the population, the subject of Part 3. The fourth and final part deals with the study of relationships between variables.

To achieve a complete and thorough understanding of statistics, you would have to know a great deal of mathematics. To understand this text, you need only know algebra, preferably to the intermediate level (i.e., the equivalent of two years of high school algebra). Just as it is possible to learn to drive a car without knowing anything about the inner workings of the engine, so it is possible to learn and apply the techniques of statistical theory without always knowing *why* they work. But be forewarned: when you learn to drive a car, you don't read how to do it in a book. Nor do you simply watch someone else do it. *You* need to get into the driver's seat and try it. And not just once, but as many times as it takes until you get it right. The same is true of learning statistics. It will not be sufficient for you to curl up by the fire and read this book. Nor will it be sufficient for you only to go to class and listen and observe your teacher. You have to get into the driver's seat and actually *do* statistics.

You will find, as you read and follow this book, that the explanation of each statistical technique is followed by one or more examples and that each section concludes with an exercise set to test your understanding. Follow the examples and do the exercises (more than once, if necessary), and you will find yourself beginning to think like a statistician. Statistical thinking is not difficult, but it is different and sometimes requires a little effort to grasp it fully.

To summarize, then, statistics is a branch of mathematics, applicable to many fields, dealing with the presentation, analysis, and interpretation of data. In the following chapters you will examine many ways in which this can be accomplished. With a little effort it is not hard to get into its way of thinking, and you will find that this "mysterious" field will bring new insights to a variety of problems. Good luck in your study of statistics!

LEON F. MARZILLIER

Elementary Statistics

Descriptive Statistics

"Let us sit on this log at the roadside," says I, "and forget the inhumanity and ribaldry of the poets. It is in the glorious columns of ascertained facts and legalized measures that beauty is to be found. In this very log we sit upon, Mrs. Sampson," says I, "is statistics more wonderful than any poem. The rings show it was sixty years old. At the depth of two thousand feet it would become coal in three thousand years. The deepest coal mine in the world is at Killingworth, near Newcastle. A box four feet long, three feet wide, and two feet eight inches deep will hold one ton of coal. If an artery is cut, compress it above the wound. A man's leg contains thirty bones. The Tower of London was burned in 1841."

"Go on, Mr. Pratt," says Mrs. Sampson. "Them ideas is so original and soothing. I think statistics are just as lovely as they can be."

The Handbook of Hymen   O. HENRY

Presentation of Data

"A few honest men are better than numbers."

Table 1–1　What the World Eats

Calories Per Day

	Wheat	Rice	Maize	Roots/Tubers	Sugars	Nuts	Vegetables	Fruit	Meat	Eggs	Fish	Oils and Fats	Milk	Alcohol	Total Calories	% of Daily Requirements
Afghan.	1004	130	285	27	38	22	14	33	63	3	—	59	66	—	1896	77.7
Argen.	925	36	44	149	388	26	50	117	708	25	9	382	218	236	3346	126.3
Australia	762	40	21	95	577	24	52	99	744	48	24	334	372	179	3415	128.9
Austria	498	31	20	106	459	22	49	134	625	54	14	658	366	247	3535	134.4
Banglad.	107	1414	—	36	63	2	8	20	14	2	19	54	16	—	1796	85.6
Bolivia	386	159	250	266	308	26	56	146	168	9	4	154	34	31	2070	89.4
Brazil	307	399	207	241	482	18	19	125	187	14	11	184	133	45	2559	107.2
Canada	574	31	21	124	507	56	62	103	685	50	22	522	353	143	3368	126.8
Chile	1199	68	47	96	338	2	49	45	185	17	13	220	150	119	2657	108.8
China	367	823	172	205	42	62	41	8	182	15	13	97	11	43	2386	104.5
Cuba	554	479	—	138	533	13	22	61	177	31	30	236	201	46	2635	114.3
E. Germ.	535	19	—	280	416	21	69	74	579	59	27	668	248	273	3644	139.0
Egypt	883	308	530	37	238	25	87	91	63	6	7	230	42	2	2787	108.4
France	702	29	8	183	431	20	73	78	583	51	32	587	316	284	3434	136.3
Guatem.	176	37	938	13	338	13	18	64	49	19	2	134	73	28	2020	98.5
Hungary	1114	40	—	127	373	20	66	95	536	63	9	539	198	248	3520	133.9
India	336	587	70	41	181	27	33	29	6	—	5	136	57	—	1919	91.5
Indon.	41	1196	186	205	142	120	9	21	20	2	21	123	4	6	2118	98.0
Iraq	941	300	1	10	333	10	64	71	84	7	5	128	70	8	2155	88.6
Ireland	743	11	55	250	571	6	35	61	592	47	24	442	437	164	3551	141.0
Italy	1166	39	40	75	307	40	88	129	386	42	23	519	245	273	3434	136.0
Japan	317	960	80	65	283	124	69	62	166	63	181	284	88	145	2946	126.0
Kenya	110	17	958	193	177	31	14	49	100	3	5	76	107	52	2166	90.3
Libya	888	137	11	43	464	103	92	214	133	4	17	444	219	—	2985	126.5
Mali	35	210	102	38	63	90	10	2	69	2	18	123	45	17	2117	90.1
Mexico	314	52	978	22	436	20	14	83	145	23	9	206	154	69	2655	113.9
Netherl.	534	32	19	153	506	25	36	90	585	40	20	642	388	188	3372	124.1
Peru	409	272	209	237	356	13	38	115	93	10	30	218	93	48	2270	96.7
Philip.	105	865	380	148	216	30	18	85	95	15	71	108	22	32	2216	96.8
S. Africa	512	41	975	42	430	19	38	38	206	23	17	196	142	129	2921	119.2
Sweden	426	26	14	159	475	21	32	104	525	47	58	488	464	164	3221	119.7
Syria	1322	93	5	29	244	93	111	133	89	17	3	261	94	7	2685	108.2
Thailand	15	1384	23	63	213	65	38	78	59	15	55	41	8	11	2098	86.9
Tunisia	1359	5	13	32	260	40	79	72	102	13	11	436	100	10	2674	111.9
Turkey	1391	44	92	93	270	92	80	139	102	16	14	282	99	10	2913	115.4
U.K.	628	18	17	165	554	36	52	61	496	52	23	563	406	199	3343	132.4
U.S.A.	470	34	77	113	579	70	64	125	758	61	21	583	374	175	3578	135.5
U.S.S.R.	1055	67	2	237	451	21	55	60	349	48	61	357	310	105	3460	135.2
W. Germ.	447	14	41	142	401	37	43	116	657	67	25	618	253	320	3382	126.6
Yugosl.	1424	24	212	128	342	16	70	93	278	34	9	371	215	138	3446	135.7
WORLD	465	505	146	161	234	48	41	59	205	21	25	225	116	70	2571	—

First printed in GEO magazine. Copyright © 1981 Knapp Communications Corporation. Reprinted with permission.

....

Chapter Overview
What Do You Do with All These Numbers?

In this opening chapter you will learn some statistical terms and how to present data in various ways so that more information can be learned from it.

....

Section 1–1 Statistical Experiments

Experiments is used here in the broad sense defined later in chapter 1.

What do you think of when you hear that someone is going to perform an experiment: a person in a white coat in a laboratory mixing chemicals? Statistical theory borrows words from the English language and assigns special meanings to them. When you use these words during the course of your study of statistics, you must forget, for the time being, the various other associations you have for these words and learn the specialized meanings that statistics has for them. As Humpty Dumpty says in *Through the Looking Glass,* "When I use a word, it means just what I choose it to mean—neither more nor less."

What then is an experiment? In statistics, an **experiment** (or sometimes random experiment) is any study that can yield one of several possible outcomes. If there is only one possible outcome each time you perform the experiment, it is not really an experiment—the results are predetermined. List the possible outcomes for the following experiments:

EXAMPLE 1–1   Roll a die and note the number of spots on the uppermost face.

Solution   1, 2, 3, 4, 5, 6   ●

Downstown   **By Tim Downs**

EXAMPLE 1–2   Ask high school students whether they currently smoke cigarettes or not.

Solution   Yes, no   ●

EXAMPLE 1–3   Measure adult males for height.

Solution   5 ft 9 in., 6 ft 0½ in., and many more depending on the accuracy of your measuring tool   ●

Experiment on What?

On whom or what do you perform experiments? A target group will be either a population or a sample. What are these to statisticians?

> A **population** is typically a large group of objects, words, numbers, or people.
> A **sample** is a part of a larger population.

Samples are always smaller than the population of which they are a part but may be larger than other populations.

All the desks in your classrooms (maybe 30) is a population. But 100 desks from all over the school is a sample of all the desks at the school.

Infinite means "without end" and is the opposite of finite, which means "having an end."

Notice that population is not restricted to people. You could talk about the population of all rats in the United States or the population of all people currently living in Iowa. The population of interest should be specified in advance. If you perform an experiment that allows you to draw conclusions about a certain population, you may not be allowed to draw similar conclusions about a larger population of which the specified population may be a part. For instance, if you determine the average family income in San Francisco, this tells you very little about the average family income in California, and probably less about the average family income in the United States. A population may be infinite, very large, or relatively small. In describing it, the word "all" is usually used. *All* rolls of a die is an infinite population since you could (theoretically) go on rolling it forever. All people in the United States at a certain point in time is large but not infinite. All female schoolchildren in Des Moines, Iowa, could still be considered a population but is relatively small compared to all people in the United States.

At the beginning of any experiment, you should state just what it is that you want to find out about the population. Take the example of the population of all rats in the United States. You may be interested in their body weights, the lengths of their tails, or how well they negotiate their way through a maze. This "thing" of interest is called the **response variable.** It is a *vari*able since it will *vary* (i.e., change) from subject to subject in a population. It is a response since it is an answer to a question you are asking about the population. The collection of all the responses comprises the *data,* or more precisely, the **raw data.**

San Francisco families may not be representative of all California families, whereas a sample of 100 families from all over the state has a better chance of being representative.

More often than not, because the population either is very large or is infinite, it is impossible or impractical to gather all the data in the population. What statisticians do, then, is gather data from a sample. They attempt to make the sample *representative* of the population of which the sample is a part. How they ensure that the sample represents the population is the subject of chapter 4 in part 2 of this book, which introduces you to inferential statistics. For now, in your study of descriptive statistics, you need only be aware of the distinction between a population and a sample.

Raw Data

The word *data* was originally the plural of the word *datum,* meaning a fact, although today it is quite common to see the word *data* used as a singular noun. The raw data is the set of facts gathered by the statistician from the population or sample under investigation *before* anything has been done to the data. It might be said, at the risk of a groan, that the raw data are the facts before the statistician has had a chance to "cook" them. There are two types of raw data: **qualitative** and **quantitative.** As you might guess, qualitative data is produced when the response variable is not a number, and quantitative data is produced when it is. You should be careful with this—remember that what decides the issue is what the response variable is, nothing else. For instance, consider example 1-2. Suppose 50 high school students were interviewed, and 21 said they smoked and 29 said they didn't. You might be tempted to conclude that this is quantitative data since numbers are involved; but look at the response variable: it could be yes or no, which is nonnumeric. This is an example of qualitative data. The 21 and 29 are not part of the raw data—something has been done to the data to arrive at those figures, and that is counting.

Another example of qualitative data would be if you went to a parking lot at your campus and collected a sample of colors of automobiles. If your sample contained 5 red cars, the 5 does not make it quantitative—the response variable is red for this category.

Momma

By Mell Lazarus

If in the sample of automobiles, you had recorded the number of passenger seats or the size of the engine, your data would have been quantitative, since the response variables would be numeric.

Quantitative data, itself, is divided into two types: **discrete** and **continuous.** What is the distinction between these? Any time you are collecting quantitative data, of either type, the possible values assumed by the response variable will be restricted to a certain interval of values. For example, when rolling one die, nothing below a one or above a six is possible. Therefore, the **interval** of values is one to six, inclusive of the endpoints. When the response variable is adult male height, according to the *Guinness Book of World Records* "9 feet may be regarded as the limit toward which the tallest giants tend, so 23 inches must be regarded as the limit toward which the shortest mature dwarfs tend." Therefore, the interval of values is 23 inches to 108 inches (which is 9 feet). Now, the distinction between discrete and continuous data is decided by what can occur within this interval of values. If there is (theoretically) no restriction placed on a possible value of the response variable, then the data is continuous. If there is some kind of restriction, then it is discrete. The number of passenger seats in an automobile is discrete, since the value of the response variable is restricted to be a whole number.

Height, on the other hand, is an example of a continuous variable. You might talk about someone being 5 feet 7.3826495... inches tall (where the decimal is taken out forever). There is no *theoretical* restriction on the accuracy with which you can measure a person's height. The restriction imposed on you is determined by a practical consideration: how accurate is your measuring device? If you are using a tape measure divided into sixteenths of an inch, you can be no more accurate than that. Can you see, then, that even continuous variables are *recorded* discretely?

The experiment of rolling one die produces discrete data: you can't roll a $3\frac{1}{2}$ for instance. If a college administration wanted to know how many students were in each class meeting from 10 A.M. to 11 A.M. on Mondays, it might ask the teachers of those classes to report this. The responses they would get would be discrete since, it is hoped, there wouldn't be any fractions of students attending class! Although there are exceptions, such as stock prices, the most common restriction that results in discrete data is limiting potential values to whole numbers.

Most of the time, continuous data is a result of *measurement,* while discrete data is a result of *counting.*

Continuous variables are variables falling in a certain interval on which no theoretical restrictions are placed.
Discrete variables have a restriction placed on them.

There is not always general agreement about which variables are discrete and which are continuous. Suppose you were performing an experiment in which the response variable is money expressed in dollars and cents. You might decide that you are dealing with discrete data, since, usually, money is not expressed in fractions of a cent. But because the gap between consecutive numbers of cents is so small, some statisticians will consider the gaps to have no practical significance and call the data continuous data.

Figure 1-1 shows in the form of a diagram what has been discussed so far.

Figure 1-1

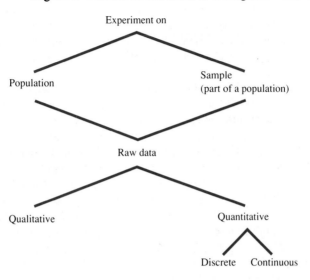

Are the statistical meanings of all the words in the diagram now clear to you? If not, review the material on the preceding pages.

Exercise Set 1-1

Exercises 1-4 describe statistical experiments. List the possible outcomes in each case.

 1. Tossing a coin and noting what is shown on the uppermost facing.

 2. Rolling two dice and counting the total spots on the uppermost faces.

 3. Asking a sample of students how many hours they study per week.

 4. Going to a parking lot at your school and noting the colors of the cars there.

In exercises 5-8, state whether each reference is to a sample or to a population and, if to a population, whether it is finite or infinite.

 5. The titles of all novels in print as of January 1 this year.

 6. All rolls of a roulette wheel, noting the number produced each time.

7. Standing outside an entrance to your school's cafeteria and asking fifty students whether they have ever smoked marijuana.

 8. The results of an opinion poll asking how people will vote in an upcoming presidential election.

In exercises 9–12, state whether the problem deals with qualitative or quantitative data.

 9. The results of tossing a coin fifty times.

 10. The results of rolling a die 100 times.

 11. The results of weighing a sample of the contents of 20 jars of instant coffee.

 12. The results of asking 25 of your fellow students whether they have ever smoked marijuana.

In exercises 13–16, all data is quantitative. State whether each is an example of discrete or continuous data.

 13. The total number of people who get married each week, during 10 successive weeks, in your home town.

 14. The lengths of tails of a sample of 80 newborn rats.

 15. The temperature on each day of August last year.

 16. The lengths of all the words on this page, where the length of a word is given as the number of letters contained in it.

Section 1–2 Frequency Distributions

The subject of this section will be explained using example 1–4.

EXAMPLE 1–4   Suppose a sample of 38 female university students was asked their weights in pounds. This was actually done, with the following results:

130	108	135	120	97	110
130	112	123	117	170	124
120	133	87	130	160	128
110	135	115	127	102	130
89	135	87	135	115	110
105	130	115	100	125	120
120	120				

What information can be derived from these numbers?

Solution   Even with such a relatively small group of numbers, it is difficult to gain any information from them while they are disorganized like this. Imagine what it would be like if there were 238!   ●

A start can be made by searching for the smallest and largest values. In this case, they are 87 and 170. The difference between these is $170 - 87 = 83$. This value is the **range** of the data.

$$\text{Range} = \text{Highest value} - \text{Lowest value}$$

To organize the data, you could list each weight from 87 to 170 and indicate how many in the sample (or population) you have at each weight. However, as you can see in this example, this would involve a very long list and the result would not be very helpful. For other types of data with a short range, this may be possible, as you will see later.

Better, here, is to group the data into intervals. How many intervals should there be? A very rough guideline is 5 to 10 intervals. The range in example 1–4 is 83. Suppose you wanted 10 intervals. The number of values in each interval would be $83 \div 10 \approx 8$. With 5 intervals, it would be $83 \div 5 \approx 17$. Suppose you decide to have 10 values in each interval (10 is between 8 and 17). The first interval, which needs to include the low value of 87, might be chosen to be

80–89 or 85–94 or 87–96

You could not use 85–94 and then 94–103. Do you see why not?

or a number of others. Each of these intervals has 10 values covered by it (*not* 9, count them!), and each includes the value 87. How, then, do you decide which interval to use? There is no easy answer to this question. Any of them might be chosen. Suppose, arbitrarily, 85–94 is chosen. The next interval would have to start at 95 and would, therefore, be 95–104. Continuing in this way, the complete list would become

There are 10 values covered by each interval. This number 10 is called the **width** of the interval.

85–94
95–104
105–114
115–124
125–134
135–144
145–154
155–164
165–174.

This gives nine intervals or **classes.** The numbers that actually appear in this list are called the **class limits.**

The next step is to decide how many values in the sample are in each interval. You could go through the whole sample counting how many fall in the interval 85–94 and then go through the whole sample again counting how many fall in the interval 95–104, and so on. This method is tedious and prone to error; better is to use the **tally** method.

List the intervals again and tally the first row (reading across) of the sample as shown in table 1–2.

Table 1–2

Class	Tally
85–94	
95–104	I
105–114	I I
115–124	I
125–134	I
135–144	I
145–154	
155–164	
165–174	

Do you see how this works? Table 1–3 shows how to do the complete sample.

Table 1–3

Class	Tally	f
85–94	I I I	3
95–104	I I I	3
105–114	⊤⊢⊢ I	6
115–124	⊤⊢⊢ ⊤⊢⊢ I	11
125–134	⊤⊢⊢ I I I I	9
135–144	I I I I	4
145–154		0
155–164	I	1
165–174	I	1

Verify that this has been done correctly by doing the tally yourself.

Notice that another column has been added, headed by f, which is the conversion of each tally to a number. f stands for **frequency** (or **simple frequency**). Frequency means how many. Thus, the frequency column tells you how many are in each class. If you add the numbers in this column, it should total the number in the sample, in this case 38. Check this.

By looking at the column of classes together with the frequency column, you can see how the sample is distributed. For this reason, it is called a **frequency distribution** or sometimes a **frequency table.**

Table 1–4 shows the frequency distribution again with another column of interest.

Table 1–4

Class	f	F
85–94	3	3
95–104	3	6
105–114	6	12
115–124	11	23
125–134	9	32
135–144	4	36
145–154	0	36
155–164	1	37
165–174	1	38

$3 + 3 = 6$
$3 + 3 + 6 = 12$
$3 + 3 + 6 + 11 = 23$
etc.

The column headed by F is the **cumulative frequency** column. It tells how many values are in that interval and all intervals less than it. The first numbers in both columns are the same. To get the succeeding numbers in the F column, accumulate (i.e., add) the numbers in the f column. The last number in the F column should be the size of the sample, since the entire sample is in the highest interval *and* all the intervals less than it.

The data has now been organized and it is much easier to see what is going on. Some information that can be inferred from this organized presentation of the data, which would not be so easy to do from the raw data, is as follows:

Over half the data falls in the two intervals 115–124 and 125–134.
Over two-thirds of the data falls in the three intervals running from 105–134.
Half the values in the sample are at some value in 115–124 or below it.

The first two pieces of information were arrived at by adding figures in the f column. The third piece of information is arrived at as follows:

Since the size of the sample is 38, half is 19. Concentrating on the F column, can you see that only 12 are in 105–114 or below it, whereas 23 are in 115–124 or below it? Therefore, the nineteenth value from the lowest value is among the 11 values in the interval 115–124.

The average implied here is the median, as you will realize after completing chapter 2.

Summing up, although the range of the students' weights in this sample is 83 lb, two-thirds of the students weighed between 105 lb and 134 lb, a range of only 29 lb, and the average weight is, very approximately, 120 lb. More (and more precise) information will be derived from this sample in later chapters.

The foregoing example dealt with quantitative data, and it illustrated how quantitative data is typically organized into a frequency distribution. Weight is a continuous variable since there is no theoretical restriction on which numbers can be a weight, so the weights are presumably rounded to the nearest pound. If this is so, the six students who weighed between 105 lb and 114 lb might actually

In the extremely rare event that a student weighed *exactly* 104.5 lb, she would be recorded as 104 lb, rounding to the *even* number.

weigh anything from 104.5 lb to 114.5 lb. This is so because if students actually weighed 104.3 lb or 114.8 lb, they would be recorded as 104 lb and 115 lb, respectively, and they wouldn't be classed in the given interval. On the other hand, if a student weighed 104.51, 104.7, 114.1, 114.499, or anything between 105 and 114 pounds, she would appear in the interval. The values 104.5 and 114.5 lb are called the **true class limits** (or boundaries) of the interval 105–114. The true class limits of the interval 155–164 are 154.5 and 164.5. All the intervals of a frequency distribution dealing with continuous data have true class limits. True class limits are found by finding the number halfway between the upper limit of one interval and the lower limit of the next. They are always taken to one more decimal place than the data. The interval after 155–164 is 165–174, and 164.5 is halfway between 164 and 165. Also, 164.5 has one decimal place, whereas the data was in whole numbers.

Discrete Data

Here is an example involving discrete data.

EXAMPLE 1–5   Examine the frequency of the final digits of the weights in example 1–4.

Solution   A width = 1 is going to be used since there are only ten digits possible.

Table 1–5

Class	Tally	f	F
0	⊦⊦⊦ ⊦⊦⊦ ⊦⊦⊦		16	16
1		0	16
2			2	18
3			2	20
4			1	21
5	⊦⊦⊦				9	30
6		0	30
7	⊦⊦⊦	5	35
8			2	37
9			1	38

●

There is something striking about table 1–5, that is, the predominance of 0s and 5s. If these students had been accurately weighed, approximately an equal number of each digit should appear in the final place. There is absolutely no reason why a particular digit should appear in the last place more frequently than any other. There are 10 digits and 38 weights. Now, 38 ÷ 10 = 3.8. Therefore, you should expect 3 or 4 weights to end in each of the 10 digits. The fact that there are sixteen 0s and nine 5s and no 1s or 6s indicates that something is wrong. What is wrong can be found in the statement of example 1–4. The students were *asked* their weights. Do you know your weight to the nearest pound? Most people don't and will tend to round it to the nearest five pounds and probably downwards! This might explain the above discrepancies.

What has been done is a useful device when you are presented with continuous data and you are not sure how it was collected. If it were important to have these students' weights accurate to the nearest pound, after examining the frequency table of final digits, you could be pretty certain that they were not weighed and the above data would be useless. Reliable data is essential when attempting to draw conclusions from it. If the data is unreliable, it matters little how sophisticated the statistical technique applied to it is, the results will be unreliable. For that reason, it is important how you gather data, and if you didn't gather it, that you examine it carefully.

Exercise Set 1–2

1. Classify this continuous data into a frequency distribution, using 5–9 as the first interval.

    20 18   6 24 33   9 10 19 27 33 22 17 19 31
    25 21 28 13 21 12 33 23 18 13   7 16   7 26

2. Classify this continuous data into a frequency distribution, using 131–133 as the first interval.

    132 135 136 140 131 134 139 133 132 135 131 132
    150 144 131 138 132 134 132 135 138 140 133

During the 1970s, the thirty most populous countries in the world were consulted and the life expectancies at birth determined for each sex in each country. Exercise 3 gives the results for males, while exercise 4 gives them for females. Classify each of these into frequency distributions, using an interval width of 5 in each case.

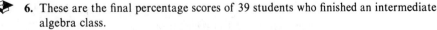

3.   69 65 46 58 49 60 59 52 36 69 68 42 48 58 69
        72 63 63 37 54 57 67 50 70 54 54 64 68 43 42

4.   76 71 47 61 52 63 63 54 40 77 75 41 48 57 75
        77 67 67 37 49 60 75 53 75 59 54 74 74 46 45

5. In a psychological experiment, 25 rats were made to run through a maze. The number of errors made by each was counted. Here are the results.

    22 17 18 21 19 12 15 23 25 11 17 22 24
    14 20 14 20 16 19 17 25 19 20 19 21

Classify these into a frequency distribution of width 3.

6. These are the final percentage scores of 39 students who finished an intermediate algebra class.

    49 64 52 80 77 43 45 59 40 50 73 62 61
    46 91 40 76 74 59 71 67 69 60 71 49 85
    75 59 77 69 60 72 98 78 77 83 45 64 67

Classify them into a frequency distribution of width 9.

7. The following is a frequency distribution of female students' heights. Add the cumulative frequency column to it, study it, and answer the questions following it.

Height	f	F
59	1	
60	2	
61	1	
62	1	
63	4	
64	5	
65	5	
66	4	
67	0	
68	4	

a. What is the class width used here?

b. Find the shortest interval that contains at least two-thirds of the sample. (E.g., the interval 60–62 contains $2 + 1 + 1 = 4$ heights. Obviously, this is not two-thirds of the sample, however.)

c. Approximately, what is an average of this sample?

 8. Repeat all parts of exercise 7 for the following frequency distribution of 100 rolls of a pair of dice, having recorded the sum of faces on each roll.

Interval	f
2–3	11
4–5	15
6–7	30
8–9	28
10–11	11
12–13	5

Section 1–3 Qualitative Data and Relative Frequency

Qualitative Data

Another term that is sometimes used for qualitative data is **categorical data.** The reason for this is that it can be divided into categories. These categories can then be used to form a frequency distribution.

EXAMPLE 1–6   Twenty-five people were examined for the color of their eyes. Here are the results.

Brown	Brown	Brown	Brown	Brown
Brown	Blue	Brown	Brown	Blue
Blue	Brown	Brown	Brown	Brown
Hazel	Brown	Brown	Hazel	Blue
Black	Brown	Blue	Blue	Green

Construct the frequency distribution.

Solution

Category	f
Brown	15
Blue	6
Hazel	2
Green	1
Black	1
Gray	0

This is a little easier than with quantitative data. A few points are worth mentioning.

Because there are relatively few distinct values in most categorical data, tallying is not absolutely necessary since we can simply count each category.

Cumulative frequency, or F, would be meaningless.

The order of the categories is unimportant. Here they have been placed in descending order of frequency.

Class limits and true class limits are also meaningless here.

Relative Frequency

Suppose you had a sample of 55 students' weights from another college. How could you compare the distribution of their weights with that of the sample of 38 in example 1–4? It would be difficult to make a straight comparison because of the different sample sizes. Having 6 out of 38 within a class is different than having 6 out of 55 in a class. For this and other reasons, samples or populations are often presented as *relative frequency distributions*. To get the **relative frequency** of a class, take the frequency of that class and figure it as a **proportion** of the size of the sample or population. For instance, a class with a frequency of 6 within a sample of size 38 would have a relative frequency of

$$\frac{6}{38} \approx 0.158.$$

This is often converted to a percentage by multiplying by 100. Answer: 15.8% or approximately 16%.

If the sample size were 55, a frequency of 6 becomes a relative frequency of

$$\frac{6}{55} \approx 0.109 \approx 11\%.$$

Thus, a class with a frequency of 6 in a sample of 38 would have a higher *relative* frequency than a class with the same frequency in a sample of 55.

Tables 1–6, 1–7, and 1–8 on page 20 show the three frequency distributions discussed in this chapter expressed as relative frequency distributions. Relative frequencies are denoted by $f\%$ and $F\%$ rather than f and F, respectively.

There are a couple of minor discrepancies. If you accumulate 29.0 with 31.6, you get 60.6, not 60.5, which was obtained by converting 32 to a percent. This is called *round-off* error. It would happen too if you rounded the *f%* column to whole numbers. The total would be 102% instead of 100%, as it should be.

Table 1–6   Weights

Class	$f\%$	$F\%$
85–94	7.9	7.9
95–104	7.9	15.8
105–114	15.8	31.6
115–124	29.0	60.5
125–134	23.7	84.2
135–144	10.5	94.7
145–154	0.0	94.7
155–164	2.6	97.4
165–174	2.6	100.0

Table 1–7   Last Digits

Class	$f\%$	$F\%$
0	42.1	42.1
1	0.0	42.1
2	5.3	47.4
3	5.3	52.6
4	2.6	55.3
5	23.7	79.0
6	0.0	79.0
7	13.2	92.1
8	5.3	97.4
9	2.6	100.0

Table 1–8   Eye Color

Category	$f\%$
Brown	60
Blue	24
Hazel	8
Green	4
Black	4
Gray	0

EXAMPLE 1–7   Using the data given at the beginning of the chapter in table 1–1 on page 6, construct frequency and relative frequency distributions of the daily calorie intake in meat for the 40 countries given.

Solution   The highest is the United States with 758, while the lowest is India with 6. Hence, the range is 752. Using the rough guideline, you should use an interval width between 75 and 150. Suppose 100 is used. Table 1–9 shows the results.

Table 1-9

Interval	Tally	f	F	$f\%$	$F\%$
1–100	THL THL |||	13	13	32.5	32.5
101–200	THL THL	10	23	25.0	57.5
201–300	||	2	25	5.0	62.5
301–400	||	2	27	5.0	67.5
401–500			1	28	2.5	70.0
501–600	THL		6	34	15.0	85.0
601–700	|||	3	37	7.5	92.5
701–800	|||	3	40	7.5	100.0

●

Exercise Set 1-3

1. A sample of investment "experts" was asked what, in their opinion, money should be invested in. Here are their responses, where S = stocks/shares, C = commodities, R = real estate, P = precious metals, A = art, F = foreign money.

S R P S C R A P F S P R C C R R

P A S R S R P C S S R P F S R P F

Classify these recommendations into a frequency distribution.

2. A sample of gamblers in New York was interviewed and asked what their favorite form of legal gambling was. Here are their responses, where H = betting on horses at the track, O = off-track betting, C = casino gambling, B = bingo, L = lotteries.

H H C C B O B C C C H H H H H

C H L C C C H H C C H H C H H

H C O B B

Classify these preferences into a frequency distribution.

3. During a campaign for a city council seat, a random sample of voters was asked if they supported J = Jones, K = Kelly, L = Latimer, U = undecided. Here are the results.

J K L U J J J U L K K L L L L

K J J K K U U J U K L J U K J

Classify these opinions into a frequency distribution.

4. In the same election as exercise 3, a proposition is to be on the ballot. The same sample of voters was asked whether they intended to vote Y = Yes, N = No, U = undecided.

Y N N U Y Y N N Y N Y N N Y U

N Y U N U Y N U N Y U N Y N Y

Classify these opinions into a frequency distribution.

In exercises 5–8, change the given frequency distribution into a relative frequency distribution.

5. Interval	**f**	**F**
12–15	1	1
16–19	3	4
20–23	8	12
24–27	7	19
28–31	4	23
32–35	2	25

6. Interval	**f**	**F**
80–84	8	8
85–89	12	20
90–94	6	26
95–99	2	28
100–104	1	29
105–109	0	29
110–114	1	30

7. Race	**f**
White	28
Hispanic	20
Black	9
Asian	2
Native American	1

8. Opinion	**f**
Yes	42
No	36
Declined to say	7
Don't know	5

9. Using the data given at the beginning of the chapter, construct frequency and relative frequency distributions of the daily calorie intake in nuts for the 40 countries given.

10. Repeat exercise 9 using the data for eggs. (*Note:* Ignore India, leaving 39 countries.)

Section 1–4   Graphs

Lewis Carroll, in his book *Alice's Adventures in Wonderland,* had Alice express the opinion, "What is the use of a book without pictures or conversations?" This book does not have any of the latter but it does have plenty of the former! The old saying goes, "A picture is worth a thousand words." With data collection, it might be said that a picture is worth a thousand numbers! This section introduces you to "pictures" used in statistics, which are referred to as graphs.

Figure 1–2

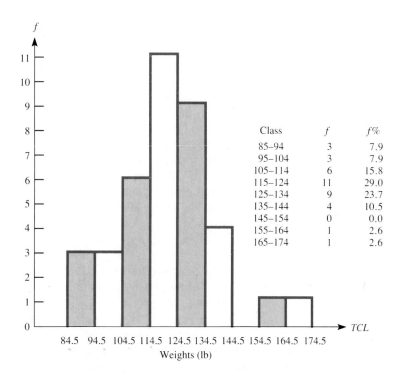

Class	f	f%
85–94	3	7.9
95–104	3	7.9
105–114	6	15.8
115–124	11	29.0
125–134	9	23.7
135–144	4	10.5
145–154	0	0.0
155–164	1	2.6
165–174	1	2.6

The Histogram

TCL = true class limits.

A graph using rectangles is called a **histogram.** The histogram for the weights in example 1–4 is shown in figure 1–2. Notice the data is represented by rectangles. Since the width of each class is the same, the width of each rectangle is the same while the height of each rectangle is proportional to the simple frequency of each class. The vertical axis, therefore, has been labeled f, the simple frequency. The horizontal axis is labeled TCL. This is an abbreviation for true class limits, introduced in section 1–2. Remember a weight in the interval 105–114, for instance, could be any value from 104.5 to 114.5. Since no number is theoretically impossible for weight, it being a continuous variable, there are no gaps between the rectangles. The gap that you see is not really a gap, but rather a rectangle of zero height. This corresponds to the fact that the interval 145–154 had zero simple frequency. For consistency in data handling, discrete data is usually graphed in the same way, even though it may not be possible to have intermediate values. You will see an example of this later in this chapter.

The width of each rectangle in the figure 1–2 histogram is 10 lb, corresponding to the width of the intervals. The area of a rectangle is given by multiplying length times width. Therefore, the total area covered by the rectangles is 380, found by adding the areas of all the separate rectangles. The area of the rectangle over the 115–124 interval is 110. Therefore, the proportion of the total area that is above this interval is

$$\frac{110}{380} = 0.2895 \approx 29\%,$$

which is exactly the relative frequency for this interval. Thus, the relative simple frequencies can be found from the *area* of the individual rectangles as a proportion of the total area covered by the histogram. This is true of all histograms, even if, as sometimes happens, you have rectangles of unequal widths. In fact, quite often, the vertical scale is omitted, since it is not the heights of the rectangles that is important, but the areas.

For instance, it is possible to regroup a frequency distribution with *unequal* intervals. Then careful attention must be paid to the areas of the bars of the corresponding histogram.

How do you decide which, of all the different groupings possible, to use so that the histogram will give us a good "picture" of the sample or population? This is not an easy question to answer. Generally what happens is this: when the intervals are narrow, requiring more of them, the histogram will tend to be "jagged." When the intervals are grouped more, fewer intervals result and the histogram tends to "smooth" out. However, if you group too much, valuable information may be lost. To take an extreme example, all 38 weights could be grouped into one interval, 95–174, and the histogram would consist of one rectangle—very smooth, but next to useless in giving us any additional information about the data. The answer to the above question is to strike a balance between the two extremes. How to do this comes mainly from experience in handling data.

The Simple Frequency Polygon

Another way to graph is to use points and straight line segments in what is called a **simple frequency polygon.** For the weights, grouped as they were originally, the simple frequency polygon is shown in figure 1–3.

79.5 is the midpoint of 75–84 and 179.5 is the midpoint of 175–184.

The method for constructing this graph is as follows: the points are placed directly above the **midpoints** of the intervals at a height equal to the height of the rectangles of the histogram. The points are then joined by straight line segments. Finally, the polygon is continued down to the horizontal axis, joining this axis at midpoints of intervals immediately beyond the range of intervals of the frequency table. An easy way to construct it is to join the midpoints of the tops of the rectangles of the histogram. See figure 1–4.

In point of fact, the simple frequency polygon is just another way of presenting the same information as is presented in the histogram.

Figure 1–3

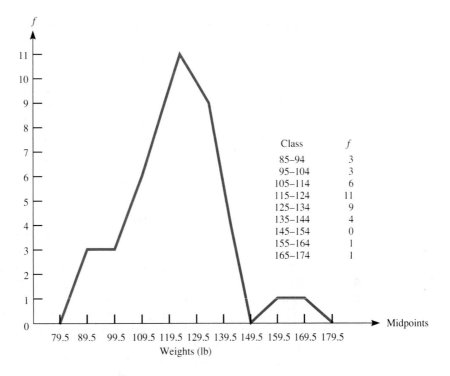

Class	f
85–94	3
95–104	3
105–114	6
115–124	11
125–134	9
135–144	4
145–154	0
155–164	1
165–174	1

Figure 1–4

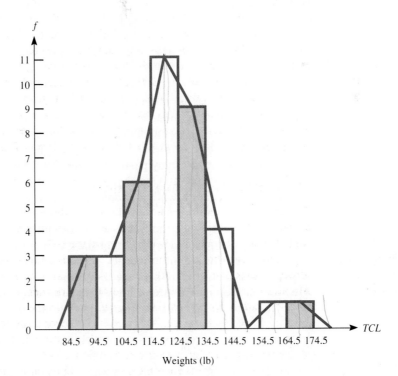

The Ogive

A graph that is constructed similarly to the simple frequency polygon, but gives different information, is the **ogive.** Another name for this graph is the **cumulative frequency polygon,** which should give you a clue as to what it is. One is shown for the weights data in figure 1–5.

Figure 1–5

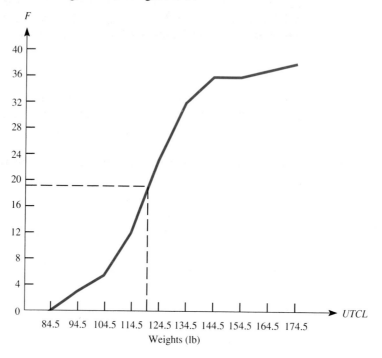

This vertical axis is labeled F, the cumulative frequency. Otherwise, the graph's construction is similar to that of the simple frequency polygon. However, the points here are not placed above the midpoints. Why not? Remember what the cumulative frequency means: the number of pieces of data at or below that interval. The midpoint of any particular interval is not relevant to those values that are below this interval. One characteristic that *all* values (at or below) have in common is that they are all *less than* the upper true class limit of the interval. The points are, therefore, plotted above the upper true class limits (UTCL) on the horizontal axis. Since the frequency is accumulated, the graph rises to the right. It would be meaningless to bring the graph down to the axis on the right (as is done with the simple frequency polygon), but it is brought down to the horizontal axis on the *left.* There are *zero* pieces of data less than the upper true class limit of the interval immediately below the lowest interval covered by the data. Therefore, the graph is continued down to the axis (or zero) at this point.

Think of the ogive as the "less-than" curve.

As in section 1-3, this "average" is the median, which is discussed in detail in chapter 2.

The ogive can be used to read off the "average" graphically. Find half the sample size (in this case, 19) on the vertical axis and then go across horizontally to the ogive and down vertically to the horizontal axis. This value is the approximate average. In this example, you will arrive at a point approximately halfway between 114.5 and 124.5, which is 119.5. You can deduce, then, that 19 (or half) of the sample values are at 119.5 or less. The average weight of the sample weights is about 119.5 lb. The appropriate lines are marked in on the graph. More information that can be obtained from the ogive will be discussed later.

Curves

Sometimes, when there is a large amount of data, such as with a population, the simple frequency polygon or the ogive are approximated by the smooth lines of a curve. Doing this enables you to compare the graph to theoretical mathematical curves.

Some of the more common shapes that can arise are shown in figure 1-6.

Figure 1-6

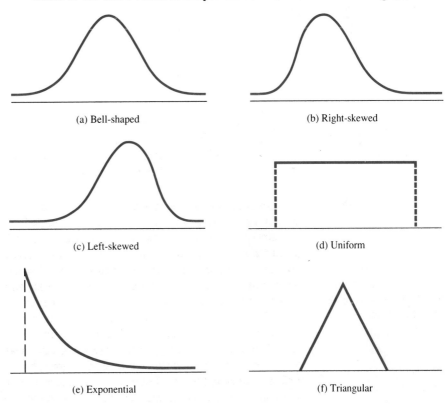

(a) Bell-shaped

(b) Right-skewed

(c) Left-skewed

(d) Uniform

(e) Exponential

(f) Triangular

If you compare the simple frequency polygon of the sample of weights to these theoretical curves, you would probably say that the weights have a distribution which is slightly right-skewed. Thus, different samples can be given a commonality by referring them to the standard shapes.

Discrete Data

As promised earlier, you will now be introduced to graphs of discrete data, using the final digits data presented in example 1–5. See figures 1–7, 1–8, and 1–9. The frequency table is reproduced in table 1–10.

Table 1–10

Class	f	F
0	16	16
1	0	16
2	2	18
3	2	20
4	1	21
5	9	30
6	0	30
7	5	35
8	2	37
9	1	38

Figure 1–7

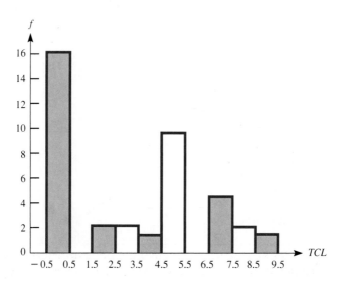

Figure 1–8

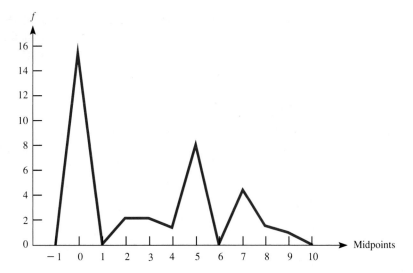

Figure 1–9

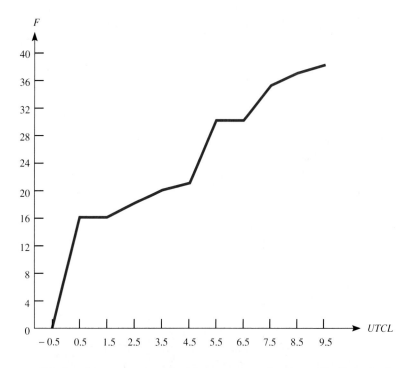

Notice that, even though this is an example about the digits 0–9, to be consistent, true class limits have been used although, in reality, they could not occur. The important information that you want from the graphs, the *shape* of the data, is there.

Qualitative Data

Since categories, rather than numbers, are involved with qualitative data, this is graphed differently than with quantitative data. Two types that are commonly used are the **bar chart** and the **pie chart.** Take a look at these for eye color in example 1–6. The frequency table together with the relative frequencies is reproduced in table 1–11.

Table 1–11

Category	f	$f\%$
Brown	15	60
Blue	6	24
Hazel	2	8
Green	1	4
Black	1	4
Gray	0	0

The construction of the graph in figure 1–10 is similar to that of the histogram; but, unlike the histogram, there *are* gaps between the bars here and the order of the bars is immaterial. The bar for each category has the same width as any other category, while its height is proportional to its frequency.

Figure 1–10

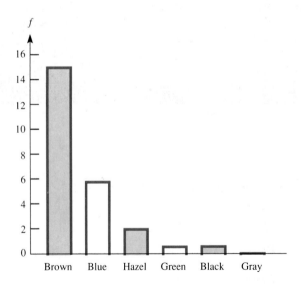

In figure 1–11, a circle represents the "pie." The areas of the slices of the pie are proportional to the relative frequencies of the different categories.

Figure 1–11

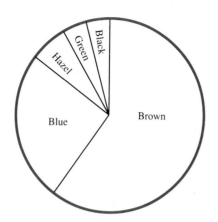

There are 360° in a complete revolution. The brown data was 60%, and 60% of 360° is 216°. Therefore, there should be 216° between the lines that border the "brown" slice. Similarly for the other colors.

THE WIZARD OF ID   by Brant parker and Johnny hart

By permission of Johnny Hart
and NAS, Inc.

Exercise Set 1-4

1. Construct a histogram from this frequency table.

Interval	f
0–4	2
5–9	3
10–14	8
15–19	7
20–24	5
25–29	1

2. Construct a histogram from this frequency table.

Interval	f
80–91	13
92–103	18
104–115	10
116–127	7
128–139	3
140–151	0
152–163	1

3. Construct a histogram from this data, using 30–32 as the first interval.

30	50	46	42	33	50	43	51	38	50	56	36	36
44	51	55	40	46	48	34	51	46	40	30	35	
48	52	31	46	37	41	50	34	54	42	34	32	

4. Construct a histogram from this data, using 251–252 as the first interval.

257	255	253	256	258	260	256	258	257	260	255	253
256	258	259	253	257	260	251	254	257	257	260	255

5. A sample of recently born babies was studied and the hour of birth of each determined to the nearest hour. Here, 1 = 1 A.M., 2 = 2 A.M., . . . , 12 = noon, 13 = 1 P.M., . . . , 23 = 11 P.M., 24 = midnight. Construct a histogram from this frequency table of results.

Class	f
1–4	4
5–8	3
9–12	6
13–16	5
17–20	9
21–24	3

6. A sample of college students was asked how many hours per week they worked on a job. Construct a histogram from this frequency distribution of their responses.

Class	f
0–4	20
5–9	1
10–14	4
15–19	5
20–24	5
25–29	0
30–34	0
35–39	1
40–44	3

7. A study of divorced couples was undertaken. The following data is a collection of the number of years each couple was married. Construct a simple frequency polygon for this data.

3	5	12	11	20	2	25	7	15	14	8	10	9
10	26	4	23	6	18	7	13	8	12	9	

8. The following data shows the scores of a high school class on a college aptitude test in which the maximum possible score was 800. Construct a simple frequency polygon for this data.

356	481	612	204	318	187	598	707	514	555	450	390	421
602	684	512	572	441	324	212	222	371	371	526	534
592	466	471	621	597	669	245	783	649	487	550	401

 9. (Optional) The American Council on Education at UCLA studied national norms of American freshman in the fall of 1974. One of the questions asked concerned the parental income of these freshmen. Here are the results.

Estimated parental income	Percent
Less than $3,000	3.4
$3,000–3,999	2.6
4,000–5,999	4.6
6,000–7,999	5.7
8,000–9,999	8.0
10,000–12,499	15.5
12,500–14,999	13.5
15,000–19,999	16.6
20,000–24,999	12.0
25,000–29,999	5.9
30,000–34,999	3.9
35,000–39,999	2.4
40,000–49,999	2.2
50,000 or more	3.8

Construct a histogram for this data, noting that the intervals are of unequal width. You may assume that the first interval has a width of $3,000 and the last a width of $10,000. This, of course, is not entirely accurate; but little accuracy will be lost.

 10. (optional) Data was collected for pupil/teacher ratio in each of 20 cities' schools. Here is the data grouped into a frequency distribution with unequal intervals. Construct a histogram from it.

Ratio	f
17	1
18–20	9
21–22	7
23–26	2
27	1

11. Construct a simple frequency polygon from this frequency table.

Interval	f
51–55	2
56–60	5
61–65	9
66–70	20
71–75	17
76–80	7
81–85	4
86–90	2
91–95	0
96–100	1

12. Construct a simple frequency polygon from this frequency table.

Interval	f
142–143	1
144–145	4
146–147	7
148–149	10
150–151	7
152–153	4
154–155	1

13. Construct an ogive from this frequency table.

Interval	f
12	2
13	5
14	8
15	6
16	3
17	1

14. Construct an ogive from this frequency table.

Interval	f
25–28	3
29–32	1
33–36	2
37–40	5
41–44	2
45–48	6
49–52	1

For exercises 15–18, (a) construct a simple frequency polygon, (b) construct an ogive, and (c) decide which of the more common shapes your graph in (a) most resembles.

 15. The life expectancies at birth for the 30 most populous countries in the 1970s, classified as follows:

Class	f
36–40	2
41–45	3
46–50	4
51–55	4
56–60	5
61–65	4
66–70	7
71–75	1

 16. In a study of a particularly dangerous section of highway, 300 cars were clocked for speed. Here are the results.

Speed (mph)	f
40–45	18
46–51	48
52–57	75
58–63	129
64–69	21
70–75	9

17. A sample of students at a certain college were asked how many siblings (total number of brothers and sisters) they each had. Here are the responses.

```
3   0   6   4   4   1   3   3   0   6   3   3   3   3   1   3
2   2   4   2   2   2   2   1   1   2   4   2   2   2   0   1
3   1   7   2   5   6   0   1
```

 18. A sample of 50 light bulbs was tested to determine their lifetime. Here are the results.

```
803   641     702   926   1,014     943   807   826   837   914
777   908     846   727     891   1,006   941   809   622   808
779   964   1,035   914     889     928   638   697   747   778
810   910   1,002   934     642     699   683   672   721   754
608   762     888   872     831     652   818   671   824   794
```

 19. The American Council on Education at UCLA studied the national norms of American freshmen in the fall of 1974. This is a distribution of average grades earned by all freshmen while in high school.

Grade	Percent
A or A+	7.5
A−	11.3
B+	19.1
B	26.7
B−	13.6
C+	12.7
C	8.7
D	0.5

Construct a bar chart for this data.

 20. This data gives the number of murder victims in the United States in 1978 classified by the type of weapon used on them.

Weapon	f
Gun	11,910
Cutting/Stabbing	3,526
Blunt object	896
Strangulation/Beating	1,422
Arson	255
All others	705

Construct a bar chart for this data.

 21. A sample of people from a certain city was asked their marital status. Here are the results.

Status	f	$f\%$
Single	15	37.5
Married	12	30.0
Separated	1	2.5
Divorced	10	25.0
Widowed	2	5.0

Construct a pie chart for this data.

 22. In the United States, approximately 45% of the population has blood type O; 40% type A; 11% type B; and 4% type AB. Illustrate this distribution of blood types with a pie chart.

Section 1–5   Summary of Terms

bar chart 30	ogive 26
bell-shaped 27	pie chart 30
categorical data 18	population 8
class 13	proportion 19
class limits 13	qualitative 9
continuous 10	quantitative 9
cumulative frequency 15	range 13
cumulative frequency polygon 26	raw data 8
curves 27	relative frequency 19
discrete 10	response variable 8
experiment 7	right-skewed 27
exponential 27	sample 8
frequency 14	simple frequency 14
frequency distribution 14	simple frequency polygon 24
frequency table 14	tally 13
histogram 23	triangular 27
interval 10	true class limits 16
left-skewed 27	uniform 27
midpoints 24	width 13

Chapter 1 Review Exercises

 1. List the possible outcomes of the experiment to determine at what hour of the day, to the nearest hour, a sample of people was born.

 2. List the possible outcomes of the experiment of going to a city's hall of records each week for a sample of weeks and finding out how many people married each week.

In exercises 3 and 4, consult a copy of the *Guinness Book of World Records* to help you get the answers.

 3. Give an interval of values possible for the birth weight of human babies.

 4. Give an interval of values possible for the weight of mature humans.

 5. According to the *1985 Britannica Book of the Year* (which gives facts for the year 1984), the country with the highest number of full-time students is China, with 187,902,000, and the country with the lowest number of full-time students is Pitcairn Island, with 16. If the total number of full-time students in each country in 1984 was listed, what would be the range of this data? What limitations, if any, are there in the accuracy of your answer?

 6. These figures give the populations of all fifty states of the United States and of the District of Columbia as of 1977, to the nearest 100,000 and given in hundred thousands.

```
10    8    4   58    9   31  179   73  118  107   54  112   91
46   40   29   48    6    7   16   23    6   41    7   51   19
55   29   50   85   35   43   37   24   22   39   28  128    8
 9    4   26   12   23   13    6   37   24  219    5    9
```

This means that the first two states listed, for instance, had populations of 1,000,000 and 800,000, respectively. What is the range of this data? If you wanted to classify these numbers into a frequency distribution with 8 intervals, list the 8 intervals that you would use.

7. Twenty cities were selected and the average pupil/teacher ratio in each of the cities' elementary and secondary schools was calculated in 1970. Here are the results.

20	22	24	23	27	29	27	25	22	24	26	27	18
25	23	28	23	23	19	20						

Classify these ratios into a frequency distribution with 18–19 as the first interval.

8. The same twenty cities as in exercise 7 were selected in 1978 and the corresponding ratios obtained. Here are the results.

19	19	22	23	21	27	22	20	20	20	20	21
20	18	23	22	21	22	17	20			

Classify these ratios into a frequency distribution with 16–17 as the first interval. Compare your answer to the answer to exercise 7.

9. The land areas of the fifty states of the United States are given in millions of acres.

33	365	73	34	100	66	3	1	35	37	4	53	36	23
36	53	26	29	20	6	5	36	51	30	44	93	49	70
6	5	78	31	34	44	26	44	62	29	1	19	49	26
168	53	6	25	43	15	35	62						

Using a width of 25, classify these areas into a frequency distribution. *Note:* Because of the state with 365 million acres (guess which one that is), you will find many intervals with zero frequency. It may be easier to omit these with a line of dots.

10. A random sample of students at a certain university was asked how many miles they traveled each way to school. Here are the results.

4	30	1	1	0	1	35	0	0	0	10	0	0	0	3	4	10
3	27	15	6	5	4	0	6	3	1	3	6	1	0	2	1	4
6	15	10	1	2	1										

A distance of 0 represents a student living on campus or less than half a mile from campus. Classify these distances into a frequency distribution with interval width 5.

11. This is a sample of male students' weights.

170	125	139	135	160	190	143	170	135	165	140	160
135	180	195	195	205	190	165	145	145	173	150	165
168	145	138	123	153	168	175	150	160	142	155	155
160	165	150	145	195	150	170	170	155			

Classify these weights into a relative frequency distribution, showing both $f\%$ and $F\%$ columns. Do you think these weights are the result of measurement or were the students asked? Why?

12. It was determined one year that a large university had 26 doctoral candidates in the social sciences. These 26 students were interviewed and, among other things, asked which branch of the social sciences they were concentrating in. Here are the responses.

Psychology	Political Science	History	Economics
Economics	Economics	History	Political Science
Psychology	Psychology	Sociology	Economics
Sociology	Sociology	Political Science	Psychology
History	History	Psychology	Psychology
Psychology	Psychology	Psychology	

Three others named fields that you can classify under "other." Classify these fields into a relative frequency distribution.

In exercises 13–18, (a) construct a histogram, (b) construct a simple frequency polygon, (c) construct an ogive from the given data, and (d) decide what shape the data has in comparison to the shapes given in figure 1–6.

13.

Interval	f
31–33	2
34–36	6
37–39	7
40–42	11
43–45	12
46–48	8
49–51	3
52–54	1

14.

Interval	f
180–186	32
187–193	64
194–200	96
201–207	90
208–214	60
215–221	30

 15. In a study of an author's writing, an analysis was made of the lengths of words he used. Here are the results.

Word length	f
1–2	224
3–4	440
5–6	180
7–8	102
9–10	38
11–12	11
13 and longer	5

16. A study was made of marriages. Here is a frequency distribution of a sample of them classified according to the age of the groom.

Interval	f
18–24	72
25–31	130
32–38	54
39–45	23
46–52	11
53–59	6
60–66	3
67–73	1

 17. In a recent professional men's golf tournament, these scores were recorded after four rounds.

Total score	f
280–282	6
283–285	8
286–288	8
289–291	18
292–294	16
295–297	9
298–300	5
301–303	1

 18. Here is the same as exercise 17 for a women's professional golf tournament.

Total score	f
275–281	1
282–288	12
289–295	23
296–302	24
303–309	9
310–316	1

In exercises 19 and 20, construct (a) a bar chart and (b) a pie chart to illustrate the data.

 19. The following represents a certain city's "average" family's monthly expenditures:

Type	Amount
Housing	$450
Utilities	90
Medical Expenses	25
Food	180
Transportation	160
Clothing	20
Savings	35
Miscellaneous	50

20. An analysis of a sample of movies recently running in the theaters in the Los Angeles area produced the following results:

Rating	Number
G	4
PG	12
R	16
X	2

(Optional) In exercises 21 and 22, construct a histogram, noting the unequal intervals.

 21. The average weekly temperatures in a certain city over a one-year period were computed to the nearest fahrenheit degree, producing the following results:

Interval	f
15–24	13
25–44	8
45–84	26
85–95	5

 22. The following data gives the number of milligrams of tar in a sample of 36 cigarette brands:

Interval	f
8	1
9–10	1
11–14	7
15–17	11
18–19	15
20–21	1

Hands-on Class/Student Project

Listening to your instructor lecture about statistics, reading this text, following the examples, and doing the exercises can all help you to understand statistics better. However, to get a real "feel" for the subject, the most important learning tool might be to gather your own data and to work with it. After all, presumably you are going to apply the ideas learned in this course to your major field someday. Getting hands-on practice now may well cement into your mind the methods you require for the future. The project introduced here is designed to be a continuous one. At the end of each chapter, you will be given some exercises that you can apply to your own data to further illustrate the ideas introduced in that particular chapter.

The structure of this project could take many forms, from a student working entirely on his or her own, to a whole class, under the leadership of the instructor, working together or in parallel. Groups of two or three students might be a particularly useful way to participate in this activity, since a great deal can be learned by discussing with fellow students what steps to take next and what the results tell you.

With another student or two from the class, interview thirty or more students at your college, determining the answers to the questions.

1. What is your height?
2. What is your birth date?
3. How many semesters have you attended this college, including this semester?
4. How many units have you completed?
5. What is your Grade Point Average?
6. What was your Scholastic Aptitude Test (SAT) score?
7. What is your gender?
8. What is the color of your hair?
9. What is the color of your eyes?
10. What is your blood type?
11. How many different cars have you owned, including your present one?
12. How much do you expect your annual income to be in your first job after you have completed your formal education?

You may wish to ask all of the questions or only some of them, although in later chapters the above variables will be referred to in the project exercises. You may wish to substitute different questions, ones that will produce responses of more interest to you, or you may wish to add to the list. Some questions (such as 7, 8, 9) might be answered by observation rather than by direct questioning. You may wish to forget question 7 altogether and interview approximately half male and half female students, so that later you can make comparisons.

After you have collected the data, perform the following exercises:

1. Which of the variables 1–12 are qualitative data?
2. Which are discrete quantitative data?
3. Which are continuous quantitative data?
4. Construct frequency distributions for each of the variables 1–12.
5. Convert each of the distributions found in 4 to relative frequency distributions.

6. For the quantitative data, draw histograms (or simple frequency polygons) and ogives.

7. For the qualitative data, draw bar graphs or pie charts.

8. Compare your answers to 4 and 5 with those of other students or student groups in the class. Which do you think gives a more meaningful comparison, 4 or 5?

9. Compare your answers to 6 and 7 with those of other students or student groups in the class.

10. If there seem to be big differences between the shape of your data and the shape of the corresponding data of other students, try to come up with reasons why this might be so. Compare how you and the others collected the data and what precautions, if any, you took to try to get samples of students representative of the student body as a whole.

> Save all your data so that you can use it for similar projects at the end of later chapters.

CHAPTER

2

Measures of Central Tendency and Dispersion

....

In small proportions we just beauties see; And in short measures, life may perfect be.

....

BEN JONSON

What Is Wrong Here?

Read this chapter and find out why something in this picture is out in left field!

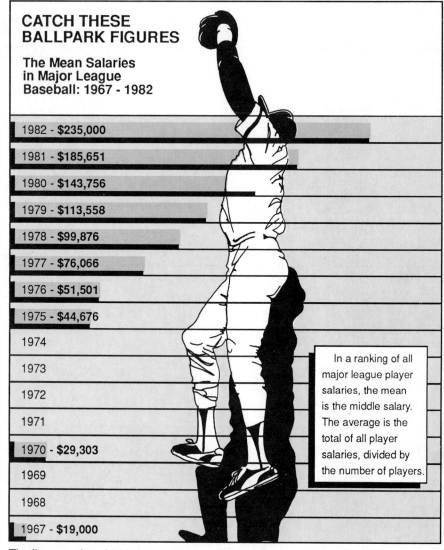

CATCH THESE BALLPARK FIGURES

The Mean Salaries in Major League Baseball: 1967 - 1982

1982 - $235,000

1981 - $185,651

1980 - $143,756

1979 - $113,558

1978 - $99,876

1977 - $76,066

1976 - $51,501

1975 - $44,676

1974

1973

1972

1971

1970 - $29,303

1969

1968

1967 - $19,000

In a ranking of all major league player salaries, the mean is the middle salary. The average is the total of all player salaries, divided by the number of players.

THE MEDIAN IS THE MIDDLE SALARY

Pete Bentovoja and Bob Allen. Copyright 1982 Los Angeles Times. Reprinted by permission.

The first year for which mean salaries were available was 1967. The first year of free agency was 1976. The 1982 mean is an estimate. All figures were supplied by the players' association.

Chapter Overview
Summary Statistics

It is often desirable when studying a sample or a population to obtain some summary statistics from it. Some of these summary values are studied in this chapter together with explanations of what information these summary values give you.

Section 2–1   Mean, Median, and Mode for Ungrouped Data

Years into the future, space exploration may advance so far that life on another planet is discovered. Suppose you are alive at that time and you read that the females on this new planet have an average weight of 42 kg. What would this mean to you? Could it be that more females weigh this than any other weight? Could it be that half the females weigh more than 42 kg and half less? Or could it be that if you took all the females' weights and added them together and divided by the number of females on the planet, the answer would come out to be 42 kg? These three ways of considering an "average" are all legitimate. The fact is that the word *average,* although used quite extensively, is an imprecise word. Statisticians generally refer to the concept of **average** as **a measure of central tendency,** that is, a single number which describes the tendency of a sample or population toward its center.

As your study of statistics progresses, you will discover several quantities that describe different characteristics of data sets. These numbers are classified as follows:

> **Statistic**   A statistic is a single number that describes a characteristic of a sample.
>
> **Parameter**   A parameter is a single number that describes a characteristic of a population.

It is not hard to remember which is which since sample and statistic both begin with the letter *s;* while population and parameter begin with the letter *p.*

You have already come across the range, which tells you a little about how much a data set is spread out. Since it is a single number, the range when applied to a sample is a statistic, and when applied to a population, a parameter.

The measures of central tendency, which you are about to study, are also considered examples of statistics or parameters. What are they?

Mode	The mode is the value (or values) in a data set that occur(s) most frequently, or that is most likely to occur.
Mean	The mean is the value found by adding all the numbers in a data set and then dividing by the number of values in the data set.
Median	The median is the value such that at least half the values in the data set are at or below the median and at least half the values in the data set are at or above the median.

Now, to clarify these definitions, examine the next few simple examples.

EXAMPLE 2–1   Find the mode, mean, and median of the data set.

$$2$$
$$3$$
$$3$$
$$5$$
$$8$$
$$9$$
$$12$$

Solution

Mode $=$ 3, since this occurs twice while all others occur only once.

$$\text{Mean} = \frac{2 + 3 + 3 + 5 + 8 + 9 + 12}{7} = \frac{42}{7} = 6.$$

Median $=$ 5, since four of the seven values are 5 or less and four are 5 or more. ●

The Mode

Why, you might ask, is the mode considered a measure of *central* tendency? It may not be anywhere near the center. Although this is true, many data sets tend to pile up at or near the center. In these cases, the mode will be a good indication of the central tendency of the data.

It is possible that a data set has no mode. If, in example 2–1, one of the 3s were a 4, there would be no mode.

It is possible that a data set has more than one mode. If, in example 2–1, the 9 were an 8, there would be two modes, namely 3 and 8. In that case, the data would be described as being **bimodal,** as opposed to **unimodal** as it is now.

It is also possible for a data set to have three, or even more, modes (for example, your grades on the first test you take in this class, because of the probable variety in your backgrounds).

The mode is the least important of the three measures of central tendency considered here. It is very rarely used, although there are situations where it might be useful. For instance, a men's clothing store may well find it useful to know the modal waist size of men living in the community that it serves, and the central tendency of a sample of incomes might be better described by the mode or the median than the mean, which would be unduly affected by a few very large incomes.

The Mean

This is the measure of central tendency most people think of when the word *average* is used. It is a very important measure, and you will be working with it extensively in your study of statistics. You are now asked to examine some symbols and formulas associated with the mean.

Σ This is the capital Greek letter corresponding to S. It is called *sigma*. In mathematics and statistics, sigma is often used to mean "sum of."

x This letter is used to represent the data values. If more than one data set is being considered at the same time, y or z or other letters may be used.

n This letter is used to represent the size of a sample.

N This capital letter is used to represent the size of a population.

$\bar{x}$ This is read "x bar" and is used to represent the mean of a sample.

μ This lowercase Greek letter is called *mu* (pronounced *mew*) and is used to represent the mean of a population.

If x is used for the data set of example 2–1,

$$
\begin{array}{c}
x \\
2 \\
3 \\
3 \\
5 \\
8 \\
9 \\
12
\end{array}
$$

then Σx means "the sum of x" and $\Sigma x = 42$ and, since this is more likely to be a sample, $n = 7$. Thus, $\bar{x} = \frac{42}{7} = 6$.

For a sample, the formula for the mean is

$$\bar{x} = \frac{\Sigma x}{n}.$$ (1)

For a population, the formula for the mean is

$$\mu = \frac{\Sigma x}{N}.$$ (2)

In mathematics, there are various means that can be calculated from a set of numbers, such as the arithmetic mean, the geometric mean, and the harmonic mean. When the word mean is used alone in statistics, the *arithmetic* mean is always the one intended and that is the one defined here.

Both formulas are essentially the same. In words, they state that: "the mean of the data set is the sum of all the values in the data set divided by the number of values." The formulas presented here are used when dealing with raw data— different formulas are used when calculating the mean from a grouped frequency distribution, as will be seen later.

There are two properties of the mean that are of interest.

1. The sum of the deviations of all the data from the mean is zero, where deviation is a directed distance, positive if it is larger and negative if it is smaller.

The letter d is used for deviation from the mean. If x represents a particular value in the data set, then its deviation from the mean, $d = x - \bar{x}$. Thus, for example 2–1, with $\bar{x} = 6$:

The sum of the negative deviations is -11, while the sum of the positive deviations is 11. Since $-11 + 11 = 0$, $\Sigma d = 0$.

x	d
2	-4
3	-3
3	-3
5	-1
8	2
9	3
12	6.

The fact that, here, $\Sigma d = 0$ is not true just for this example; it is true for the deviations from the mean for *any* data set. Try it for yourself: make up any set of numbers, calculate the mean, figure out all the deviations from the mean, and then sum these deviations. Your answer will be zero, if you don't do any rounding off.

It can be proved algebraically, and this has been done at the end of the chapter. If you don't follow the argument, don't worry; it is not critical to your understanding of statistics.

If you were told that for some data $\Sigma(x - 9) = 0$, you could conclude that the mean of this data is 9, without ever seeing the data.

The mean is the only number for which the above property is true. If you found a number for which the sum of the deviations of all the data from this number is zero, then you can be sure that the number you have found is the mean.

It is because of this property that the mean is often considered to be the balance point of the distribution. To illustrate this, suppose you have a board with weights placed on it, as in figure 2–1.

Figure 2–1

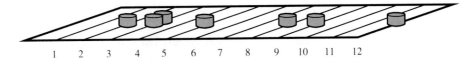

Single weights have been placed at 2, 5, 8, 9, and 12 and two weights at 3, to illustrate example 2–1. If you moved a fulcrum under this board until the board balanced, you would find the balance point at 6, the mean.

2. The mean is a sensitive measure of central tendency.

Looking again at example 2–1, suppose the 12 were changed to 21, then $\Sigma x = 51$ and $\overline{x} = \frac{51}{7} \approx 7.3$. The mode and median, however, would remain unchanged. A single change in a set of data will always alter the mean, but not always the median or mode. It is for this reason that the mean is said to be sensitive.

The Median

Although probably not as important a measure of central tendency as the mean, the median can be very useful in certain situations. As with the mean, symbols have been devised to refer to the median.

> $\tilde{x}$ This is read "x tilde" and is used to represent the median of a sample. The tilde above the x is the same symbol as the one used above the letter n in Spanish in such words as *mañana*.
>
> $\tilde{\mu}$ This is read "mu tilde" and is used to represent the median of a population. It is a combination of the Greek letter used for the population mean and the tilde.

In example 2–1, there are an odd number of pieces of data: $n = 7$. Thus, the median was the "one in the middle," namely 5, it having an equal number of values above and below it. What if n is even? Take a look.

EXAMPLE 2–2   Find the median of this data set.

$$
\begin{array}{c}
x \\
1 \\
2 \\
3 \\
5 \\
8 \\
9
\end{array}
$$

Solution   The value 3 cannot be chosen since there are three values at 3 or below, whereas there are four values at 3 or above. For a similar reason, 5 can't be the median either. The median here could be any number between 3 and 5. Take any number between 3 and 5 as the median and there will be three numbers in the data set at or below this number and three values at or above it. To be consistent, an infinite number of choices for the median cannot be allowed, so the rule is to choose the mean of the middle two numbers as the median. Thus,

$$\tilde{x} = \frac{3 + 5}{2} = \frac{8}{2} = 4.$$

Notice that 4 is halfway between 3 and 5. It is always true that the mean of two numbers is halfway between the two numbers. ●

There is no simple formula to find the median of raw data, but you can use the following rule:

> Arrange the data in order of size. If the sample (or population) size is odd, then the median is the one in the middle. If the sample (or population) size is even, then the median is found by calculating the mean of the two in the middle.

Here are two more examples. Find the median of each.

EXAMPLE 2-3   5, 7, 2, 10, 3, 5, 6, 1, 12

EXAMPLE 2-4   14, 18, 13, 10

Solutions   In example 2–3, 3 is not the median. Although it is presently the one in the middle, the data has not yet been arranged according to size. Arranged, it becomes

1, 2, 3, 5, 5, 6, 7, 10, 12

and $n = 9$, which is odd. Thus, $\tilde{x} = 5$. You might think, "Wait a minute! There are only three values less than 5 and four greater than it. How can 5 be the median?" Look back at the definition of the median and notice the phrases "at or below" and "at or above." In this example, there are five values at or below the 5 in the middle and five values at or above the 5 in the middle. The two halves are

{1, 2, 3, 5, 5} and {5, 6, 7, 10, 12}.

If the second 5 were also included in the high set, the definition of the median would not be violated. This is why the phrase "at least" is included in the definition. See also example 2–5.

Example 2–4, arranged, becomes 10, 13, 14, 18 and $n = 4$, which is even. Thus,

$$\tilde{x} = \frac{13 + 14}{2} = \frac{27}{2} = 13.5.$$

Notice the median does not have to be a whole number.

As with the mode, it is possible the median turns out to be nowhere near what you intuitively think of as the center. The median of the set $\{2, 2, 2, 2, 31, 87\}$ is 2. Here are some more samples with their medians.

EXAMPLE 2–5   **Solution**

Sample	**Median ($\tilde{x}$)**
5, 8, 11, 13, 15	11
0, 2, 5, 7, 10, 19, 23, 25	8.5
1, 2, 3, 3, 7, 9	3
1, 2, 3, 3, 7	3
2, 2, 2, 2, 31, 87, 93	2
2, 2, 2, 2, 31, 87, 930	2

Exercise Set 2–1

1. Find the mode or modes for the data.

     18   19   20   21   22   21   20   19   17   17   16   16   16   20
     21   21   22

2. Find the mode or modes for the data.

     36   36   37   38   36   37   38   37   36   37   36   38   38   37

3. Find the median of the data.

     5   3   6   9   7   2   10   8

4. Find the median of the data.

     18   16   19   22   20   15   23   21   21

5. Find the mean of the data.

     30   40   50   35   45   55   33   37   43   47   53   57   32   42
     52   81

6. Find the mean of the data.

     76   69   63   82   29   83   64   71   76

7. For the data set 8, 3, 9, 12, 6, 2, 4, 12, find $\Sigma(x - c)$ for $c = 7$ and for $c = 8$. Does this show that either 7 or 8 is the mean?

8. For the data set 13, 24, 17, 20, 15, 12, 18, find $\Sigma(x - c)$ for $c = 18$ and for $c = 19$. Does this show that either 18 or 19 is the mean?

9. Use a calculator to find the mean of the data.
  12.83   7.16   9.45   3   2.7   14.614   3.21   8.6   9.713   8.24   5.616
  13.0   12.2

10. Use a calculator to find the mean of the data.
  0.31   0.04   0.25   0.003   0.0004   0.201   0.020   0.501   0.046
  0.013

Measures of Central Tendency from Frequency Distributions

When data has been grouped into a frequency distribution, the original values have been lost. In the frequency table of weights, table 2–1, the interval 105–114 has a simple frequency of 6; but it is no longer known where those 6 weights are. They may have all been at 105, all at 114, all at some intermediate value, or spread out at various values in the interval. The calculations of the mean, median, and mode from the frequency distribution, rather than from the raw data, will necessarily be prone to error. Methods have been developed, however, to minimize errors. Before attention is turned to these methods, the "true" values of the mean, median, and mode have been calculated from the raw data, so that they can be compared to the values calculated from the frequency distribution.

$$\text{Modes} = 120, 130$$
$$\overline{x} = 119.97 \approx 120.0$$
$$\tilde{x} = 120$$

Section 2–2   The Mode and Mean of a Frequency Distribution

The Modal Interval

Table 2–1

Class	f
85–94	3
95–104	3
105–114	6
115–124	11
125–134	9
135–144	4
145–154	0
155–164	1
165–174	1

The actual mode of a grouped frequency distribution is impossible to determine. However, the interval with the highest simple frequency can be found. This is called the **modal interval.** The modal interval for the weights is 115–124, since it has $f = 11$ and 11 is the largest number in the f column. Notice that this interval does include one of the modes found from the raw data. The modal interval can also be found from the histogram by looking for the peak. Distributions with histograms having two peaks, even though the peaks are different heights, are called *bimodal.* The distributions of last digits is bimodal, since there are peaks at 0 and 5.

The Mean of a Frequency Distribution

As is stated above, it is no longer known where the 6 weights in the 105–114 interval were. In the long run, the least amount of error occurs in the calculation of the mean, when it is assumed that all values in an interval are concentrated at the midpoint of the interval. The logic behind this reasoning arises from the assumption that half the values in an interval are *liable* to be less than the midpoint and half more, so that errors will tend to balance out. The letter m will be used for midpoint. To find the midpoint of an interval, find the mean of the class limits of the interval. Thus, for the interval 105–114,

$$m = \frac{105 + 114}{2} = \frac{219}{2} = 109.5.$$

If all 6 values in this interval were concentrated at 109.5, the sum of the values in this interval would be

$$6 \times 109.5 = 657.0.$$

The sum of the actual weights in this interval is 655. Not bad! Now to do this with the whole frequency distribution.

EXAMPLE 2-6   Find the mean of the frequency distribution of weights.

Solution

Table 2–2

Class	m	f	fm
85–94	89.5	3	268.5
95–104	99.5	3	298.5
105–114	109.5	6	657.0
115–124	119.5	11	1,314.5
125–134	129.5	9	1,165.5
135–144	139.5	4	558.0
145–154	149.5	0	0
155–164	159.5	1	159.5
165–174	169.5	1	169.5
		$\Sigma f = 38$	$\Sigma fm = 4{,}591.0$

The sum of the f column in table 2–2 is, as it should be, the size of the sample. The sum of the fm column is the total of all the weights were they concentrated at the midpoints of the intervals. Thus,

$$\bar{x} = \frac{4{,}591.0}{38} = 120.82 \approx 120.8.$$

Again, not bad! This is quite close to the true sample mean, 120. So, the formula for $\bar{x}$ becomes

$$\bar{x} = \frac{\Sigma fm}{\Sigma f},\qquad(3)$$

and exactly the same formula is used for μ, if you are dealing with a population instead of a sample.

Instead of having to add 38 numbers in the original data, it is necessary to add only 9 numbers in the *fm* column. True, some accuracy has been lost, but not much. The 9 numbers, however, are unpleasant. A coding method has been devised to make life easier.

Warning: The following method works only if the intervals are all of the same width.

Suppose any midpoint is chosen, say 119.5, and it is subtracted from all the midpoints. These numbers result: $-30, -20, -10, 0, 10, 20, 30, 40, 50$. Notice that all these numbers are multiples of 10, which is the class width. The letter *w* is used for the class width. If the resulting numbers are divided by $w = 10$, these numbers result: $-3, -2, -1, 0, 1, 2, 3, 4, 5$. These numbers are much smaller and more convenient than the unpleasant *m*'s. The original data was labeled *x*. The coded midpoints are usually labeled x' (read "x prime"). Now, redo the calculations using coding.

EXAMPLE 2–7   Use coding to find the mean of the frequency distribution of weights.

Solution

Table 2–3

Class	x'	f	fx'	
85–94	-3	3	-9	
95–104	-2	3	-6	-21
105–114	-1	6	-6	
115–124	0	11	0	
125–134	1	9	9	
135–144	2	4	8	26
145–154	3	0	0	
155–164	4	1	4	
165–174	5	1	5	
		$\Sigma f = 38$	$\Sigma fx' = 5$	

The coded mean is calculated similarly to the way the mean using midpoints was calculated.

$$\overline{x}' = \frac{\Sigma fx'}{\Sigma f} = \frac{5}{38} \approx 0.132$$

To achieve the coding, the midpoint, 119.5, was subtracted and the result divided by $w = 10$. To "get back," it is necessary to multiply by w and add the 119.5, thus:

$$\overline{x} = 119.5 + 10(0.132)$$

$$= 119.5 + 1.32$$

$$= 120.82 \approx 120.8,$$

which is exactly the same answer obtained from the frequency distribution without the use of coding. There is no loss of accuracy in coding. There is a gain in simplification. ●

Instead of adding 38 fairly large numbers to calculate the mean, you need add only 9 small numbers.

To summarize the method:

1. Choose any interval, preferably near the middle, and put a zero opposite it in the x' column.
2. Fill the rest of the x' column by putting -1, -2, etc., for the intervals smaller than the chosen interval, and 1, 2, etc., for the intervals larger than it.
3. Compute the fx' column and sum it and the f column.
4. If the letter C is used for the midpoint of the interval opposite which the zero was placed and w for the class width, the formulas for the mean become

$$\overline{x}' = \frac{\Sigma fx'}{\Sigma f}$$

and   (4)

$$\overline{x} = C + w\overline{x}'.$$

Important note: It is not necessary to compute all the midpoints and do all the subtractions and divisions. Just follow the steps above. Any interval can be chosen in step 1, the same answer should result. Observe the calculations again, using a different interval midpoint as the zero value in the coded data.

EXAMPLE 2-8 Redo example 2-7.

Solution

Table 2-4

Class	x'	f	fx'
85–94	−4	3	−12
95–104	−3	3	− 9
105–114	−2	6	−12
115–124	−1	11	−11
125–134	0	9	0
135–144	1	4	4
145–154	2	0	0
155–164	3	1	3
165–174	4	1	4
		$\Sigma f = 38$	$\Sigma fx' = -33$

$w = 10$ and, this time, $C = 129.5$ (the midpoint of the 125–134 interval) and so,

$$\overline{x}' = \frac{-33}{38} \approx -0.868$$

$$\overline{x} = 129.5 + 10(-0.868)$$

$$= 129.5 - 8.68$$

$$\approx 120.8,$$

again, exactly the same answer. ●

The formulas given at (4) may seem complicated, but if you follow the method with the examples given, you should find that it is not too difficult.

Exercise Set 2-2

1. State in your own words why it is impossible to find the mode of data already grouped into a frequency distribution with width greater than 1. What is done instead?

2. If the width of a frequency distribution is $w = 1$, then the mode *can* be found. Find the mode of the data.

x	f
8	2
9	13
10	9
11	7
12	3
13	1

3. Find the modal interval of the frequency distribution.

Interval	f
20–24	2
25–29	3
30–34	8
35–39	9
40–44	6
45–49	3

4. Find the modal interval of the following frequency distribution. Here, you are asked to reconstruct the f column using the F column.

Interval	f	F
3–5	2	2
6–8	3	5
9–11	7	12
12–14	8	20

In exercises 5–8, find the mean of the given frequency distributions.

5.

Interval	f
21–25	2
26–30	4
31–35	8
36–40	7
41–45	5
46–50	4

6.

Interval	f
60–79	1
80–99	0
100–119	3
120–139	10
140–159	8

7. This frequency distribution was constructed using the data of exercise 6 of the chapter 1 review exercises, giving the sizes of the populations of all fifty states and the District of Columbia as of 1977, to the nearest 100,000 and given in hundred thousands.

Interval	f
1–30	27
31–60	15
61–90	2
91–120	4
121–150	1
151–180	1
181–210	0
211–240	1

8. This frequency distribution was constructed using the data of exercise 9 of the chapter 1 review exercises, giving the land areas of the fifty states in millions of acres.

Interval	f
1–25	14
26–50	22
51–75	9
76–100	3
101–125	0
126–150	0
151–175	1
.	.
.	.
.	.
351–375	1

Note: The line of dots indicates missing intervals, all with frequency 0. Take care when coding!

9. In the calculation of a mean from a frequency distribution, 0 was placed opposite the interval 18–24, in preparation for the coding process. It was then found that $\bar{x}'$ = 0.34. Find $\bar{x}$.

10. In the calculation of a mean from a frequency distribution of a population, 0 was placed opposite the interval 83–90, in preparation for the coding process. It was then found that $\mu' = 1.71$. Find μ.

Section 2–3   The Median of a Frequency Distribution

As with the mean, it is no longer known where the 6 weights in the 105–114 interval were. In the long run, the least amount of error to the median results when it is assumed that all the values in an interval are equally spaced in the interval. Remember, weight is a continuous variable, so these 6 weights could be anywhere from 104.5 to 114.5. The 6 weights are placed interior to this interval, as in figure 2–2. Notice that the six marks divide the interval into *seven* equal spaces. There will always be one more space than the value of f for the interval.

Figure 2–2

104.5   114.5

The distance from 104.5 to 114.5 is 10, which is w for this frequency distribution. The width of each space must be $\frac{10}{7} \approx 1.43$.

The space-width can be found from $\dfrac{w}{f+1}$.

Now, for example, the third value in this interval would be

$$104.5 + 3(\text{space-widths}) = 104.5 + 3(1.43) \approx 108.8.$$

If $LTCL$ is used to abbreviate *lower true class limit* of the interval,

$$\text{the } r\text{th value in an interval} = LTCL + r\left(\frac{w}{f+1}\right). \tag{5}$$

Now, reexamine the frequency distribution of weights in order to use these ideas to find the median.

EXAMPLE 2-9   Find the median from the frequency distribution of weights.

Solution

Table 2-5

Class	f	F	
85–94	3	3	
95–104	3	6	
105–114	6	12	
115–124	11	23	← $\tilde{x}$ is in here.
125–134	9	32	
135–144	4	36	
145–154	0	36	
155–164	1	37	
165–174	1	38	

Since $n = 38$, $\tilde{x}$ is found by calculating the mean of the 19th and 20th values from the lowest value when the data is arranged in order. Examining the F column of table 2–5, notice that 12 values have been accumulated by the end of the 105–114 interval and 23 by the end of the 115–124 interval. Since 12 is smaller than 19, and 23 is larger than 20, the 19th and 20th must be among the 11 values in the 115–124 interval. Since there are 12 immediately below this interval, to get the 19th and 20th in the data set, the 7th and 8th values in the interval are needed ($19 - 12 = 7$ and $20 - 12 = 8$). Use formula (5), remembering $w = 10$ and $f + 1 = 11 + 1 = 12$, to get

$$\text{7th value} = 114.5 + 7(\tfrac{10}{12}) \approx 120.33$$

$$\text{8th value} = 114.5 + 8(\tfrac{10}{12}) \approx 121.17$$

and finally,

$$\tilde{x} = \frac{120.33 + 121.17}{2}$$

$$= \frac{241.50}{2}$$

$$= 120.75.$$
●

Look back at the "true value" of $\tilde{x}$, found from the raw data. It was 120. As before, not bad!

To summarize this method:

1. Look at the sample size and decide which value or values you need to look for to compute the median. Remember, if n is odd, only one value is necessary.
2. Examine the F column to decide which interval must contain the median. *Note:* There is a possibility that two intervals are involved. Suppose $n = 28$ and the relevant part of the F column looks like this:

$$F$$

.

.

.

11

14

18

.

.

.

The 14th and 15th values are needed here and the 14th is in one interval and the 15th in the next.
3. Use formula (5) to calculate the value or values you need.
4. If n is odd, you only need to calculate one value in step 3, and this is the median. If n is even, the mean of the two values you calculated in step 3 is the median.

Graphic Approximation

When performing the calculations for the mean and median, mistakes are possible. Even if you are using a calculator, you might press the wrong button! It is therefore a good habit to check your answers graphically. What is meant by this is: find the mean and median on the histogram, for instance, and see if those values are plausible. What is plausible?

The median, recall, is the value that divides the data set in half. On the histogram, therefore, 50% of the area covered by the histogram should be to the left of the median and 50% to the right. Don't measure the area accurately; rather eyeball it to see if it looks right.

For the mean, imagine the histogram made of solid material. Would this solid balance at the mean? Recall that the mean is the balance point of the distribution.

The histogram for the weights is shown in figure 2-3. The dotted line placed at $\tilde{x}$ does seem to divide the area in half and the "fulcrum" placed at $\bar{x}$ would seem to balance the histogram.

Figure 2-3

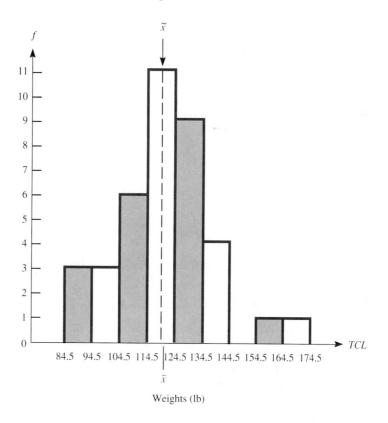

Weights (lb)

If, by some arithmetic error, you had calculated the mean or median at around 160, you could see quite clearly from the histogram that something was wrong.

Another check for the median can be made by using the ogive. This was discussed in chapter 1.

Quick graphic checks are useful for pointing out gross errors here and in numerous other instances you will encounter in your introduction to statistics.

Exercise Set 2–3

In exercises 1–4, find the median of the given frequency distributions.

1.
Interval	f
1–5	1
6–10	3
11–15	8
16–20	10
21–25	7
26–30	6

2.
Interval	f
200–203	12
204–207	15
208–211	10
212–215	8

3. This is a frequency distribution of the number of hours spent studying per week claimed by a sample of university students.

Class	f
1–6	6
7–12	7
13–18	5
19–24	6
25–30	3
31–36	1
37–42	1
43–48	1

4. This is a frequency distribution of the number of miles driven in a given week by a sample of traveling salesmen attached to a large corporation.

Class	f
400–599	3
600–799	7
800–999	12
1,000–1,199	11
1,200–1,399	5
1,400–1,599	2

Exercises 5 and 6 give the data (from chapter 1) on average pupil/teacher ratio in the schools of 20 selected cities for 1970 and 1978, respectively. Construct frequency distributions, using 16–17 as the first interval, calculate the medians of each of these frequency distributions, draw histograms, and verify your answers graphically.

5. 20 22 24 23 27 29 27 25 22 24 26 27 18 25 23 28
   23 23 19 20

6. 19 19 22 23 21 27 22 20 20 20 20 21 20 18 23 22
   21 22 17 20

Section 2–4 The Weighted Mean

Sometimes it is necessary to find the mean of a set of values, each of which has a different weight (or importance). The formula for finding the mean becomes

$$\bar{x} = \frac{\Sigma xw}{\Sigma w}, \tag{6}$$

where the w's are the weights given to each data value. This is called the **weighted mean.** For example, if you wish to calculate your GPA, you must weight the grade in each of your classes by the number of units (or credits) in the class.

EXAMPLE 2–10 If you take one each of 2-, 3-, 4-, and 5-unit classes, receiving grades of A, C, B, and B, respectively, A = 4 points, B = 3 points, C = 2 points, D = 1 point, F = 0 points, find your GPA.

Solution

$$\bar{x} = \frac{4 \times 2 + 2 \times 3 + 3 \times 4 + 3 \times 5}{2 + 3 + 4 + 5}$$

$$= \frac{41}{14}$$

$$\approx 2.93 \qquad\qquad \bullet$$

The formula for the weighted mean can also be used to find the mean of two or more means of samples of different sizes, the weights then being the different sample sizes.

EXAMPLE 2–11 Find the overall mean for the sample results.

Sample	$\bar{x}$	n
1	36	52
2	39	150
3	38	24

Solution

$$\bar{x} = \frac{36 \times 52 + 39 \times 150 + 38 \times 24}{52 + 150 + 24}$$

$$= \frac{8,634}{226}$$

$$\approx 38.2 \qquad\qquad \bullet$$

If you simply added the means and divided by 3, you would get 37.7, the *wrong* answer, since you wouldn't be taking the different sample sizes into account.

Exercise Set 2–4

1. Find the GPA of a student who has taken two 5-unit classes, earning A and B; three 4-unit classes, earning A, B, and C; six 3-unit classes, earning 3 A's, 1 B, and 2 C's; and two 2-unit classes, earning an A and a D.

2. Find the GPA of a student who has taken three 5-unit classes, earning 2 A's and a B; one 4-unit class earning a C; and four 3-unit classes earning 2 B's, a D, and an F.

3. Find the overall mean given these sample results.

Sample	$\bar{x}$	n
1	18.3	10
2	20.1	25
3	17.6	15
4	15.8	30

4. Find the overall mean given these sample results.

Sample	$\bar{x}$	n
1	51.6	20
2	72.3	200

5. If your blood pressure is 130 over 60, a nurse or doctor finds your mean blood pressure by giving twice the weight to the higher value than to the lower. Find the mean blood pressure for the figures quoted.

6. Find the mean blood pressure of a person whose blood pressure is quoted as 125 over 65.

Mean or Median?

The mode is a little-used measure of central tendency; but when should you use the mean and when the median, or does it matter?

THE WIZARD OF ID   by Brant parker and Johnny hart

By permission of Johnny Hart and NAS, Inc.

It was pointed out earlier that the mean is a sensitive measure, and correspondingly, the median is insensitive. When computing an average, however, sensitivity is not always desirable. Here is a typical, but simplified, situation. A company has 100 employees making around $10,000 per year, five executives making around $80,000, and a president making $120,000. The company president could boast that the average income of the employees of this company is about $14,000 per year, which no one makes! The "average" that the president is using is, of course, the mean. The relatively few large salaries affect the mean considerably. The median salary would be $10,000, right in the center of the majority of employees' salaries, unaffected by the few large salaries. In this situation, which do you think reflects the central tendency of the data better, mean or median?

Generally speaking, the mean is more accurate, since *all* the data is used in its calculation. However, the median may reflect the central tendency better in certain situations.

When collecting data, gross errors may be made. In a sample, all but one of the data may be clustered together, while the one is way off. The researcher looking at this data might suspect that this one value is a measurement error. Should the researcher throw this value out? It is always a bad idea to throw out data, if you are not *sure* measurement error is involved. However, the researcher has the option of using the median, which will not be affected by the extreme value, rather than the mean, which will.

A final point to be made is the usefulness of calculating *both* the mean and the median. A comparison of the relative sizes of the mean and median can give you a clue to the shape of the distribution. In a perfectly symmetrical distribution, the mean and median will be equal. In a skewed distribution, the mean will be pulled toward the long tail of the distribution. In other words, for a right-skewed distribution, the mean will be larger than the median, and for a left-skewed distribution, the mean will be smaller than the median.

In figure 2–4, imagine μ at the balance point. To balance, the fulcrum has to move toward the tail.

Figure 2–4

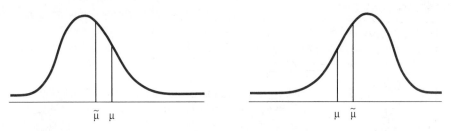

$\tilde{\mu}$ μ μ $\tilde{\mu}$

For the frequency distribution of weights, $\bar{x}$ turned out to be slightly larger than $\tilde{x}$, indicating slight right-skewness. A distribution of weights is usually more right-skewed than this one. Possibly, students are less likely to be overeaters!

> Next to the baseball player's leg at the beginning of the chapter, the newspaper states: ". . . the mean is the middle salary. The average is the total of all player salaries divided by the number of players." You should now be able to see that what they refer to as the mean is actually the median, while what they refer to as the average is actually the mean.

Is the Average Enough?

Examine these three samples together with the three measures of central tendency calculated for each.

	Mode	**Median**	**Mean**
a. 20, 40, 50, 50, 60, 80	50	50	50
b. 10, 50, 50, 50, 50, 90	50	50	50
c. 50, 50, 50, 50, 50, 50	50	50	50

Notice how three entirely different samples can have exactly the same values for the three measures of central tendency discussed in the foregoing sections. So, the answer to the question, "Is the average enough?", must be a resounding no! For these three samples, something else is needed to describe the differences among them.

Here is another example in which knowing the average alone may let you down. Suppose while in your twenties you win a competition, the first prize being a cruise of your choice. You visit the travel agent and look through available cruises, narrowing your choice down to two. To help your decision, you ask the agent what the average age of the passengers is. On one, the average age turns out to be 49 and on the other 24. You choose the second and embark on the cruise only to find that half the passengers are in their forties and fifties and half are small children! (The mean age of a father of 42 and his 6-year-old child is 24.) After having rather less than a wonderful time, you head for the nearest college to sign up for Statistics I to find out what other information would have been helpful!

Section 2–5   Dispersion and How to Measure It

Dispersion

A key concept, which is needed in addition to the central tendency of the data, is the **dispersion** of the data. Another word for dispersion is **spread.** Knowing how spread out, or dispersed, a sample (or population) is can give you valuable information about it. For instance, take the cruise with passengers whose average age is 24. If all the passengers are in their twenties, the dispersion of these ages will be much less than the ages of the passengers on the cruise you went on. So, knowing the dispersion of the ages in advance would have forewarned you.

A very simple **measure of dispersion,** which you have already met, is the *range*. The letter R is used to designate the range.

$$R = \text{Highest value in the data set} - \text{Lowest value in the data set}$$

For the three samples above,

    a. $R = 60$
    b. $R = 80$
    c. $R = 0$.

Notice that, although the measures of central tendency in all three samples are 50, the range is different for each of the three. It is true, therefore, that the range does give additional, helpful information, but is it enough? The answer to this question must be an unqualified "no!"

The range uses only two pieces of the data, the high and the low—it tells you nothing about what is going on in between. On the cruise that you were on, suppose the oldest passenger was 57 and the youngest 5, then $R = 52$. To show you that this number is not very helpful, imagine another cruise where the overwhelming majority of the passengers were in their twenties, but there is a 65-year-old and a 13-year-old, who are the oldest and youngest aboard. The range would still be 52, but knowing that would not help you in differentiating between the two sets of passengers. In some cases, the range might be helpful. If, for instance, you were told that the average age is 24 and $R = 10$, you would know that you were pretty safe. In many cases, however, the range is not very helpful in telling you about the dispersion of the data. Historically, therefore, statisticians searched for another measure of dispersion, preferably one that used *all* of the data.

The Mean Deviation

To measure spread, you might ask, spread about what? The logical answer to this is to find out how spread out the data are about their center. The most important measure of central tendency is the mean. So, when statisticians first started to tackle this problem, they investigated how spread out data sets were about their means. In section 2–1, deviations from the mean were considered with the following example:

EXAMPLE 2–12   Find the deviations from the mean and the sum of these deviations for this data: 2, 3, 3, 5, 8, 9, 12.

Solution

Recall that $d = x - \bar{x}$.

x	d
2	-4
3	-3
3	-3
5	-1
8	2
9	3
12	6
$\Sigma x = 42$	$\Sigma d = 0$
$n = 7$	
$\bar{x} = 6$	

●

As was pointed out then, the sum of the deviations around the mean is *always* zero. Calculating Σd, then, tells you nothing about the dispersion of the data. Something that occurs to many people at this point is as follows:

You are interested in the spread of the data, that is, the distances of the individual pieces of data from the mean. You are not really interested in whether they are above or below the mean. So, why not just drop the minus signs? Why not, indeed? After all, if you do that, their sum would not be zero and would be different for different data sets. The process of "dropping the minus signs" has been given a name by mathematicians. It is called taking the **absolute value.** The absolute value of a number n is written $|n|$, that is, the number with vertical bars on either side of it. The absolute value of a number is the value of the number without regard to sign, or, put another way, it is the distance between the number and zero.

For example: $|3| = 3, |-5| = 5, |0| = 0.$

Notice that the absolute value of a positive number or zero leaves the number unchanged; but that the absolute value of a negative number is the corresponding positive number.

EXAMPLE 2–13   With the data of example 2–12, find the absolute
deviations from the mean and the sum of these.

Solution

x	d	$\lvert d \rvert$
2	-4	4
3	-3	3
3	-3	3
5	-1	1
8	2	2
9	3	3
12	6	6
$\Sigma x = 42$		$\Sigma \lvert d \rvert = 22$
$n = 7$		
$\bar{x} = 6$		

●

For this sample, the sum of the absolute deviations is 22. If you had another
sample with $n = 7$ and $\Sigma \lvert d \rvert = 12$, would you not conclude that this second
sample is less spread out than the first? The 7 numbers would necessarily have
to be closer together for the sum of their absolute deviations to come out that
much less. On the other hand, suppose you had a sample with $n = 50$ and $\Sigma \lvert d \rvert$
$= 22$. Imagine that! You have 50 numbers and the sum of their absolute deviations is no larger than the sum of the absolute deviations of the above sample of
size 7. Again, would that not mean that those 50 numbers are far less spread out
than the above 7? Looking at $\Sigma \lvert d \rvert$, alone, is not enough to tell you about the
dispersion of the data. $\Sigma \lvert d \rvert$ should be compared to n, the sample size. This is
what is known as the **mean deviation** (or, sometimes, the **absolute deviation**).

$$\text{Mean deviation} = \frac{\Sigma \lvert d \rvert}{n}$$

For example 2–13, mean deviation $= \frac{22}{7} \approx 3.14$.

Notice that if $\Sigma \lvert d \rvert = 12$, mean deviation $= \frac{12}{7} \approx 1.7$, a smaller number than
3.14, and if $\Sigma \lvert d \rvert = 22$, $n = 50$, mean deviation $= \frac{22}{50} = 0.44$, a much smaller
number than 3.14. Thus, the mean deviation is a measure of dispersion. The larger
the mean deviation becomes, the more dispersed the data is; and correspondingly,
the smaller the mean deviation, the less dispersed the data. The mean deviation
is a better measure of dispersion than the range, because, unlike the range, all
the data is used in the calculation of the mean deviation. Despite this fact, the
mean deviation is a rarely used measure of dispersion. Why this is so is difficult
to justify to nonmathematicians. Although absolute value is a fairly simple concept to grasp, its use in formulas is difficult to deal with algebraically. Statisticians wanted to avoid it in formulas and therefore searched for another way to
measure dispersion.

The Variance

Mathematically, there is another way of removing negative signs, other than taking the absolute value. That is by squaring; for example, $3^2 = 9$, $(-5)^2 = 25$, $0^2 = 0$. The reason for this, if you recall your beginning algebra, is that when you multiply two numbers with like signs the answer is positive, and if they have unlike signs the answer is negative. Squaring a number is equivalent to multiplying the number by itself, and therefore, the "two" numbers that are being multiplied must have like signs. They are really the same number. The answer, therefore, must be positive or zero.

A column can be created that contains the squares of the deviations, d. This is referred to as the **squared deviations** and is headed by d^2.

EXAMPLE 2–14   For the data of example 2–12, find the squared deviations from the mean and the sum of these.

Solution

x	d	d^2
2	-4	16
3	-3	9
3	-3	9
5	-1	1
8	2	4
9	3	9
12	6	36
$\Sigma x = 42$		$\Sigma d^2 = 84$
$n = 7$		
$\bar{x} = 6$		

For this sample, the sum of the squared deviations is 84. At this point, you are confronted with the same issue as you were with the mean deviation. The 84 should be compared to the sample size. When this is done, you get what is known as the **variance.**

$$\text{Variance} = \frac{\Sigma d^2}{n}$$

For example 2–14, variance $= \frac{84}{7} = 12$. As with the mean deviation, for a more dispersed data set, the variance would be larger than 12 and, for a less dispersed data set, smaller than 12. For this reason, the variance is a valid measure of dispersion. However, it does have a weakness. Suppose the original data were measurements expressed in inches. Adding inches to inches gives inches; so, Σx must be in inches. The sample size n is a pure number. Dividing inches by a pure number gives inches; so, $\bar{x}$ is also in inches. Subtracting inches from inches gives inches; so, the *d* column must be in inches. Squaring inches, however, gives square inches. Adding square inches to square inches gives square inches, which means Σd^2 must be in square inches. Dividing square inches by a pure number gives square inches. Thus, the variance is expressed in square inches, *different* units from the original data. This will always be true. If the original data are expressed in certain units, the variance will be in the square of those units. The mean, $\bar{x}$, is always in the *same* units as the original data. It is desirable that a measure of dispersion has this property too. As you can see, the variance fails this test. The search continues!

The Standard Deviation

In mathematics, an inverse operation of a given operation is the operation that "undoes" the given one. For instance, the inverse operation of addition is subtraction, and the inverse operation of multiplication is division. If you add 3 to a certain number, you return to the original number by subtracting 3. The inverse operation of squaring is taking the **square root.** The symbol used for square root is $\sqrt{}$. For instance, $3^2 = 9$, whereas $\sqrt{9} = 3$. Also, if a quantity is expressed in square inches, by taking the square root, you arrive at a quantity expressed in inches. This is the key that leads to the most important measure of dispersion, the **standard deviation.**

$$\text{Standard deviation} = \sqrt{\text{Variance}}$$

For example 2–14, standard deviation $= \sqrt{12} \approx 3.46$. Notice that the size of the standard deviation is comparable to that of the mean deviation for the same data, which was 3.14.

Because of the importance of the standard deviation, symbolism has been devised for it.

> s This is used to represent the standard deviation of a sample.
>
> σ This is the lowercase Greek letter *sigma* and is used for the standard deviation of a population.

Since the standard deviation was calculated by taking the square root of the variance, s^2 can be used to represent the variance of a sample and σ^2 to represent the variance of a population. Now, to present the formulas.

$$s^2 = \frac{\Sigma d^2}{n} \tag{7}$$

$$s = \sqrt{\frac{\Sigma d^2}{n}} \tag{8}$$

These formulas are for a sample. The equivalent formulas for a population can be obtained simply by taking formulas (7) and (8) and replacing s by σ and n by N. Some statisticians give the formulas

$$s = \sqrt{\frac{\Sigma d^2}{n-1}}$$

and

$$\sigma = \sqrt{\frac{\Sigma d^2}{N}}.$$

The reasons why some statisticians prefer $n-1$ to n for the *sample* standard deviation will be presented later. This book will use n for the sample and N for the population standard deviations.

Another Formula

There is an alternative formula for the variance and standard deviation, which is often easier to use. It is

$$s = \sqrt{\frac{\Sigma x^2}{n} - \left(\frac{\Sigma x}{n}\right)^2}. \qquad (9)$$

This formula, when applied to the same data, will always give the same answer as formula (8), since it is mathematically equivalent to it. For those interested, the proof of this is given at the end of the chapter; but it is not vital that you follow it. You should understand, however, the meaning of Σx^2. It means: square each of the x's and then add them up. It is different from $(\Sigma x)^2$, which means add all the x's and then square the result.

EXAMPLE 2–15   For the data of example 2–14, use formula (9) to calculate the standard deviation.

Solution

$\Sigma x^2 = 336$, whereas
$(\Sigma x)^2 = (42)^2 = 1{,}764$.

x	x^2
2	4
3	9
3	9
5	25
8	64
9	81
12	144
$\Sigma x = 42$	$\Sigma x^2 = 336$
$n = 7$	
$\bar{x} = 6$	

$$s = \sqrt{\frac{336}{7} - \left(\frac{42}{7}\right)^2}$$

$$= \sqrt{48 - 36}$$

$$= \sqrt{12}$$

$$\approx 3.46$$

exactly the same answer as before.   ●

"Why burden us with another formula?" you might ask. Although formula (9) looks more complicated than formula (8), it is often easier to apply. In examples 2-14 and 2-15, both formulas looked equally easy to apply. However, this is because the data was contrived to make $\bar{x}$ come out as a nice whole number. This rarely happens with real data. Observe example 2-16.

EXAMPLE 2-16 For the data 1, 4, 6, 8, 9, find the standard deviation using both formulas (8) and (9).

Solution First, use formula (8).

x	d	d^2
1	−4.6	21.16
4	−1.6	2.56
6	0.4	0.16
8	2.4	5.76
9	3.4	11.56
$\Sigma x = 28$		$\Sigma d^2 = 41.20$
$n = 5$		
$\bar{x} = 5.6$		

$$s = \sqrt{\frac{41.20}{5}}$$
$$= \sqrt{8.24}$$
$$\approx 2.87$$

Now, use formula (9).

x	x^2
1	1
4	16
6	36
8	64
9	81
$\Sigma x = 28$	$\Sigma x^2 = 198$
$n = 5$	
$\bar{x} = 5.6$	

$$s = \sqrt{\frac{198}{5} - \left(\frac{28}{5}\right)^2}$$
$$= \sqrt{39.6 - 31.36}$$
$$= \sqrt{8.24}$$
$$\approx 2.87$$

Both formulas give the same answer, but which is easier? Another reason for introducing formula (9) is that it is most often applied to the calculation of the standard deviation of a frequency distribution, the subject of the next section.

Exercise Set 2-5

1. Find the mean deviation of the sample.

    8   12   3   8   17   6   8   11   14   13

2. Find the mean deviation of the sample.

    36   42   43   33   48   37   36   44   39   45   33   32

3. Find the variance of the sample.

    15   12   9   13   19   22   20   18   17

4. Find the variance of the sample in exercise 3 by first replacing the data value, 22, by 52.

 5. A sample of 20 women who had their first child at a city hospital were reported to have the following ages:

22   24   23   20   15   19   27   26   17   27   21   20   16   24
23   22   25   21   23   29.

Find the variance of these ages.

 6. A sample of members of a health club were monitored to see how many sit-ups they performed. Here are the results.

30   35   20   55   100   15   40   32   25   50   30   30   41   12
28   35

Find the variance of these numbers.

7. Find the standard deviation of the sample.

7   13   2   9   16   7   7   12   13   14

8. Find the standard deviation of the sample.

54   48   63   65   47   51   55   54   62   65   51   45

 9. The population of all graduate students in psychology at a small university scored the following scores on a statistics examination:

92   84   93   81   72   95   80   75   81   99   100   82   89   90
76   94   90   81   88.

Find the mean and standard deviation of these scores considering them to be a population.

 10. In 1985, the top 10 leading money winners in the Professional Golf Association won, to the nearest $10,000:

540,000   450,000   380,000   380,000   380,000   370,000   370,000
360,000   340,000   340,000.

Find the mean and standard deviation of these amounts.

11. Take the set of numbers in exercise 7 and add 12 to each number. Find the mean and standard deviation of this new set of numbers. Compare your answers to the mean and standard deviation of the original set. What do you notice?

12. Again, take the set of numbers in exercise 7, but this time multiply each number by 3. Find the mean and standard deviation of this new set and compare your answers to the mean and standard deviation of the original set. What do you notice? How could you have used this idea to simplify your work in answering exercise 10?

Section 2–6 The Standard Deviation of a Frequency Distribution

The calculation of s from a frequency distribution is similar to the calculation of the mean. The same assumption is made: all values within each interval are concentrated at the midpoint of the interval. Formula (9) can be adapted to include frequencies giving

$$s = \sqrt{\frac{\Sigma fm^2}{\Sigma f} - \left(\frac{\Sigma fm}{\Sigma f}\right)^2}. \tag{10}$$

EXAMPLE 2–17 Use formula (10) to calculate the standard deviation of the frequency distribution of weights.

Solution

Table 2–6

Class	m	f	fm	fm^2
85–94	89.5	3	268.5	24,030.75
95–104	99.5	3	298.5	29,700.75
105–114	109.5	6	657.0	71,941.50
115–124	119.5	11	1,314.5	157,082.75
125–134	129.5	9	1,165.5	150,932.25
135–144	139.5	4	558.0	77,841.00
145–154	149.5	0	0	0
155–164	159.5	1	159.5	25,440.25
165–174	169.5	1	169.5	28,730.25
		$\Sigma f = 38$	$\Sigma fm = 4,591.0$	$\Sigma fm^2 = 565,699.50$

$$\bar{x} = \frac{\Sigma fm}{\Sigma f} = \frac{4,591.0}{38} = 120.8$$

$$s = \sqrt{\frac{\Sigma fm^2}{\Sigma f} - \left(\frac{\Sigma fm}{\Sigma f}\right)^2} = \sqrt{\frac{565,699.50}{38} - \left(\frac{4,591.0}{38}\right)^2}$$

$$= \sqrt{14,886.829 - 14,596.455}$$

$$= \sqrt{290.374}$$

$$= 17.0 \qquad \bullet$$

Coding can also be used, with s' (s prime) used for the coded standard deviation. The formula is

$$s' = \sqrt{\frac{\Sigma f(x')^2}{\Sigma f} - \left(\frac{\Sigma fx'}{\Sigma f}\right)^2}. \tag{11}$$

$\Sigma f(x')^2$ means square each x', multiply by the corresponding f, and add up the results. The f's are *not* squared.

Recall that to get from m to x', C is subtracted and the result divided by w. How is s obtained from s'? Subtracting C from *all* the midpoints has the effect of moving *all* the data down the scale. Would this alter the spread of the data? No. However, dividing by w will move the data closer together. Therefore,

$$s = ws'.$$

Consider the data 12, 14, 16, 18, 20. Subtract 10 to get 2, 4, 6, 8, 10. Is it not true that these two sets have the same spread? Now, divide by 2 to get 1, 2, 3, 4, 5. These new data points are each one unit apart instead of two, meaning they are half as spread out as the original data.

EXAMPLE 2–18   Use coding to calculate the standard deviation of the frequency distribution of weights.

Solution

Table 2–7

Class	x'	f	fx'	$f(x')^2$
85–94	−3	3	−9	27
95–104	−2	3	−6	12
105–114	−1	6	−6	6
115–124	0	11	0	0
125–134	1	9	9	9
135–144	2	4	8	16
145–154	3	0	0	0
155–164	4	1	4	16
165–174	5	1	5	25
		$\Sigma f = 38$	$\Sigma fx' = 5$	$\Sigma f(x')^2 = 111$

$C = 119.5$, $w = 10$

$$\bar{x}' = \frac{\Sigma fx'}{\Sigma f} = \frac{5}{38} = 0.132$$

$$\bar{x} = C + w\bar{x}' = 119.5 + 10(0.132) \approx 120.8$$

$$s' = \sqrt{\frac{\Sigma f(x')^2}{\Sigma f} - \left(\frac{\Sigma fx'}{\Sigma f}\right)^2} = \sqrt{\frac{111}{38} - \left(\frac{5}{38}\right)^2}$$

$$= \sqrt{2.92 - 0.017}$$

$$= \sqrt{2.903}$$

$$\approx 1.70$$

$$s = ws' = 17.0$$

The calculations may look a bit forbidding, but remember all but the last column of table 2–7 and the last few lines of the calculations have been done before, when the mean was calculated.

Out of interest, *s* was calculated, using a calculator, directly from the original 38 weights. The result was 17.045585. Including more decimal places in the above calculation from the frequency distribution, the result is 17.040363. Extraordinarily good agreement! Much less work too! Even with a calculator, punching 38 numbers into it is no fun. Imagine if the sample size were larger!

Exercise Set 2–6

1. Find the standard deviation of the frequency distribution.

Interval	f	m	Fm	Fm^2
30–36	2			
37–43	5			
44–50	9			
51–57	6			
58–64	3			

2. Find the standard deviation of the frequency distribution.

Interval	f
1–5	1
6–10	2
11–15	3
16–20	5
21–25	6
26–30	13

3. This is a random sample of adult women's heights.

Height	f
60–61	2
62–63	5
64–65	10
66–67	8
68–69	6
70–71	2

Find the mean and standard deviation of this sample.

4. This is a random sample of adult women's weights.

Weight	f
100–119	6
120–139	12
140–159	6
160–179	4
180–199	3
200–219	2

Find the mean and standard deviation of this sample.

5. Find the mean and standard deviation of the area in square miles of the 18 districts in Uganda.

Class	f
500–2,499	5
2,500–4,499	3
4,500–6,499	5
6,500–8,499	2
8,500–10,499	1
10,500–12,499	2

Note: Since this comprises all districts in Uganda, it is considered to be a population.

 6. Another population that can be considered is the populations (as of 1978) of all 52 independent African states. The smallest was Seychelles with 63,000, while the largest was Nigeria with 80,862,000. Here are the complete results in millions.

Population (in millions)	f
0–4	26
5–9	13
10–14	3
15–19	5
20–24	1
25–29	1
30–34	1
35–39	1
.	.
.	.
.	.
80–84	1

Find the mean and standard deviation of this population frequency distribution.

In exercises 7–10, data in various random samples is given. In each case, find the mean and standard deviation in two ways: first, directly with the use of a calculator and, second, by grouping the data into a frequency distribution and using formula (10) or the coding method.

 7. A random sample of houses in a housing development were monitored for energy consumption over a one-week period. Here are the results in kilowatt hours.

230	242	238	237	236	245	215	214	221	235	218	202
196	238	250	224	242	207	252	205	238	185	227	224
226	218	211	232	258	235	213	207	224	219	218	212
248	230	240	264	192	256	239	244	210			

8. An aptitude test was given to a random sample of applicants for a certain job. The maximum possible score was 250. Here are the results.

184	161	172	178	218	195	209	165	204	213	241	189
180	233	204	211	181	204	221	229	162	183	177	174
216	198	208	163	206	210	240	190	176	237	202	210
185	205	222	227								

9. Fifty guinea pigs were fed a special diet and their weight gains were recorded after four weeks. Their weight gains in grams are given here.

16	10	7	12	22	9	16	11	8	14	4	9	8	12	7
18	12	10	11	5	11	7	17	5	12	10	21	8	16	7
17	8	13	7	15	21	18	2	6	10	19	14	10	16	19
8	11	14	17	13										

10. Forty patients were admitted to a city hospital and tested for blood sugar. Here are the results.

86	77	82	76	80	72	68	86	78	79	84	66	78	75
67	84	71	67	81	65	84	73	80	76	74	85	78	81
77	81	70	65	88	77	80	82	68	79	74	66		

In exercises 11 and 12, find the *variance* of the frequency distributions.

11.

Interval	f
30–32	12
33–35	18
36–38	31
39–41	54
42–44	30
45–47	16
48–50	5

12.

Interval	f
18	52
19	108
20	146
21	92
22	71
23	42
24	12
25	3

Section 2–7   An Empirical Rule and Chebyshev's Theorem

What Does All This Mean?

It has now been discovered that the standard deviation of the sample of 38 weights is 17.0 lb. What does that mean to you? Probably not very much at this point. People generally have a more intuitive notion of an average than they do of a measure of spread. One piece of information that you can derive from it is by comparing it to the standard deviation of another sample of weights. If the other sample has a standard deviation that is larger, it is more dispersed than this sample; and if its standard deviation is smaller, less dispersed. However, there is an empirical rule that you can use to give you still more information. In samples and populations that pile up in the middle, such as the bell-shaped distribution, about 68% of the sample values will be within one standard deviation of the mean, and about 95% of them will be within two standard deviations of the mean. (Where the figures 68% and 95% come from will be explained in chapter 3). See figure 2–5. What is meant by "within one standard deviation" and "within two standard deviations"? Remember that both the mean and the standard deviation are expressed in the same units as the original data. If the standard deviation is subtracted from the mean, a value, call it A, is obtained; and if the standard deviation is added to the mean, another value, call it B, is obtained. A and B will also be in the same units as the original data. To say that "about 68% of the sample values will be within one standard deviation of the mean" is to say that about 68% of the raw data will be between A and B.

Figure 2–5

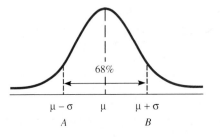

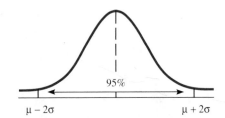

EXAMPLE 2-19   Find what percentage of the weights are within one standard deviation of the mean and what percentage are within two standard deviations of the mean.

Solution   Since $\bar{x} = 120.8$ and $s = 17.0$,

$$A = \bar{x} - s = 103.8 \text{ and } B = \bar{x} + s = 137.8.$$

Now, how many of the original data are between these two values? Below, the original data has been reproduced with those values falling between 103.8 and 137.8 circled.

(130)	(108)	(135)	(120)	97	(110)
(130)	(112)	(123)	(117)	170	(124)
(120)	(133)	87	(130)	160	(128)
(110)	(135)	(115)	(127)	102	(130)
89	(135)	87	(135)	(115)	(110)
(105)	(130)	(115)	100	(125)	(120)
(120)	(120)				

There are 30 (out of 38) values between $\bar{x} - s$ and $\bar{x} + s$ and 30 is 79% of 38, quite a bit more than the empirical rule predicted. The reason for this is that the empirical rule works best for distributions that pile up in the center (especially those close to the bell shape), and this one, if you remember, was slightly right-skewed. The second part of the empirical rule works quite well here. Observe:

$$\bar{x} - 2s = 86.8 \text{ and } \bar{x} + 2s = 154.8.$$

All but the two weights 160 and 170 lie between these two values. Since all but two out of 38 is 36,

$$\frac{36}{38} = 0.947 = 94.7\% \approx 95\%.$$

            •

The **empirical rule** states:

> For distributions close to the bell shape: about 68% of the data lie between $\bar{x} - s$ and $\bar{x} + s$, and about 95% of the data lie between $\bar{x} - 2s$ and $\bar{x} + 2s$.

Chebyshev's Theorem

There is another result that, unlike the empirical rule, is true for *all* distributional shapes. It is stated as follows:

Chebyshev's Theorem

The fraction of any data set that is within C standard deviations of the mean is at least $1 - \dfrac{1}{C^2}$. Here, C is any number larger than 1.

EXAMPLE 2-20   Show that the sample of weights is consistent with the conclusion of Chebyshev's theorem for $C = 1.5$ and $C = 2$.

Solution

$$\text{For } C = 1.5 = \frac{3}{2}, \ 1 - \frac{1}{C^2} = 1 - \frac{1}{\frac{9}{4}} = 1 - \frac{4}{9} = \frac{5}{9} \approx 0.56 = 56\%.$$

Thus, at least $\frac{5}{9}$ or 56% of the weights should be between $\overline{x} - 1.5s$ and $\overline{x} + 1.5s$. Since $\overline{x} = 120.8$ and $s = 17.0$,

$$\overline{x} - 1.5s = 120.8 - (1.5)(17.0) = 120.8 - 25.5 = 95.3$$

and

$$\overline{x} + 1.5s = 120.8 + (1.5)(17.0) = 120.8 + 25.5 = 146.3.$$

All but 5 of the weights are between these two figures, and 33 out of 38 is about 87%. Since 87% is more than 56%, it is *at least* 56% and the result is thus consistent with the conclusion of Chebyshev's theorem.

$$\text{For } C = 2, \ 1 - \frac{1}{C^2} = 1 - \frac{1}{4} = \frac{3}{4} = 0.75 = 75\%.$$

You saw above that 95% of the weights are within two standard deviations of the mean, and since 95% is more than 75%, this result is again consistent with the conclusion of Chebyshev's theorem.   ●

Both the empirical rule and Chebyshev's theorem are intended to give you only rough guidelines. Their purpose is simply to give you a better feel for the data and what the standard deviation is telling you about it.

Suppose you visit another travel agent and, disgusted after your cruise, wish to take your next vacation on land. The agent tells you of a nice spot with an average temperature of 74° F. This sounds pleasant, but you have become suspicious of averages and ask the agent what the standard deviation is. Assuming that the agent knows, what will the answer tell you? Suppose the answer is that $s = 3°$ F. You can conclude that about 68% of the time, the temperature will be between 71° and 77° and about 95% of the time it will be between 68° and 80°. Sounds good! But suppose, instead, the answer came back, $s = 20°$ F. Now, all you can say is that about 68% of the time the temperature will be between 54° and 94° and about 95% of the time it will be between 34° and 114°. Could be a place with very hot days and very cold nights, with "average" temperature at 74° F. Sounds like a place to avoid!

Had you known the standard deviation of the ages of the passengers, your disastrous cruise might have been avoided. To show you this, consider that 10 numbers in the twenties have a standard deviation of about 3, whereas five numbers in the forties together with five numbers less than 10 have a standard deviation of about 18. The means of the two sets, however, may be the same at around 24.

It is to be hoped that these last examples have given you a better feeling for standard deviation. If the purpose of standard deviation is still vague, however, do not worry. Standard deviation is used a great deal in the concepts to come. As you use it, its meaning should become increasingly clear.

Exercise Set 2-7

1. In a random sample, it is found that $\bar{x} = 12.6$, $s = 2.1$, $n = 50$. How many in the sample would you expect to lie between 10.5 and 14.7, according to the empirical rule?

2. In a random sample, it is found that $\bar{x} = 34.2$, $s = 6.6$, $n = 36$. How many in the sample would you expect to be between 24.3 and 44.1, according to Chebyshev's theorem?

3. In a random sample, it is found that $\bar{x} = 109.1$, $s = 10.4$, $n = 75$. How many in the sample would you expect to lie between 88.3 and 129.9, according to Chevyshev's theorem?

4. Answer exercise 3 according to the empirical rule.

In exercises 5–8, find the mean $\bar{x}$ and the standard deviation s, then test the empirical rule by finding what percentages lie in the intervals $\bar{x} - s$ to $\bar{x} + s$ and $\bar{x} - 2s$ to $\bar{x} + 2s$. Also, test Chebyshev's theorem for the second of these intervals.

5. The pulse rates of a sample of adult males:
　　68　60　79　81　77　82　58　87　79　96　77　90　75　74
　　63　80　65　74　79　92.

6. The temperatures of the same sample of adult males:
　　98.6　98.9　99.0　98.5　98.4　99.2　99.6　98.2　98.0　98.6　98.5
　　98.6　97.9　98.8　98.6　98.6　98.9　99.1　98.6　98.7.

7. The number of errors made by a sample of rats as they negotiate their way through a maze:
　　13　15　12　11　3　12　16　6　20　7　19　15　18　11　4
　　9　11　10　22　20　12　1　7　8　13　8　12　17　20　25.

First group these into a frequency distribution with 0–5 as the first interval and then make the calculations.

8. A random sample of students' GPA's were as follows:
　　3.8　1.6　3.5　2.2　2.5　2.0　3.1　2.8　3.2　1.9　3.8　1.8　3.1
　　2.1　2.2　1.8　2.1　2.7　2.0　1.8.

First group these into a frequency distribution with 1.4–1.7 as the first interval and then make the calculations.

9. Look at the last three lines of the Chess article in figure 2–6. Verify the figures they quote. (*Note:* The computer took 0 seconds for the first move and the times for the other moves are given in parentheses after its moves.)

10. Find what percentages of Chess 4.6's moves were within one and two standard deviations of the mean.

Figure 2–6

Making Seconds Count
Hash/transposition tables give a computer even more time to think about its move

by Evan Katz

In evaluating a terminal node (the final position at the end of a tree search), a computer assumes that the position is static, or *quiescent*. Problems arise however, if there are meaningful captures, checks or pawn promotions that require further examination. We would not want our program to make what it believes is a good move, based on a seven-ply search, if it will be mated or lose its Queen in two additional moves.

A related problem is the *horizon effect*. If one side must part with some material or incur a positional loss, the computer will interpose one or more poor moves to push the greater, but inevitable, loss beyond the limits of its search (over the "horizon").

For example, let's suppose that a program, searching to eight-ply, sees that it must lose a piece in all variations on the eighth ply. However, by adding into its variations an additional pawn move—even though the pawn will be lost—the computer can push the loss of the piece beyond the "horizon" of its eight-ply search. It doesn't mind losing the pawn, because it thinks that other variations lose a piece instead.

Thus, the machine suffers additional tactical and positional disadvantages in such an instance or anticipates faulty moves on the part of its opponent. Not until the loss of the piece is realized will the machine get back on the right track.

To compensate for the computer's lack of foresight, most programmers have added a *capture/quiescent search* to their programs. Extensions are made to the tree for the examination of captures and checking sequences, greatly reducing tactical mishaps. Usually, some limit is placed on the examination of check variations that don't produce mate or a win of material—perhaps four plies—so that the program won't extend "forever."

The capture/quiescent search is naturally very crucial in speed chess. With only five to seven seconds per move, the computer must extend beyond its limited five-ply exhaustive search in order to enhance its tactical prowess.

Below is an example of what happens when even the best of humans attempts to out-tactic Chess 4.6 in speed chess. England's Michael Stean became the first grandmaster to lose to a computer. Bear in mind that this contest is several years old and that chess programs now search much more deeply and have more intelligent evaluative functions. The time, in seconds, that 4.6 used in considering its choice is given after each move.

continued

Evan Katz is a leading writer and speaker on computer chess programming theory and commercially-marketed chess computers. He has served as a Contributing Editor with *Personal Computing* and *Chess Life* magazines, and currently sits on the United States Chess Federation's Computer Chess Committee. Mr. Katz may be contacted at: 161 West 61st Street, New York, NY 10023.

Figure 2–6   *continued*

September 18, 1977
Chess 4.6 (Cyber 176)   **Stean/**
                                  2485

1. e4		**b6**
2. d4	**(2)**	**Bb7**
3. Nc3	**(10)**	**c5**
4. dxc5?!	**(3)**	

Stean had noticed that 4.6 had a tendency to exchange its central pawns for side pawns. This positional misjudgment helps Black control the center and the b-file.

4. . . .		**bxc5**
5. Be3	**(6)**	**d6**
6. Bb5+	**(6)**	**Nd7**
7. Nf3	**(3)**	**e6**
8. 0–0	**(2)**	**a6**
9. Bxd7+	**(6)**	**Qxd7**
10. Qd3	**(8)**	**Ne7**
11. Rad1	**(8)**	**Rd8**
12. Qc4	**(14)**	**Ng6**
13. Rfe1	**(3)**	

Not 13. Bxc5, which drops a piece to **13. . . . Qc6.**

13. . . .		**Be7**
14. Qb3	**(3)**	

An unnecessary retreat which will permit Black to gain a tempo and chase the White Queen to a passive position.

14. . . .		**Qc6**
15. Kh1?!	**(9)**	

Chess 4.6 moves its King to the corner, usually a safer place. But in this case, the Black Queen and Bishop are dangerously aligned on the h1-a8 diagonal. Notice that White's pieces were placed optimally as far as the program

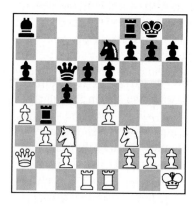

was concerned; Chess 4.6 had probably hit "the top of its hill" and was unsure of how to improve its position.

15. . . .		**0–0**
16. Bg5	**(6)**	**Ba8**
17. Bxe7	**(5)**	**Nxe7**
18. a4	**(8)**	**Rb8**
19. Qa2	**(3)**	**Rb4**
20. b3	**(5)**	

Solidifying the pawn structure but further shutting out the Queen. Black not only controls the center but puts great pressure along the critical diagonal and has an active Rook at b4. Stean has outplayed 4.6, but now enters into uncertain tactics, always a risky decision when competing against a computer.

20. . . .		**f5?!**

A better way of opening the diagonal, according to David Levy, is 20. . . . d5; for example, 21. Ne5 Qb7 22. Nd3 Rd4 23. Nxc5 Rxd1 24. Rxd1 Qc6, followed by 25. . . . dxe4.

21. Ng5	**(8)**	**fxe4**
22. Ncxe4!	**(14)**	**Rxf2**
23. Rxd6	**(6)**	

Stean: "Bloody iron monster."

23. . . .		**Qxd6?**

Stean misses the clever resources Chess 4.6 finds (moves 25–29). Better would have been 23. . . . Qb7 24. Rd8+ Rf8 25. Rxf8+ Kxf8 26. Nxe6+ Kg8 27. N/6xc5, although White still obtains a winning edge.

24. Nxd6	**(4)**	**Rxg2**
25. Nge4	**(17)**	**Rg4**
26. c4	**(3)**	**Nf5**
27. h3	**(3)**	

Stean: "This computer is a genius."

27. . . .		**Ng3+**
28. Kh2	**(4)**	**Rxe4**
29. Qf2!	**(20)**	

Chess 4.6 mercilessly forces its opponent into a Queen vs. Rook-and-pawn ending. Stean had counted on 29. Nxe4, missing White's mate threat.

29. . . .		**h6**
30. Nxe4	**(5)**	**Nxe4**
31. Qf3	**(10)**	**Rb8**
32. Rxe4	**(6)**	**Rf8**
33. Qg4	**(6)**	**Bxe4**
34. Qxe6+	**(10)**	**Kh8**
35. Qxe4	**(4)**	**Rf6**
36. Qe5	**(5)**	**Rb6**
37. Qxc5	**(9)**	**Rxb3**
38. Qc8+	**(7)**	**Kh7**
39. Qxa6	**(3)**	**Resigns**

Chess 4.6 averaged 6.5 seconds per move, with a standard deviation of 4.2 seconds.

Where Do You Stand?

Suppose you take a test in one of your college classes and score a 78, while a friend of yours takes a different test in another class and scores 83. Who has done better? On the surface, it appears that your friend has. However, suppose your score was top of your class, while your friend was fifth. Would your performance not be judged more impressive? Or perhaps the mean score in your class was 75, while the mean score in your friend's class was 85. In that case, you have performed above average, while your friend is below. Or, again, perhaps 90% of your class scored less than you, while only 80% of your friend's class scored less than 83. Does this not mean that you have a higher standing in your class than your friend has in that class?

Looking at the scores in these different ways should convince you that what appears on the surface may be hiding "a more meaningful comparison." The illustrations above begin to consider the individual scores *relative* to the sets of scores of which they are a part. Unless you and your friend took the same test under the same conditions, a straight comparison is not very meaningful. But converting the scores somehow to show their relative standings within their respective groups and then making the comparison would be more meaningful, would it not? The next two sections deal with methods of converting individual pieces of data to show their location within the data as a whole. Such conversions are grouped under the general heading of **measures of location.**

Section 2–8   Ranks and Percentile Ranks

Ranks

Ranking scores is the simplest way of converting data to show the relative positions of pieces of data. The highest value in the data set is given a **rank** of 1, the second highest, a rank of 2, and so on.

EXAMPLE 2–21   Rank the data 9, 4, 7, 2, 3.

Solution

x	Rank
9	1
4	3
7	2
2	5
3	4

●

As you can see, this is not difficult; but what happens if there are two or more values equal in the data set? Suppose, for example, your 78 on the test was not the only 78. There was one other 78 and four students who scored higher than 78 on that test. It would not be fair for the other student to be ranked 5 and you 6, or vice versa. This is resolved by taking the ranks that the tied values would have received had they all been different, calculating their mean, and giving each of the tied values this mean rank. Thus, the two 78s would each be ranked 5.5. Now, suppose the next lowest score was a 77 by itself. It would be ranked 7, *not* 6, since the two 78s would have been 5 and 6, had they been different. If there were three 77s, how would they be ranked? They would have been 7, 8, 9, if they were different, and $\dfrac{7 + 8 + 9}{3} = 8$. Thus, each of the 77s would be given a rank of 8, and the next lower score a rank of 10.

EXAMPLE 2–22   Rank the data

29, 32, 25, 33, 30, 29, 23, 34, 24, 29, 32, 25.

Solution   The following is the solution in which the data has been sorted to help you see how it's done.

x	**Rank**
34	1
33	2
32	3.5
32	3.5
30	5
29	7
29	7
29	7
25	9.5
25	9.5
24	11
23	12

A quick check to see if you have performed the ranking correctly is that the lowest value in the data set should have a rank equal to the sample size. For the above example, $n = 12$, and the lowest value has a rank of 12. The one exception to this "rule" is when there is a tie at the bottom. For instance, suppose there had been two 23s in example 2–22. Then, $n = 13$, but each of the 23s would have been given a rank of 12.5.

This is the usual method of ranking, with the highest being given a 1 and so on down. However, some statistical methods, as you will see later, require ranking the other way. In that case, the lowest value is given a rank of 1, the next lowest a 2, with the largest value receiving a rank equal to the sample size. It is no more difficult ranking this way, and ties are dealt with in the same way.

The great advantage to ranking is that it is a very simple method of converting scores. However, there is one big disadvantage. Suppose, in example 2–22, the highest score had been 94, instead of 34. It would still have a rank of 1, while 33 would still have a rank of 2. There is absolutely no indication of whether two values with neighboring ranks are close together or far apart. Other methods of converting scores, without this disadvantage, are needed.

Percentile Ranks

Figure 2–7 represents a data set by a rectangle. Here, H means the highest value and L the lowest. This data set can be divided into two halves, as in figure 2–8.

This idea has already been discussed, and, as indicated in figure 2–8, the value that achieves this is called the median, $\tilde{x}$.

Figure 2–7

Figure 2–8

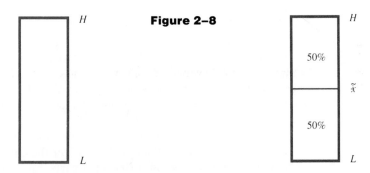

Now divide the data set into four equal parts, as in figure 2–9. Three values are required to do this, and these three values are labeled, Q_1, Q_2, Q_3. They are called the **quartiles** because they divide the data into four quarters.

One of these quartiles is the same as the median. Which is it?

Q_1, or the first quartile, comes ahead of the lowest 25% of the data. Q_2, or the second quartile, comes ahead of the lowest 50% of the data. Q_3, or the third quartile, comes ahead of the lowest 75% of the data, or just below the top 25% of the data.

Figure 2–9

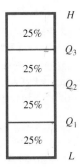

Figure 2–10

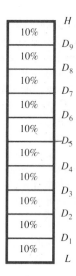

As with the quartiles, one of these deciles is the same as the median. Which is it?

Now, divide the data set into ten equal parts, as in figure 2–10. Nine values are required to do this and are labeled, $D_1, D_2, \ldots, D_9$. These are called the **deciles** because they divide the data into ten tenths. (The latin prefix deci- means ten.) D_1, or the first decile, is larger than the lowest 10% of the data. D_2, or the second decile, is larger than the lowest 20% of the data. Continuing in this manner, you would finally reach D_9, or the ninth decile, which is larger than the lowest 90% of the data, or just less than the top 10% of the data.

One of the percentiles is also the same as the median. Again, which is it?

Now, divide the data set into 100 equal parts. Imagine a similar diagram to those above. Just as it requires one value to divide the data into 2 equal parts, three values to divide it into 4 equal parts, nine values to divide it into 10 equal parts, it would require ninety-nine values to divide the data set into 100 equal parts. These values are labeled $P_1, P_2, P_3, \ldots, P_{99}$ and are called **percentiles** or **percentile ranks.** P_1, or the first percentile, is larger than the lowest 1% of the data. P_{53}, or the 53rd percentile, is larger than the lowest 53% of the data. And so on. Notice that each quartile and each decile could be expressed as a percentile. For example,

$$Q_1 = P_{25}, D_4 = P_{40}, Q_3 = P_{75}, D_9 = P_{90}.$$

Percentile ranks are used extensively to indicate relative positions of scores. Standardized test score grades are often given in percentiles. If you are given a percentile score of 88 on a particular examination, this means you did better than 88% of the students who took the examination at the same time. You did well! If, however, you received a percentile score of 15, this means you did better than only 15% of the students—not so hot!

Percentiles are often calculated by counting how many in a sample or population fall below the value of interest and converting this to a percentage of the size of the sample or population. This might often turn out to be a decimal. This decimal is generally rounded downward, like age. Just as you would not say you are 21 until your 21st birthday (even on the day before, you are still 20!), so a value is not at the 88th percentile until 88% or more of the data set falls below it. If 87.3%, 87.5%, 87.9%, or 87.99% of a data set falls below a certain value, then that value is placed at the 87th percentile.

The simplest and quickest way of approximating percentiles is from the ogive with $F\%$, or relative cumulative frequency, on the vertical axis.

Recall the sample of weights from the first chapter. Suppose you wanted the answers to the questions in examples 2–23 and 2–24.

EXAMPLE 2–23   Which weight is at the 90th percentile?

EXAMPLE 2–24   What is the percentile rank of a weight of 105 lb?

Solutions   The ogive is shown in figure 2–11.

Figure 2–11

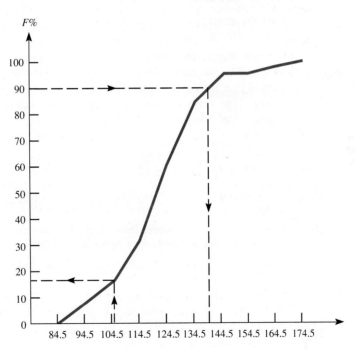

Solution 2–23   Start on the vertical axis at 90%, move horizontally across to the polygon, and then move vertically down to the axis. This reaches the axis at slightly more than halfway from 134.5 to 144.5. Thus, the weight at the 90th percentile is 140 lb. Put another way, $P_{90} = 140$.

Solution 2–24   Start on the horizontal axis at 105 lb, move vertically up to the polygon, and then move horizontally across to the axis. This reaches the axis at slightly more than 16. Rounding down, you would conclude that 105 lb has a percentile rank of 16. Put another way, $105 = P_{16}$.   ●

Can you see that, using this graph, the weight at any percentile rank, or the percentile rank of any weight, can be approximated quickly and easily? To see why it works, consider example 2–23.

By the time 140 lb has been reached, 90% of the data has been accumulated—this is the interpretation of $F\%$. This can be turned around to claim that the weight of 140 lb is greater than 90% of the sample. But this is just the interpretation of the 90th percentile rank.

Exercise Set 2–8

1. Rank the sample from highest to lowest.
   18   81   66   53   5   93   27   64   36

2. Rank the sample from highest to lowest.
   45   47   42   46   44   42   40   41   48   43

3. Rank the sample from lowest to highest.
   155   157   165   163   155   150   154   164   165   168   167   166
   159   155   152   151

4. Rank the sample from lowest to highest.
   0.12   0.18   0.17   0.10   0.14   0.15   0.16   0.15   0.10   0.11   0.19
   0.17   0.12   0.17   0.15   0.14   0.13   0.20   0.17   0.12

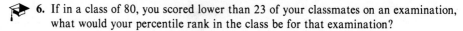

5. If in a class of 250, you scored higher than 180 of your classmates on an examination, what would your percentile rank in the class be for that examination?

6. If in a class of 80, you scored lower than 23 of your classmates on an examination, what would your percentile rank in the class be for that examination?

Exercises 7–10 refer to the ogive in figure 2–12.

Figure 2–12

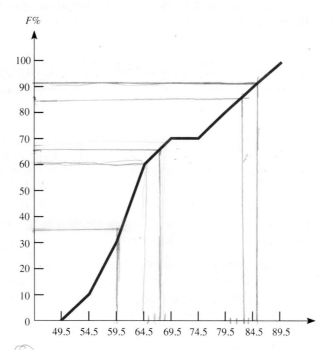

7. Find the percentile rank of 67.

8. Find the percentile rank of 85.

9. Find the value of P_{35}.

10. Find the value of the sixth decile.

{ ASK ABOUT these

Section 2–9 Standard Scores

Another method of converting scores to show their relative positions within the data set is by the use of **standard scores.** The formula that accomplishes this is

$$z = \frac{x - \mu}{\sigma}.$$

(12)

This is the formula that is used to convert x, a raw score, into z, a standard score, also called **z-score,** for obvious reasons. The numerator of the fraction in this equation, $x - \mu$, tells you how far, in raw units, x is above or below the mean. Dividing this by σ tells you how many standard deviations (or what fraction of a standard deviation) x is above or below the mean. Standardizing a score in this way takes into consideration not only the center of the population but also its dispersion. It is, therefore, a very precise method of measuring the location of a raw score within a population (but not necessarily *preferable* to a rank or percentile rank).

If only *sample* data is available, a formula very similar to formula (12) is used, but with the parameters replaced by the corresponding sample statistics. Thus,

$$z = \frac{x - \bar{x}}{s}. \tag{13}$$

To illustrate how a comparison with z-scores can be useful, consider example 2–25.

EXAMPLE 2–25   A man weighing 195 lb is to be compared to a chimpanzee weighing 115 lb. Who is heavier?

Solution   Obviously, in absolute terms, the man is; but, who is heavier in relation to their respective species? Suppose the man comes from a population with $\mu = 170$, $\sigma = 25$, and the chimpanzee comes from a population with $\mu = 100$, $\sigma = 7$. Then, for the man, $z = \dfrac{195 - 170}{25} = 1$, while for the chimp, $z = \dfrac{115 - 100}{7} = 2.14$. The man is only one standard deviation above his species' mean, while the chimp is over two standard deviations above its species' mean. Relative to their species, the chimp is heavier.

Notice that in absolute terms the man is heavier than the chimp. Even relative to their respective means, it might seem that the man is heavier, since he is 25 lb above his species' mean, while the chimp is only 15 lb above its. However, not only is the population of men generally heavier than the population of chimps, male human weights are also more dispersed (i.e., they have a larger standard deviation). The z-scores bring all this out.   ●

EXAMPLE 2–26   In the introduction to the ideas of measures of location, it was hypothesized that you scored 78 in a class, while a friend scored 83 in another class. Suppose your score of 78 came from a sample with $\bar{x} = 75$, $s = 10$ and your friend's score of 83 from a sample with $\bar{x} = 80$, $s = 9$. Who did better?

Solution   You both scored above the mean, and by the same amount, 3 points. However, for you, $z = \dfrac{78 - 75}{10} = 0.30$, and for your friend, $z = \dfrac{83 - 80}{9} = 0.33$. So, your friend would have performed better, but only slightly.   ●

EXAMPLE 2-27   Look back at example 2-22. Find the z-scores of the values that were ranked 1 and 2.

Solution   Check for yourself that $\bar{x} = 28.75$, $s = 3.56$. 34 and 33 were ranked 1 and 2.

$$\text{For } 34,\ z = \frac{34 - 28.75}{3.56} = 1.47$$

$$\text{For } 33,\ z = \frac{33 - 28.75}{3.56} = 1.19$$

EXAMPLE 2-28   Later, it was postulated that if in example 2-22, 34 were replaced by 94, 94 would be ranked 1 and 33 still ranked 2. What would their z-scores be now?

Solution   Now, if 34 is replaced by 94, $\bar{x} = 33.75$, $s = 18.44$. Thus,

$$\text{for } 94,\ z = \frac{94 - 33.75}{18.44} = 3.27$$

$$\text{for } 33,\ z = \frac{33 - 33.75}{18.44} = -0.04.$$

Notice that the ranks or percentiles did not reflect the change from 34 to 94 but that the z-scores are radically affected by it.

Exercise Set 2-9

1. In a population that has $\mu = 47$ and $\sigma = 12$, find the z-score for a data value of 55.

2. In a population that has $\mu = 19.6$ and $\sigma = 1.8$, find the z-score for a data value of 16.

3. In a sample with mean 33.2 and standard deviation 5.1, find the z-score for 25.

4. In a sample with mean 192 and standard deviation 41, find the z-score corresponding to 70.

5. Bob earns $18,000 in a country with mean income $15,000 and standard deviation $6,000, while Carl earns $16,000 in a country with a mean income $10,000 and standard deviation $4,000. Convert Bob's and Carl's incomes to z-scores. Looking at these conversions alone, who has the higher income relative to his countrymen?

6. Ann smokes 25 cigarettes per day, while the smokers of her brand smoke a mean of 30 cigarettes per day with a standard deviation of 5 cigarettes per day. Diane smokes 20 cigarettes per day, while smokers of her brand smoke a mean of 35 cigarettes per day with a standard deviation of 8.2 cigarettes. Convert Ann's and Diane's smoking habits to z-scores. Looking at these conversions alone, who is the heavier smoker relative to other smokers of her brand?

7. A random sample of novels had the following numbers of pages: 248, 195, 226, 317, 154, 298, 237, 186. Use a calculator to find the mean and standard deviation of this sample and use these to find the z-score of the novel with 195 pages.

 8. A random sample of repeats of the same chemical reaction take these times in minutes and seconds.

    2:12   4:07   3:21   4:06   4:37   3:35   3:57   4:01   3:30   4:18   2:24
    2:47   3:09   2:48   2:17   3:18

Find the z-score of the reaction that took 3 min 35 sec.

In the formula, $z = \dfrac{x - \bar{x}}{s}$, there are four quantities. If any three are known, the fourth can be calculated. Use this idea to do exercises 9 and 10.

9. In a sample, the z-score of 38 is 1.26. If the standard deviation of the sample is 3, find the mean.

10. In a sample, the z-score of 83 is 2.61. If the mean of the sample is 60, find the standard deviation.

....

Section 2–10   Summary of Terms

absolute deviation   69
absolute value   68
arithmetic mean   48
average   45
bimodal   46
deciles   89
dispersion   67
empirical rule   81
mean   46
mean deviation   69
measure of central tendency   45
measure of dispersion   67
measure of location   87
median   46
modal interval   52
mode   46

parameter   45
percentile ranks   90
percentiles   90
quartiles   89
ranks   87
spread   67
square root   71
squared deviations   70
standard deviation   71
standard scores   92
statistic   45
unimodal   46
variance   70
weighted mean   63
z-scores   92

Section 2–11   Summary of Symbols

Symbol	Meaning
C	Midpoint of the chosen interval in the coding process
d	Deviations from the mean
D_n	The nth decile, n is 1, 2, . . . , 9.
m	Midpoints of the intervals
μ	Population mean
$\tilde{\mu}$	Population median

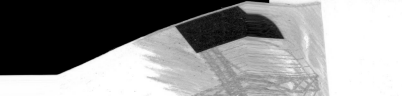

n	Sample size
N	Population size
P_n	The nth percentile rank of a data value n is 1, 2, . . . , 99.
Q_n	The nth quartile, n is 1, 2, or 3.
s	The standard deviation of a sample
s'	The coded standard deviation when calculating from a frequency distribution using the coding process
s^2	The variance of a sample, equal to the square of the standard deviation
σ	The standard deviation of a population
σ^2	The variance of a population
Σ	Sum of
w	Width of intervals; weights
x	Data values
x'	Coded data values
$\bar{x}$	Sample mean
$\bar{x}'$	Coded sample mean
$\tilde{x}$	Sample median
n	The absolute value of any number n
z	The z-score or standard score

Section 2–12   Summary of Formulas

Formula	**Use**
(1) $\bar{x} = \dfrac{\Sigma x}{n}$	To find the sample mean from raw data
(2) $\mu = \dfrac{\Sigma x}{N}$	To find the population mean from raw data
(3) $\bar{x} = \dfrac{\Sigma fm}{\Sigma f}$ or $\mu = \dfrac{\Sigma fm}{\Sigma f}$	To find the mean from a frequency distribution
(4) $\bar{x}' = \dfrac{\Sigma fx'}{\Sigma f}$ and $\bar{x} = C + w\bar{x}'$	To find the mean from a frequency distribution using coding
(5) $LTCL + r\left(\dfrac{w}{f+1}\right)$	To find the rth value in an interval, used in calculating the median from a frequency distribution
(6) $\bar{x} = \dfrac{\Sigma xw}{\Sigma w}$	To find the weighted mean
Mean deviation $= \dfrac{\Sigma \lvert d \rvert}{n}$	To find the mean deviation

(7) $s^2 = \dfrac{\Sigma d^2}{n}$

To find the variance of a sample

(8) $s = \sqrt{\dfrac{\Sigma d^2}{n}}$

To find the standard deviation of a sample

(9) $s = \sqrt{\dfrac{\Sigma x^2}{n} - \left(\dfrac{\Sigma x}{n}\right)^2}$

Alternative to formula (8), often easier to apply

(10) $s = \sqrt{\dfrac{\Sigma f m^2}{\Sigma f} - \left(\dfrac{\Sigma f m}{\Sigma f}\right)^2}$

To find the standard deviation from a frequency distribution

(11) $s' = \sqrt{\dfrac{\Sigma f (x')^2}{\Sigma f} - \left(\dfrac{\Sigma f x'}{\Sigma f}\right)^2}$

and $s = w s'$

Formulas to find the standard deviation from a frequency distribution using the coding process

(12) $z = \dfrac{x - \mu}{\sigma}$

To convert raw scores to standard scores in a population

(13) $z = \dfrac{x - \bar{x}}{s}$

To convert raw scores to standard scores in a sample

Section 2–13 Proofs

Proof About the Sum of the Deviations from the Mean

1. The sum of the deviations from the mean is zero.

$$\Sigma(x - \bar{x}) = \Sigma x - \Sigma \bar{x}$$

$$= n\left(\frac{\Sigma x}{n}\right) - \Sigma \bar{x} \qquad \text{multiplying and dividing } \Sigma x \text{ by } n$$

$$= n\bar{x} - n\bar{x} \qquad \text{since } \bar{x}, \text{ a constant, is summed } n \text{ times}$$

$$= 0$$

2. If the sum of the deviations from a number c is zero, then c is the mean.

$$\Sigma(x - c) = 0$$

$$\Sigma x - \Sigma c = 0$$

$$\Sigma x = \Sigma c$$

$$\Sigma x = nc \qquad \text{since } c, \text{ a constant, is summed } n \text{ times}$$

$$nc = \Sigma x$$

$$c = \frac{\Sigma x}{n} \qquad \text{dividing both sides by } n$$

$$c = \bar{x} \qquad \text{since } \bar{x} = \frac{\Sigma x}{n}$$

Proof of the Equivalency of Formulas (8) and (9)

Formula (8) states $s = \sqrt{\dfrac{\Sigma d^2}{n}}$ where $d = x - \bar{x}$.

Thus,

$$s = \sqrt{\frac{\Sigma(x - \bar{x})^2}{n}}$$

$$= \sqrt{\frac{\Sigma(x^2 - 2x\bar{x} + \bar{x}^2)}{n}} \quad \text{since } (a - b)^2 = a^2 - 2ab + b^2$$

$$= \sqrt{\frac{\Sigma x^2}{n} - \frac{\Sigma(2x\bar{x})}{n} + \frac{\Sigma \bar{x}^2}{n}}$$

$$= \sqrt{\frac{\Sigma x^2}{n} - \frac{2\bar{x}\Sigma x}{n} + \frac{n\bar{x}^2}{n}} \quad \text{since } \bar{x} \text{ is constant}$$

$$= \sqrt{\frac{\Sigma x^2}{n} - 2\bar{x}\,\bar{x} + \bar{x}^2} \quad \text{since } \frac{\Sigma x}{n} = \bar{x}$$

$$= \sqrt{\frac{\Sigma x^2}{n} - 2\bar{x}^2 + \bar{x}^2}$$

$$= \sqrt{\frac{\Sigma x^2}{n} - \bar{x}^2}$$

$$= \sqrt{\frac{\Sigma x^2}{n} - \left(\frac{\Sigma x}{n}\right)^2} \quad \text{which is formula (9).}$$

Chapter 2 Review Exercises

In exercises 1–4, find (a) the mean, (b) the median, and (c) the mode(s) of the given samples.

1. 5   9   3   6   12

2. 25   21   28   23   24   20   27   21   24

3. 15   13   18   16   13   19

4. 45   41   40   42   41   44   43   44   41   45   47   44

 5. In this chapter, you were introduced to some Greek letters. Here is a list of the names of *all* the Greek letters: alpha, beta, gamma, delta, epsilon, zeta, eta, theta, iota, kappa, lambda, mu, nu, xi, omicron, pi, rho, sigma, tau, upsilon, phi, chi, psi, omega. Find (a) the mean, (b) the median, and (c) the modal length of the names of these letters. Do *not* group the lengths into a frequency distribution.

6. In 1980, Ronald Reagan, president-elect of the United States, chose the nominees for his cabinet. Here are their ages.

63   63   61   67   34   54   58   49   50   56   54   58   53   42
45   59

Find (a) the mean, (b) the median, and (c) the modal age of these nominees.

In exercises 7–10, find (a) the mean, (b) the median, and (c) the modal interval for the frequency distributions of samples.

 7. A random sample of schoolchildren from a large, metropolitan school district was tested for IQ, using the Stanford-Binet IQ test. Here are the results.

Interval	f
60–69	3
70–79	6
80–89	15
90–99	26
100–109	20
110–119	18
120–129	7
130–139	4
140–149	1

 8. A random sample of subjects was tested by a psychologist to see how many words on a list were recalled incorrectly. Here are the results.

```
9  6  8  8  1  3  3  6  5  0  5  7  8  3  0  2  3  6  7
0  5  1  4  7  4  2  4  2  5  5  6  7  4  1  4
```

Group these into a frequency distribution, using 0–1 as the first interval, and then answer the questions.

 9. A sample of customers of a company producing cereal placed the following orders.

Size of order	f
$20–29.99	2
30–39.99	9
40–49.99	15
50–59.99	4
60–69.99	3
70–79.99	1

 10. A sample of salesmen for a large insurance company sold the following number of policies in a randomly selected month.

Policies sold	f
20–29	2
30–39	6
40–49	10
50–59	4
60–69	1

In exercises 11 and 12, there are population frequency distributions. Find (a) the mean and (b) the median for these.

 11. The country of Uganda, Africa, has 18 districts. This frequency distribution is a distribution of these 18 districts according to area in square miles.

Class	f
500–2,499	5
2,500–4,499	3
4,500–6,499	5
6,500–8,499	2
8,500–10,499	1
10,500–12,499	2

12. This frequency distribution is of the number of electoral votes assigned to each state and to the District of Columbia in the 1980 presidential election.

Class	f
1–6	21
7–12	18
13–18	5
19–24	1
25–30	4
31–36	0
37–42	1
43–48	1

13. Do you think two *different* samples can have exactly the same mean, median, and mode? Give a reason for your answer and then do exercise 14.

14. Find the mean, median, and mode for each of the data sets, which was an example in the text.

a. 20, 40, 50, 50, 60, 80

b. 10, 50, 50, 50, 50, 90

c. 50, 50, 50, 50, 50, 50

 15. These are the numbers of teenagers appearing in a suburban traffic court with traffic citations in each of the past 12 months.

28 28 39 14 19 20 25 31 27 21 15 34

Find the mean and standard deviation of this data.

 16. A doctor with a large practice finds herself faced with an outbreak of influenza. She prescribes a new drug and monitors the number of days to complete recovery. Here are the results for a sample of her patients.

5 4 3 6 4 5 5 4 3 6 4 5 7

Find the mean and standard deviation of this sample.

 17. In a study of plumbing practices in Los Angeles, a random sample of plumbing firms were called and asked to estimate a job. Here are the results in dollars.

265 250 168 200 45 230 142 212 155 65 140 200
250 145 200 120 240 75 100 150

Find the mean and standard deviation of these quotes.

 18. A dental research study was conducted and, among other things, the age of a child's first loss of a milk tooth was determined. Here are the results for a sample of children with the age given in months.

63 59 65 60 62 67 58 62 61 60 62 64 63 60
66 64 62 60 59 57 66 68 61 60 63 62

Find the mean ($\bar{x}$) and standard deviation (s) of this sample.

19. A sociological survey of students at a large metropolitan university includes the question, "How many members are there in your family, including yourself?" Here are the responses.

Number	f
1	2
2	10
3	24
4	52
5	41
6	23
7	13

Find the mean ($\bar{x}$) and standard deviation (s) of this frequency table.

 20. A random sample of light bulbs were tested for length of life in hours. Here are the results.

Lifetime	f
500–649	4
650–799	21
800–949	62
950–1,099	89
1,100–1,249	31
1,250–1,399	15
1,400–1,549	8

Find the mean and standard deviation of this sample.

 21. A photographic supply store conducted a survey of its customers to see how many pictures of a roll of 36 pictures came out satisfactorily. Here are the results.

Number of good pictures	f
1–6	3
7–12	2
13–18	5
19–24	11
25–30	24
31–36	22

Find the mean and standard deviation of this sample.

 22. A simple puzzle was given to a group of fourth graders to solve. A few failed to solve it; but of those who did, the following indicates the length in time required to solve the puzzle.

Time in minutes	f
0–1	4
2–3	1
4–5	7
6–7	6
8–9	2

Find the mean and standard deviation of these times.

23. Find the percentage of survey results that fall between (a) $\bar{x} - s$ and $\bar{x} + s$ and (b) $\bar{x} - 2s$ and $\bar{x} + 2s$ for the data of exercise 19.

24. Do the same as exercise 23 for the results of exercise 18.

25. The planets in our solar system have the following equatorial radii, where earth's is taken as the unit.

Mercury	0.382	Venus	0.949
Earth	1.000	Mars	0.533
Jupiter	11.200	Saturn	9.380
Uranus	4.230	Neptune	3.950
Pluto	0.500		

Rank these planets in order from smallest to largest.

26. The number of public universities and colleges in each of the fifty states and the District of Columbia is given here in alphabetical order of the states.

```
36   12   17   19   135   28   24    6    1   37   34    9    6
63   24   22   29     9   20   11   32   33   45   30   27   28
 9   16    5   10    31   16   82   73   11   60   29   21   61
 3   33    8   23    95    9    6   39   33   17   30    8
```

Rank these in order from largest to smallest.

27. For the data of exercise 26 find the percentile rank of Alabama the first of the list.

28. Find P_{33} for the data in exercise 26.

29. The following figures are the birthrates per 1,000 in a random sample of world nations.

```
16.3   20.3   16.0   12.9   32.1   27.2   14.5   37.3   28.3   34.0   18.8
13.3   19.4   15.0   15.5   17.7   13.6   10.2   19.8   17.1   43.4   44.7
20.7   40.7   38.7
```

Find the z-score of the nation with birthrate 15.0 per 1,000.

30. Using the data of exercise 29, find the z-score of the nation with birthrate 37.3 per 1,000.

Hands-on Class/Student Project

If you did this project at the end of chapter 1, retrieve your data and perform the following exercises.

1. Find the mean, median, mode, range, and standard deviation for some or all of the variables 1, 3, 4, 6, 11, and 12 *before* they were grouped into frequency distributions.
2. Find the weighted mean of variable 5, using variable 4 for the weights.
3. For each of the frequency distributions of the same variables as in 1, find the mean, median, modal interval, and standard deviation.
4. Compare your answers to 1 and 3 for corresponding variables.
5. Compare your answers to 1 and 3 to the corresponding answers of other students or student groups in the class.
6. For each of the variables in 1, find the percentage of each data set that lies within one and two standard deviations of the mean.
7. Check the percentages you found in 6 against those promised by the empirical rule and Chebyshev's theorem. (*Note:* Look at the graphs you constructed in chapter 1 for these variables—those that have a shape closest to the bell shape should provide the best agreement with the empirical rule.)
8. Rank variable 6 from highest to lowest.
9. Find the percentile rank and standard score for your *own* height in relation to your sample of variable 1. Do this again for any other variable you might be interested in. (*Note:* For percentile ranks, you can use the ogives that you constructed in chapter 1.)
10. For variables 4, 5, and 6, find which values lie at the first and third quartiles.

A Transition to Statistical Inference

"All positive reasoning is of the nature of judging the proportion of something in a whole collection by the proportion found in a sample. Accordingly, there are three things to which we can never hope to attain by reasoning, namely, absolute certainty, absolute exactitude, absolute universality."

Collected Papers, Volume 1 (Harvard University Press, 1965).

CHARLES SANDERS PIERCE (1839–1914)

Probability

....

"But to us, probability is the very guide of life."

....

BISHOP JOSEPH BUTLER

The probability machine demonstrates how a probability curve can be found by experiment. When the balls are dropped from the center top of the machine they invariably form the pattern shown in the picture. This pattern is called the "Normal Distribution Curve."

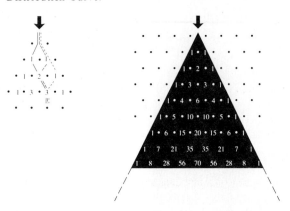

The number of possible paths to a given space in the array of pins is given by Pascal's triangle. For the number of paths to a space is the sum of the number to the two spaces above it. The probability of a ball's dropping in any box can be found by counting the number of paths to that box, and comparing it with the total number of paths. A ball can land in any box, and yet any given box fills to nearly the same height each time the experiment is repeated. This stability is due to the fact that there are many balls.

Unpredictable   More predictable

If a random event happens a great many times the average results are likely to be predictable. Probability theory is used in genetics, thermodynamics, games of chance and strategy, traffic control, insurance and queuing theory.

....

Chapter Overview
Probability and Statistics

This chapter and the next introduce you to some basic ideas that you will need to have a more complete understanding of inferential statistics.

Probability theory forms the foundation of modern statistical theory, and so it is necessary for you to learn some probability before continuing your study of statistics.

By the end of the chapter, you will have a good idea of what is meant by probability and a probability distribution. In addition, you will learn about a pair of probability distributions that are of particular importance in the study of inferential statistics.

....

Section 3–1   Introduction to Probability Theory

While you are watching the 6 o'clock news one evening, the weather forecaster might inform you that there is a 30% chance of rain tomorrow. Forecasters might reach such a conclusion as follows: they would study the weather as it is today and count all the days in the past, for which they have records, that had similar weather as today's. Then they would count all the days following those days that had rain. If they found that out of every 10 days that had weather like today's, 3 of the following days were wet, it would be reasonable for them to conclude that there are 3 chances out of 10 for us to have rain tomorrow, and 3 out of 10, expressed as a percent, is 30%.

It is possible to estimate the chance of any event occurring by repeating an experiment a large number of times and observing what proportion of the outcomes result in that event. In fact, a seventeenth-century French nobleman by the name of Chevalier de Méré found that long-run observation of dice games did not fit with his theories. He enlisted the help of Blaise Pascal and Pierre de Fermat, mathematicians who subsequently became interested in the problem, and together they founded probability theory.

Probability

Rather than use the word chance (or chances), statisticians prefer to use the word **probability.** Thus a weather forecaster who was in a more statistical frame of mind would make a prediction like, "the probability of rain tomorrow is 30%." This prediction, as explained above, is the result of long-run observation, and the probabilities of any events can be estimated by long-run observation. Consider this question: What are the chances of getting heads when you toss a coin? Many

THE WIZARD OF ID   by Brant parker and Johnny hart

By permission of Johnny Hart
and NAS, Inc.

people would answer 50–50. The statistician would say, "The probability of getting heads when a coin is tossed is $\frac{1}{2}$." (The answer 50–50 comes from the fact that $\frac{1}{2} = 50\%$, and there is a 50% probability of getting heads and a 50% probability of getting tails when a coin is tossed.)

You could arrive at the figure of $\frac{1}{2}$, or 50%, in much the same way as weather forecasters arrive at their predictions, by long-run observation. Try it. Toss a coin a large number of times and count how many times you get heads—you should find that it is about half the time. For some types of experiments, probabilities can be determined using reasoning rather than through long-run observation. Where probabilities can be determined by logical experiments rather than empirical observation, these "thought experiments" generally are much more efficient. To understand when and how you can take advantage of this streamlined method of calculating probabilities, it is necessary to study probability theory. This study is initiated by defining a number of terms.

sample space	A sample space is a list of all possible outcomes of an experiment. The list is usually enclosed in braces, as with sets in algebra.
event	An event is an **outcome** or outcomes of an experiment that form a subset of a sample space.
simple event	A simple event is an event that cannot be expressed as a list of other events.
success	A success is an outcome of an experiment in which you are interested.
failure	A failure is an outcome of an experiment that is not a success.

EXAMPLE 3–1   Find the sample space for tossing a coin.

Solution   Since the only possibilities are "heads," or *H*, and "tails," or *T*, the solution is $\{H,T\}$.

EXAMPLE 3-2   Find the sample space for tossing two coins.

Solution   If you are interested in how many heads you get when you toss two coins, then when you get heads that is a success and when you get tails that is a failure, and the solution to this problem might be $\{0H,1H,2H\}$. This solution indicates the three possible results, no heads, one head, and both heads, when the two coins are tossed. However, there is a fundamental difference between this solution and the solution to example 3-1. In example 3-1, the different outcomes in the sample space, namely, H and T, are events that are also simple events. In the solution to this example, this is not the case. The event $1H$ is not a simple event, since it could be expressed as follows:

$$1H = \{HT,TH\}.$$

HT is the simple event of obtaining heads on the first coin and tails on the second, and TH is the simple event consisting of tails on the first coin and heads on the second. In each case, there is one success (head) in the tossing of the two coins. Another possible solution to this example, using only simple events, is $\{TT,TH,HT,HH\}$.   ●

The theoretical probability of an event E is written $P(E)$. Suppose that all the simple events in S, the sample space, have an equal likelihood of occurring. Then, the probability of any particular event $P(E)$ is found by the following formula:

$$P(E) = \frac{\text{number of simple events in } E}{\text{number of simple events in } S}. \qquad \textbf{(14)}$$

EXAMPLE 3-3   Find the probability of getting heads when tossing a coin.

Solution   As was seen in example 3-1, $S = \{H,T\}$. Using a fair coin, $P(H) = \frac{1}{2}$.   ●

EXAMPLE 3-4   Find the probability of rolling a 2 or less when rolling a fair die (one of two dice).

Solution   For this experiment, $S = \{1,2,3,4,5,6\}$. Let $E = \{2 \text{ or less}\}$. Then, another way of writing E is as follows: $E = \{1,2\}$ and $P(E) = \frac{2}{6} = \frac{1}{3}$.   ●

EXAMPLE 3-5   Answer example 3-4 if the die is weighted so that faces 4, 5, and 6 come up twice as often as the other faces.

Solution   To satisfy the assumption of formula (14), the sample space can be rewritten $S = \{1,2,3,4,4,5,5,6,6\}$. Then, it becomes clear that $P(E) = \frac{2}{9}$.   ●

EXAMPLE 3–6   Find the probability of getting both heads when tossing two coins.

Solution   Look back at example 3–2 and notice that S could be taken as $\{0H,1H,2H\}$ or $\{TT,TH,HT,HH\}$. If $S = \{0H,1H,2H\}$, then is $P(2H) = \frac{1}{3}$? If $S = \{TT,TH,HT,HH\}$, then is $P(2H) = P(HH) = \frac{1}{4}$? In the first case, the event $1H$ is not a simple event since $1H = \{HT,TH\}$, and formula (14) requires that you count *simple* events. Thus, the second listing for S is the correct one to use and the solution to this problem is $P(2H) = \frac{1}{4}$.   ●

The Meaning of Probability

You can see, in example 3–6, that $P(2H) = \frac{1}{4}$. What does this mean to you? Does it mean that whenever two coins are tossed a number of times, exactly $\frac{1}{4}$ of the outcomes will be both heads? Some people think so when first thinking about probability. However, this is far from the truth. Table 3–1 shows the results of actually throwing a fair coin 4 times; then 40 times; then 400 times; then 4,000 times.

Table 3–1

Number of Times Two Coins Tossed	Expected Number of 2H	Observed Number of 2H	Absolute Difference	Observed Relative Frequency = Empirical Probability
4	1	2	1	0.5
40	10	8	2	0.2
400	100	103	3	0.2575
4,000	1,000	995	5	0.24875

According to the logical experiment, the probability of getting two heads is $\frac{1}{4}$. The expected numbers of outcomes $2H$ are shown in the second column, where it is expected that exactly $\frac{1}{4}$ of all tosses will have this result. Column three indicates what actually happened, and column four shows the absolute difference between the expected value (the best guess in advance based solely on the thought experiment) and what actually occurred.

Concentrate on the fourth column for a moment. This shows how the actual results deviate from the expected results, and it seems that, for these results, the more the experiment is repeated, the worse the agreement between the expected and the observed. Of course, this will not necessarily always happen—the agreement could actually improve. However, even given this bad agreement, look at the last column, which gives the empirical probability. The more the experiment is repeated, the closer the empirical probability gets to the theoretical probability of $\frac{1}{4}$ or 0.25.

This is what theoretical probability should mean to you. It does not allow you to predict exactly what is going to happen, but it does tell you approximately what proportion of the time an event will occur when an experiment is repeated a large number of times.

Properties of Probability

Study formula (14) for a moment. Since the number of simple events in any event E, or S for that matter, can never be negative, $P(E)$ can never be negative. Also, since the number of simple events in E can never be larger than the number of simple events in S, the fraction that appears in formula (14) can never be larger than 1. This leads to the first property of probability:

$$0 \leq P(E) \leq 1,$$

where E is any event.

Here, given an impossible event E, the statement $P(E) = 0$ would result, whereas given an event E that is certain to happen, the statement $P(E) = 1$ would result. Given an event E that may or may not happen, $P(E)$ would result in a fractional value between 0 and 1.

Probability $\begin{bmatrix} 1 = \text{certainty} \\ \quad \longrightarrow \quad \text{may or may not happen} \\ 0 = \text{impossibility} \end{bmatrix}$

EXAMPLE 3–7   Experiment: Draw a card at random from a well-shuffled 52-card, ordinary deck of playing cards.
Find the following probabilities:
P(jack of hearts), P(ace), P(club), P(red card), P(bishop).

Solution

$$S = \{2C,3C,4C,5C,6C,7C,8C,9C,10C,JC,QC,KC,AC,2D,3D,4D,$$
$$5D,6D,7D,8D,9D,10D,JD,QD,KD,AD,2H,3H,4H,5H,6H,7H,$$
$$8H,9H,10H,JH,QH,KH,AH,2S,3S,4S,5S,6S,7S,8S,9S,10S,JS,$$
$$QS,KS,AS\}$$

where C = clubs, D = diamonds, H = hearts, S = spades, J = jack, Q = queen, K = king, A = ace, diamonds and hearts are red suits, clubs and spades are black suits. In all later examples and exercises, the ace is considered the highest card in a suit, while the 2 is considered the lowest. Now, by counting simple events, the solutions are

$$P(JH) = \frac{1}{52}$$

$$P(A) = \frac{4}{52} = \frac{1}{13}$$

$$P(C) = \frac{13}{52} = \frac{1}{4}$$

$$P(\text{red card}) = \frac{26}{52} = \frac{1}{2}$$

$$P(\text{bishop}) = \frac{0}{52} = 0.$$

Notice that in example 3–7, and indeed in the previous examples, all values for probabilities satisfy the first property: none are negative; none exceed 1; all are in the interval [0,1]. In example 3–7, P(bishop) $= 0$ comes from the fact that it is impossible to draw a card that is a bishop from an ordinary deck of cards. There is no card with a bishop on it (wrong game!).

In example 3–7, you saw that $P(JH) = \frac{1}{52}$. Since there is nothing special about the jack of hearts, it is also true that $P(2C) = \frac{1}{52}, P(3C) = \frac{1}{52}, \ldots, P(AS) = \frac{1}{52}$, for all 52 cards. Adding these probabilities gives

$$P(2C) + P(3C) + \cdots + P(AS) = \frac{1}{52} + \frac{1}{52} + \cdots + \frac{1}{52}$$

$$= 52\left(\frac{1}{52}\right) \text{ (since there are 52 cards)}$$

$$= 1.$$

This illustrates the second property of probability:

$$\Sigma P(e) = 1,$$

where the summation is taken over all simple events e.

EXAMPLE 3–8   Show that the second property of probability holds for the experiment of tossing a coin.

Solution   Here, $S = \{H, T\}$ and $P(H) = \frac{1}{2}, P(T) = \frac{1}{2}$. Adding gives that $P(H) + P(T) = \frac{1}{2} + \frac{1}{2} = 1$, showing that the second property holds.   ●

The two properties of probability are

1. $0 \leq P(E) \leq 1$, where E is any event.
2. $\Sigma P(e) = 1$, where the summation is taken over all simple events e.

Odds

When thinking of the chances of an event happening, many people think in terms of **odds** rather than probability. "There are 2 to 1 odds against rain tomorrow," for instance. Odds and probability are different ways of expressing **chances,** but they are related.

Probability is calculated as the fraction, or ratio, of the number of simple events in the sample space that produce a success to the total number of simple events in the sample space.

Let $s =$ the number of simple events in the sample space that produce a success

and $f =$ the number of simple events in the sample space that produce a failure.

Then, the total number of simple events in the sample space must be given by $s + f$, and the probability of an event E is given by

$$P(E) = \frac{s}{s + f}.$$

On the other hand, the odds of event E happening are expressed as follows: "f to s against" or "s to f for."

EXAMPLE 3–9   Find both the probability and the odds of rolling a 6 with one roll of a fair die.

Solution $S = \{1,2,3,4,5,6\}$ and only one simple event produces a success, namely "6." Hence, $s = 1, f = 5$, and $s + f = 6$ and

$$P(6) = \frac{1}{6}.$$

    The odds can either be expressed as "5 to 1 against" or "1 to 5 for" getting a 6.   ●

EXAMPLE 3–10   Find the odds of getting heads in one toss of a fair coin.

Solution   In example 3–3, you saw that $P(H) = \frac{1}{2}$. This means that $s = 1$ and $s + f = 2$, which in turn implies that $f = 1$. The odds are therefore "1 to 1" to get heads. When s and f are equal like this, this is often expressed as **evens** or **even odds**.   ●

EXAMPLE 3–11   "There are '2 to 1 odds against' rain tomorrow." What is the probability of rain tomorrow?

Solution   Since the word "against" is used, $f = 2$ and $s = 1$, which means $s + f = 3$. Hence, $P(\text{rain tomorrow}) = \frac{1}{3}$.   ●

In order to avoid decimals, odds of "1.5 to 1 for" are often expressed as "3 to 2 for" (found by multiplying by 2). Check for yourself that these odds give the same answer as the solution to example 3–12.

EXAMPLE 3–12   Suppose the odds of some event happening are "1.5 to 1 for." What is the probability of this event?

Solution   Since the word "for" is used, $s = 1.5, f = 1$, and $s + f = 2.5$. So, the probability of this event is

$$\frac{1.5}{2.5} = \frac{15}{25} = \frac{3}{5}.$$ ●

With odds, the larger number is usually placed first and the appropriate word "for" or "against" placed after the odds. Did you check that the odds in example 3–12, "1.5 to 1 for," are the same odds as "3 to 2 for"? As a contrast to this, what probability corresponds to odds of "3 to 2 against"? Well, then, $f = 3$, $s = 2$, and $s + f = 5$ and the probability is $\frac{2}{5}$ instead. Notice that it does make a difference whether the word "for" or "against" is placed after the odds. If you see neither word, you can usually assume "against" is implied. For example, if you were told the odds of something happening are "5 to 1," it would generally be assumed that "5 to 1 *against*" was intended.

Exercise Set 3–1

1. A coin is suspected of being biased. It is tossed 100 times and found to show heads 42 times. What is the observed (empirical) probability of heads for this coin?

2. A die is suspected of being unfair. It is rolled 360 times and found to show the ace (one-spot) 42 times. What is the observed (empirical) probability of an ace for this die?

3. List the sample space for tossing three coins.

4. A married couple has three boys and one girl. Using B for boy and G for girl, list the sample space of possible orders of birth.

5. What is the probability of getting exactly two heads when tossing three coins?

6. If a married couple has four children, what is the probability that they have three boys and a girl?

7. In the casino game of roulette, the possible outcomes are the numbers 1, 2, 3, . . . , 36 (half of which are red and half black) and 0 and 00, both of which are green. What is the probability of an odd number coming up on a roulette wheel if 0 and 00 are considered to be neither odd nor even?

8. In roulette (see exercise 7), find the probabilities of each of the following colors coming up: black, blue, green, red.

9. If the ace of spades has been removed from an ordinary deck of cards and you draw a card at random from the remaining cards, what is the probability that you draw an ace?

10. If all cards less than 6 are removed from an ordinary deck of cards and a card is drawn at random, what is the probability that it is a jack?

11. A box has three blue marbles, five red marbles, and one yellow marble. If a marble is drawn at random from this box, what is the probability that it is blue?

12. A deck of cards is shuffled and dealt to you and three other players until the cards are exhausted. What is the probability that you hold the ace of spades?

13. If the odds against an event happening are "7 to 1," what is the probability of the event?

14. If the probability of an event is $\frac{1}{7}$, what are the odds of the event happening?

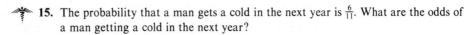
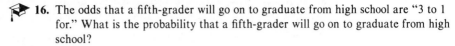

15. The probability that a man gets a cold in the next year is $\frac{6}{11}$. What are the odds of a man getting a cold in the next year?

16. The odds that a fifth-grader will go on to graduate from high school are "3 to 1 for." What is the probability that a fifth-grader will go on to graduate from high school?

17. At roulette, if you place a bet on the numbers 1–12, the casino gives you odds of "2 to 1." What are the true odds of winning this bet?

18. At roulette, if you place a bet on red, the casino gives you "even odds." What are the true odds of winning this bet?

Section 3–2 Compound Probabilities

In this section, you will learn three probability rules, two concerning the combination of two events and one concerned with the "complement" of an event. These rules, once learned, should make the solution of probability problems that much easier.

Complementary Events

Since the idea of complementary events is very likely new to you, to begin, the definition of "complement" is presented to you.

Suppose A represents an event. Then, the **complement** of A, which is written A', is the event that A does not happen. For example, suppose A is the event that you roll an even number in the roll of one die. Then A' is the event that you roll an odd number. Now, it must always be true that all the simple events in the sample space are included in either A or A'. So, by the second property of probability,

$$P(A) + P(A') = 1$$

or

$$P(A') = 1 - P(A). \tag{15}$$

EXAMPLE 3–13 Experiment: Roll one die. Find the complement of the event $E = \{$at least 3$\}$ and find the probability of the complement.

Solution $S = \{1,2,3,4,5,6\}$. $E = \{$at least 3$\} = \{3,4,5,6\}$. Hence, $E' = \{1,2\}$. By counting simple events directly, can you see that $P(E') = \frac{2}{6} = \frac{1}{3}$? If you wanted to use formula (15), first find $P(E)$, which is $\frac{4}{6}$ or $\frac{2}{3}$, by counting simple events and then,

$$P(E') = 1 - P(E) = 1 - \frac{2}{3} = \frac{1}{3}$$

as before.

From the relatively simple example 3–13, you might wonder about the advantage, if any, of using formula (15). In later examples, you will see that, when attempting to calculate a probability of an event, it is often easier to calculate the probability of the complementary event. In that case, formula (15) can be very useful.

EXAMPLE 3–14 Experiment: Ask a friend for his/her birthday. Find the complement of the event $E = \{$your friend's and your birthdays fall on the same month and day$\}$ and find the probability of E'.

Solution $E' = \{$your friend's and your birthdays don't match$\}$. Since there are 365 days in a year and only one of these that is your birthday, $P(E) = \frac{1}{365}$ and

$$P(E') = 1 - P(E) = 1 - \frac{1}{365} = \frac{364}{365}.$$
                                                                                               ●

Addition Rule

You are now presented with the first of the two rules concerned with the calculation of probabilities of a combination of events. The one immediately following deals with the rule for finding the probability of one event *or* another.

    This rule is stated as follows:

$$P(A \text{ or } B) = P(A) + P(B) - P(A \text{ and } B). \tag{16}$$

Here, the event A or B is interpreted to mean that A occurs or B occurs or that both A and B occur.

EXAMPLE 3–15 Experiment: Draw a card from an ordinary deck of cards. Find the probability that the card is either red or a king.

Solution Let $R = \{$a red card$\}$ and $K = \{$a king$\}$. Then the problem is asking for $P(R \text{ or } K)$. Since there are 26 red cards and 4 kings, $P(R) = \frac{26}{52}$ and $P(K) = \frac{4}{52}$. Also, the event R and K is the event that a card is both red and a king. There are 2 such cards in the deck, namely, the king of hearts and the king of diamonds. Thus, $P(R \text{ and } K) = \frac{2}{52}$. Thus, by formula (16),

$$P(R \text{ or } K) = \frac{26}{52} + \frac{4}{52} - \frac{2}{52} = \frac{28}{52} = \frac{7}{13}.$$
                                                                                               ●

Example 3–15 could have been solved directly by counting simple events in the sample space. Among those 52 simple events, there are 28 that are either red or a king: 26 red cards plus 4 kings would make 30, but then the king of hearts and king of diamonds would have been counted twice. That is the reason the fraction $\frac{2}{52}$ is subtracted in the solution to example 3–15. For a very similar reason, $P(A \text{ and } B)$ is subtracted in formula (16) to correct for counting simple events twice.

EXAMPLE 3-16 For the experiment of rolling two dice, find the probability of getting a total of 8 spots showing or doubles.

Solution When rolling two dice, there are six possible outcomes on each die, which means that there are a total of $6 \times 6 = 36$ simple events in the sample space for this experiment. Let $E = $ {a total of 8 spots showing} and D = {rolling doubles}. Then, since there are 5 ways of rolling an 8 [(2,6), (3,5), (4,4), (5,3), and (6,2)], $P(E) = \frac{5}{36}$ and there are 6 doubles, 1 for each of the different faces on one die, $P(D) = \frac{6}{36}$. Also, there is only one way to get E and D [(4,4)], so that $P(E \text{ and } D) = \frac{1}{36}$. Thus,

$$P(E \text{ or } D) = \frac{5}{36} + \frac{6}{36} - \frac{1}{36} = \frac{10}{36} = \frac{5}{18}.$$

EXAMPLE 3-17 For the experiment of example 3-15, find the probability of drawing an ace or a king, that is, $P(A \text{ or } K)$.

Solution Since there are 4 aces and 4 kings in a deck of cards,

$$P(A) = P(K) = \frac{4}{52}.$$

Also, a card cannot be both an ace and a king, which means that $P(A \text{ and } K) = 0$. Therefore,

$$P(A \text{ or } K) = \frac{4}{52} + \frac{4}{52} - 0 = \frac{8}{52} = \frac{2}{13}.$$

Two events A and B are said to be **mutually exclusive** if both cannot occur at the same time. If A occurs then B cannot, and vice versa.

In example 3-17, the events A, K are mutually exclusive events because a card cannot be both an ace and a king at the same time. In example 3-16, E and D are not mutually exclusive, since a roll of a pair of dice can total 8 and be a double at the same time. In the experiment of choosing a person at random, examples of mutually exclusive events are

  a mother and a nephew,
  a blond and a brunette,

whereas examples of nonmutually exclusive events might be

  an honor-roll student and a paroled ex-felon,
  a mother and a daughter.

For mutually exclusive events A and B, it is always true that $P(A$ and $B)$ = 0, since by definition mutually exclusive events cannot occur at the same time. This was demonstrated in example 3–17, where $P(A$ and $K)$ was shown to be zero for the mutually exclusive events ace and king. Therefore, when any two events A and B are mutually exclusive, formula (16) simplifies to

$$P(A \text{ or } B) = P(A) + P(B).$$

It is advisable to use formula (16) and not the one immediately above, since formula (16) applies to all situations, whether or not A and B are mutually exclusive events.

EXAMPLE 3–18   Experiment: Toss a coin once. Find $P(H$ or $T)$.

Solution   The events H, T are mutually exclusive events because if the result of a toss is heads it cannot be tails, and vice versa. Thus,

$$P(H \text{ or } T) = P(H) + P(T) = \frac{1}{2} + \frac{1}{2} = 1. \qquad \bullet$$

Earlier, you were informed that if an event was certain to happen, that event had a probability of 1. Example 3–18 illustrates this concept. When a coin is tossed once, it is a certainty that the result is heads or tails. Another point illustrated by example 3–18 is the complementary rule: since H and T are complementary events, the equation $P(H) + P(T) = 1$ is the same as $P(A) + P(A')$ = 1.

Multiplication Rule

Here, you will be introduced to the third and final rule of this section and the second concerning a combination of events. This time, the one immediately following deals with the rule for finding the probability of one event *and* another.

Before stating this rule, it is necessary to define the concept of **conditional probability.** The event written $B \mid A$ is the event that B occurs with the condition that A has already occurred.

> The **conditional probability** $P(B \mid A)$ is the probability of B occurring, given that A has occurred.

When encountering the probability $P(B \mid A)$, this is usually read "the probability of B, given A."

EXAMPLE 3-19   Experiment: Draw 2 cards from an ordinary deck of cards. If A is the event that the first card is black and B is the event that the second card is black, find $P(B \mid A)$. In words, find the probability that the second card is black, given that the first was black.

Solution   After the first card is removed from the deck and it is given that it is black, there are 51 cards remaining, 25 of which are black. So, $P(B \mid A) = \frac{25}{51}$.

The multiplication rule of probability can now be stated.

$$P(A \text{ and } B) = P(A)P(B \mid A) \qquad \textbf{(17)}$$

EXAMPLE 3-20   For the experiment of example 3-19, find the probability that both cards are black.

Solution   When the 2 cards are drawn from the deck, let the events A and B be as in example 3-19, with $P(B \mid A) = \frac{25}{51}$, as was shown there. Since 26 of the original 52 cards were black initially, in the first drawing, $P(A) = \frac{26}{52}$. Therefore,

$$P(\text{both cards are black}) = P(A \text{ and } B)$$

$$= P(A)P(B \mid A) \quad \text{from formula (17)}$$

$$= \left(\frac{26}{52}\right)\left(\frac{25}{51}\right)$$

$$= \frac{25}{102}.$$

0.245098   *650* / *2652*

EXAMPLE 3-21   Experiment: Toss a coin twice. What is the probability that it shows heads twice?

Solution   The experiment of tossing the same coin twice is equivalent to the experiment of tossing two coins. So this question was answered in example 3-6: it was $\frac{1}{4}$. But this can be done another way, using formula (17). Let the events H_1, H_2 be the occurrences of heads on the first and second tosses, respectively. Then,

$$P(\text{both tosses show heads}) = P(H_1 \text{ and } H_2)$$

$$= P(H_1)P(H_2 \mid H_1),$$

but what is $P(H_2 \mid H_1)$? Contrary to many gamblers' popular beliefs, coins and dice have no memories, and the probability of getting heads on the second toss is exactly the same as the probability of getting heads on the first, namely $\frac{1}{2}$. Thus,

$$P(H_1 \text{ and } H_2) = \left(\frac{1}{2}\right) \times \left(\frac{1}{2}\right) = \frac{1}{4}.$$

Notice that, in example 3–21, $P(H_2 | H_1) = P(H_2)$. What this implies, in words, is that the probability of heads on the second toss is the same regardless of the result of the first toss. When, as in this case, the probability of an event is unaffected by the occurrence of another event, the events are said to be **independent**.

Two events A, B are said to be **independent** if the occurrence of one has no effect on the probability of the other. In symbols: $P(A | B) = P(A)$ and $P(B | A) = P(B)$.

When A and B are independent events, formula (17) simplifies to

$$P(A \text{ and } B) = P(A)P(B).$$

As with the addition rule, it is advisable to use formula (17), which applies to all situations, whether or not A and B are independent events.

EXAMPLE 3–22   Experiment: Roll two dice. What is the probability of rolling "snake-eyes," that is double-ones?

Solution   Put another way: What is $P(\text{one and one})$? Since dice cannot communicate, the outcome on one die cannot possibly affect the probability of an outcome on the other. This means that the individual eyes of the snake-eyes are independent of each other. Recall that $P(\text{one}) = \frac{1}{6}$. Thus,

$$P(\text{one and one}) = P(\text{one})P(\text{one}) = \left(\frac{1}{6}\right) \times \left(\frac{1}{6}\right) = \frac{1}{36}. \qquad \bullet$$

The two concepts of mutual exclusivity and independence are often confused. To differentiate between them, note the following: To say that two events are mutually exclusive, you need never refer to probability—the idea simply means that the two events cannot occur at the same time. However, look back at the definition of independence, it cannot be defined *without* using the concept of probability.

Looking back at example 3–17, notice that there are 36 simple events in the sample space. Note also that only one of these gives you snake-eyes, which is another reason why the answer to example 3–22 is $\frac{1}{36}$.

EXAMPLE 3–23   For the experiment of example 3–22, find the probabilities of rolling the following sums of the uppermost faces: 4, 7, 11. Also, find the odds of rolling these sums.

Solution   Remember that there are 36 possible simple events when two dice are rolled. There are 3 ways of rolling a sum of dots on the two dice equal to 4: (1,3), (2,2), and (3,1). Hence:

$$P(4) = \frac{3}{36} = \frac{1}{12}.$$

There are 6 different ways of rolling a sum of 7, but only 2 ways of rolling a sum of 11. Check these results for yourself. Thus,

$$P(7) = \frac{6}{36} = \frac{1}{6}$$

$$P(11) = \frac{2}{36} = \frac{1}{18}.$$

For rolling a 4, there are 3 simple events giving success and 33 giving failure. Thus, the odds of rolling a 4 are "33 to 3 against" or "11 to 1 against." Similarly, the odds of rolling 7 are "5 to 1 against," while the odds of rolling 11 are "17 to 1 against." ●

Applying the Three Probability Rules Together

Three rules of probability have been introduced in this section: the rules for complementary events, occurrence of either of two events, and occurrence of two events simultaneously. To complete this section, here are two examples that use all three of these formulas.

EXAMPLE 3–24   For the experiment of example 3–22, find the probability that the sum is larger than 3.

Solution   A possible sample space for this experiment is

$$S = \{2,3,4,5,6,7,8,9,10,11,12\}.$$

However, you should be aware that this is not a list of simple events. For instance, as you saw in example 3–23, the event 4 can be broken down to {(1,3), (2,2), (3,1)}. Since the events listed in S are mutually exclusive, this problem could be solved by finding each of the probabilities of the events 4, 5, 6, 7, 8, 9, 10, 11, 12 and adding. This is a considerable amount of work. Instead, find the probability of the complementary event "3 or less," and use formula (15). The event "3 or less" consists merely of 2 or 3. Using formulas (16) and (17) and the ideas of examples 3–22 and 3–23,

$$P(2 \text{ or } 3) = P(2) + P(3) = \frac{1}{36} + \frac{2}{36} = \frac{3}{36} = \frac{1}{12}$$

and

$$P(\text{roll larger than } 3) = 1 - P(2 \text{ or } 3)$$

$$= 1 - \frac{1}{12}$$

$$= \frac{11}{12}.$$

●

EXAMPLE 3–25   A hat has ten poker chips in it, one blue, four red, and five white. If three poker chips are chosen at random, find the probability that all three are white if the drawing is done (a) with replacement, (b) without replacement.

Solution (a)   The phrase "with replacement" means that the chips are drawn one at a time, the color noted, and each chip replaced before the next is drawn. Thus, the three drawings of the three chips are independent of each other. If W_1, W_2, and W_3 represent a white chip on each of the three drawings, then

$$P(\text{all 3 white}) = P(W_1 \text{ and } W_2 \text{ and } W_3)$$

$$= P(W_1)P(W_2)P(W_3)$$

$$= \left(\frac{5}{10}\right)\left(\frac{5}{10}\right)\left(\frac{5}{10}\right) = \frac{1}{8}.$$

(b) When the drawing is made "without replacement," the events W_1, W_2, and W_3 are dependent, since the probability of drawing a white chip changes as the chips are removed. Thus:

$$P(\text{all 3 white}) = P(W_1 \text{ and } W_2 \text{ and } W_3)$$

$$= P(W_1)P(W_2 | W_1)P(W_3 | W_1 \text{ and } W_2)$$

$$= \left(\frac{5}{10}\right)\left(\frac{4}{9}\right)\left(\frac{3}{8}\right) = \frac{1}{12}.$$

●

Exercise Set 3-2

1. The probability of an event Q is $\frac{3}{4}$. What is the probability that Q does not occur?

2. The probability of an event X is 0.519. What is the probability that X does not occur?

3. A ball is drawn at random from a box containing seven blue balls, eight white balls, and five green balls. What is the probability that it is blue or green?

4. A card is drawn from an ordinary deck of cards. What is the probability that it is either a queen or a spade?

5. A coin is tossed and a die is rolled. What is the probability that both heads and a 6 are obtained?

6. Two cards are drawn at random from an ordinary deck of cards. What is the probability that they are both hearts?

7. The game of craps is a dice game played in casinos. A pair of dice is rolled. One bet that can be made is on the field, which consists of the numbers 2, 3, 4, 9, 10, 11, and 12. If you bet on the field, you win if any one of those numbers comes up. What is your probability of winning a bet on the field?

8. There is another bet in craps called, "craps-eleven." A roll of 2, 3, 12, or 11 is a win with this bet. What are the odds of winning a craps-eleven bet?

9. What is the probability that two people, chosen at random, do not have the same birthday (day not year)? What is the probability that three people do not have the same birthday? (*Note:* In this problem and the next, ignore 29 February [leap year] birthdays.)

10. (Calculator problem) What is the smallest number of people gathered together such that the probability that any two of them *do* have the same birthday is greater than one-half? (If you do not already know the answer to this question, the answer may surprise you.)

11. If the king's advisor in this cartoon is assuming that the event of one army surrounding a castle is independent of the event of any other army surrounding the same castle, and that these two events have equal probability, what must he be assuming is the probability of one army surrounding a castle?

THE WIZARD OF ID   by Brant parker and Johnny hart

By permission of Johnny Hart and NAS, Inc.

12. If the odds that you are helped by any psychiatrist are 3 to 1 against and the event of being helped by one psychiatrist is mutually exclusive of the event of being helped by any other psychiatrist, what are the odds that you are helped by going to two psychiatrists?

13. For the box of exercise 3, if two balls are drawn at random without replacement, what is the probability that they are both green?

14. Repeat exercise 13 but "with replacement."

15. A whole number is chosen at random from the whole numbers from 1 to 50. What is the probability that it is an even square?

16. In exercise 15, find the probability that the number is a square or even.

17. A packet of twenty seeds contains 15 that will germinate and five that won't. If three seeds are chosen at random, what is the probability that at least 1 will germinate?

18. Five people are chosen at random from a group of 100 people. If 20 of the 100 have cancer, what is the probability that at least 2 of the 5 chosen have cancer?

19. A number between 0 and 1 is to be chosen at random. The resulting number might be 0.348935. . . , where the decimal expansion goes on forever. Since there are an infinity of such numbers, the probability that any particular one, such as $\frac{1}{3}$ = 0.3333. . . , is chosen is 0. Which of the following statements is true and which is false?

a. If $P(E) = 0$, then E is an impossible event.

b. If E is an impossible event, then $P(E) = 0$.

20. True or false?

a. If $P(E) = 1$, then event E is certain to happen.

b. If event E is certain to happen, then $P(E) = 1$.

Make up an example, such as the one in exercise 19, to illustrate this point.

Section 3–3 Probability Distributions

When performing an experiment, the results obtained may or may not be numerical quantities. Whether or not the results are numeric, however, it is often possible to assign a number to each possible outcome. For example, when a coin is tossed a number of times, the outcomes are sets of heads and tails, nonnumeric results. Viewed another way, however, the results could be expressed as the number of heads resulting from the tosses. A number such as this, which is assigned to the outcome of a random experiment, is called a **random variable.**

> A **random variable** is a numerical quantity that is assigned to the outcome of a random experiment.

Usually the letter x is used to denote a random variable.

EXAMPLE 3-26   While playing roulette, you bet on red. If red comes up, you get your bet back plus an equal amount from the casino. If anything else comes up, you lose your bet. Assign random variables for this experiment.

Solution   The solution to this problem depends, of course, on the size of your bet. Let us assume you bet $1. If red comes up, you get your dollar back plus $1, which means you are up $1. If black or green comes up, you lose your $1, which means you are down $1. So, for the event red, $x = +1$, and for the event black or green, $x = -1$.

Once all possible results of the experiment are identified, the **probability distribution** of these events can be determined. A probability distribution is defined as follows:

> A **probability distribution** is a list of all possible values of a random variable associated with an experiment, together with the probability of each value.

EXAMPLE 3-27   Form the probability distribution for the experiment of example 3-26.

Solution   Since there are 18 numbers on a roulette wheel that are red and 20 that are black or green, the solution is

x	$P(x)$
$+1$	$\frac{18}{38}$
-1	$\frac{20}{38}$.

EXAMPLE 3-28   For the experiment of rolling two dice, form the probability distribution using the sum of the two uppermost faces as the random variable.

Solution   A particular roll of two dice, such as (4,3), would be assigned the value $x = 7$. You know from example 3-22 that any particular sequence of outcomes of rolling two dice has a probability of $(\frac{1}{6}) \times (\frac{1}{6}) = \frac{1}{36}$. There are 36 such possible sequences. To form the probability distribution of the sum of values on the two uppermost faces of the dice, it is merely necessary to determine how many ways each sum can be obtained. This is shown in table 3-2.

Table 3–2

Sum	Possible Sequences of Dice Forming Sum					
2	1–1					
3	1–2	2–1				
4	1–3	2–2	3–1			
5	1–4	2–3	3–2	4–1		
6	1–5	2–4	3–3	4–2	5–1	
7	1–6	2–5	3–4	4–3	5–2	6–1
8		2–6	3–5	4–4	5–3	6–2
9			3–6	4–5	5–4	6–3
10				4–6	5–5	6–4
11					5–6	6–5
12						6–6

Thus, the probability distribution for the sum is

x	$P(x)$
2	1/36
3	2/36
4	3/36
5	4/36
6	5/36
7	6/36
8	5/36
9	4/36
10	3/36
11	2/36
12	1/36.

●

Properties of Probability Distributions

Probability distributions are similar to relative frequency distributions discussed in chapter 2.

In a relative frequency distribution, intervals of values are listed together with a relative frequency corresponding to each. The relative frequency of any particular interval is computed by dividing the frequency (f), or number of values occurring in that interval, by the sum of the frequencies (Σf), or the size of the sample n.

In a probability distribution, in place of intervals of observed values, values (or intervals of values) of the *random variable* are listed together with the *probabilities* (theoretical or empirical) associated with the random variable. Since a probability is in fact a relative frequency (observed or theoretical), you can imagine that there are various tie-ins between relative frequency distributions and probability distributions, as you will now see.

Figure 3-1

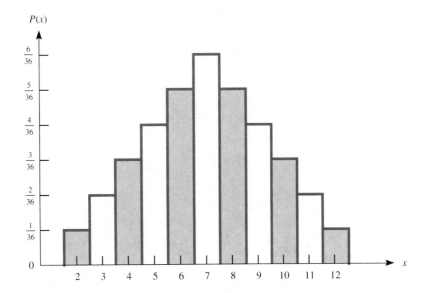

In a frequency distribution, the sum of the frequency column (Σf) was always equal to the sample size n, while the sum of the relative frequencies was always equal to 100% or 1. In a probability distribution, the sum of the probability column is always equal to 1. Check for yourself that the sum of each of the probability columns in examples 3-27 and 3-28 is 1.

Graphs

Two graphs that depicted frequency distributions were the histogram and the simple frequency polygon. These graphs can also be used for probability distributions.

EXAMPLE 3-29   Draw a histogram and polygon for the probability distribution of example 3-28.

Solution   The histogram is shown in figure 3-1.
The polygon is shown in figure 3-2 on page 132.   ●

As with histograms for frequency distributions, the histogram is drawn so that the sides of the rectangles come at the true class limits of the intervals. Usually, though, with probability distributions, the values of the random variable are put at the middle of each of the bases of the rectangles.
In a polygon, points are placed directly above the values of the random variable at a height equal to the probability. These points are joined to form the polygon and, as with a polygon for a frequency distribution, the polygon is brought down to the horizontal axis at both ends. Note that the values of 1 and 13 for x in example 3-29 have probabilities equal to 0, since it is impossible to get these values.

Figure 3–2

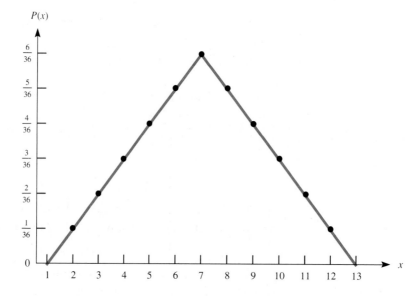

The total area covered by the bars of a histogram or the interior of a polygon corresponding to a probability distribution is always 1. This property corresponds to the fact that the sum of the probability column of a probability distribution is always 1. For example, the polygon in the solution to example 3–29 is a triangle with base $= 12$ and height $= \frac{6}{36}$ or $\frac{1}{6}$. The area of a triangle is given by $\frac{1}{2}$ $\times$ base $\times$ height, and note that $\frac{1}{2} \times 12 \times \frac{1}{6} = 1$. Check for yourself that the sum of the areas of the bars of the histogram in the solution to example 3–29 is 1.

Mean and Standard Deviation of a Probability Distribution

Consider another method of finding the mean from a frequency distribution with this example.

EXAMPLE 3–30   Find the mean of the frequency distribution in table 3–3.

Table 3–3

Interval	m	f	$f\%$	fm	$mf\%$
3–5	4	2	10	8	40
6–8	7	5	25	35	175
9–11	10	8	40	80	400
12–14	13	4	20	52	260
15–17	16	1	5	16	80
		$n = \Sigma f = 20$	$\Sigma f\% = 100$	$\Sigma fm = 191$	$\Sigma mf\% = 955$

Solution   The totals of the columns have been calculated. So,

$$\bar{x} = \frac{\Sigma fm}{\Sigma f} = \frac{191}{20} = 9.55.$$
But notice that $\Sigma mf\% = 955$ percent, which is also 9.55.

The letters μ and σ are used for probability distributions since they are theoretical and not samples.

A probability distribution, like a frequency distribution, will always have a mean and a standard deviation.

Formula (3), presented in chapter 2, gave the mean of a frequency distribution as

$$\mu = \frac{\Sigma fm}{\Sigma f}.$$

Now, any individual frequency of a class f divided by Σf gives the relative frequency for that class. Thus, formula (3) could be rewritten as

$$\mu = \Sigma mf\%.$$

For probability distributions, in place of intervals, midpoints (m), and relative frequencies, there are values for the random variable (x) and probabilities. So, formula (3) becomes

$$\mu = \Sigma x \cdot P(x). \tag{18}$$

Formulas (9) and (10), presented in chapter 2, can also be adapted to give the standard deviation of a probability distribution.

$$\sigma = \sqrt{\Sigma x^2 \cdot P(x) - [\Sigma x \cdot P(x)]^2} \tag{19}$$

EXAMPLE 3-31   Find the mean and standard deviation of the probability distribution found in example 3-27.

Solution

x	$P(x)$	$x \cdot P(x)$	$x^2 \cdot P(x)$
$+1$	$\dfrac{18}{38}$	$\dfrac{18}{38}$	$\dfrac{18}{38}$
-1	$\dfrac{20}{38}$	$-\dfrac{20}{38}$	$\dfrac{20}{38}$
		$\Sigma x \cdot P(x) = -\dfrac{2}{38}$	$\Sigma x^2 \cdot P(x) = \dfrac{38}{38} = 1$

From formulas (18) and (19),

$$\mu = -\frac{2}{38} \qquad\qquad \sigma = \sqrt{1 - \left(-\frac{2}{38}\right)^2}$$

$$\approx -0.053 \qquad\qquad \approx 0.9986.$$

The mean of a probability distribution is the theoretical value you would expect if the experiment were performed a large number of times and you then found the mean of all the results, whereas the standard deviation is an indication of the average spread you would expect in the results.

In example 3–31, the experiment was betting $1 on red at roulette. The mean −0.05 represents a loss to you of 5 cents. At first, that might seem ridiculous, since you can only win or lose $1. However, what you must realize is that the mean is a long-run average. If you were to sit down at the roulette table hour after hour simply betting $1 on red each time the wheel is spun, what the above mean is telling you is that you should expect to be down 5 cents for every dollar you bet. If you bet 100 times, you should expect to be down $5, and for 1,000 bets you should expect to be down $50, and so on. From the casino's point of view, they can expect to rake in 5 cents for every $1 bet. Some gamblers might win and others will lose, but as long as there are sufficient numbers of bets being made, the casino always wins—5 cents on average for each dollar bet in roulette.

EXAMPLE 3–32   Find the mean and standard deviation of the probability distribution found in example 3–28.

Solution

Table 3–4

x	$P(x)$	$x \cdot P(x)$	$x^2 \cdot P(x)$
2	1/36	2/36	4/36
3	2/36	6/36	18/36
4	3/36	12/36	48/36
5	4/36	20/36	100/36
6	5/36	30/36	180/36
7	6/36	42/36	294/36
8	5/36	40/36	320/36
9	4/36	36/36	324/36
10	3/36	30/36	300/36
11	2/36	22/36	242/36
12	1/36	12/36	144/36
		$\Sigma x \cdot P(x) = \dfrac{252}{36}$	$\Sigma x^2 \cdot P(x) = \dfrac{1{,}974}{36}$

From formulas (18) and (19),

$$\mu = \frac{252}{36} \qquad\qquad \sigma = \sqrt{\frac{1{,}974}{36} - 7^2}$$
$$= 7 \qquad\qquad\qquad \approx \sqrt{5.833}$$
$$\approx 2.42.$$

Here, these results indicate that if two dice were rolled a large number of times, the mean of all the rolls is expected to be 7 with a standard deviation of 2.42. If you look back at figures 3–1 and 3–2, do you see that 7 is indeed a reasonable value for the mean?

Exercise Set 3–3

 1. Construct a probability distribution for the experiment of tossing three coins, using the number of heads as the random variable.

 2. Toss three coins 80 times, counting the number of heads each time. Construct a relative frequency distribution for the observed number of heads. Compare your answer to the answer to exercise 1.

3. Identical pieces of paper having each of the integers 1, 2, 3, 4, 5 on them are placed into a hat. Two pieces of paper are drawn at random and the sum of the numbers on them is taken to be the random variable. Construct the probability distribution for this experiment.

4. Draw the histogram and polygon for the probability distribution constructed in exercise 3.

5. A wallet contains three $1 bills, two $5 bills, four $10 bills, and a $20 bill. If a bill is selected at random and its value used as the random variable, construct the probability distribution.

6. If two bills are selected at random from the wallet in exercise 5 and the random variable is their total value, construct the probability distribution.

7. A probability distribution must satisfy the two basic properties of probability: $0 \leq P(x) \leq 1$ for all x, and $\Sigma P(x) = 1$ over all x. For the mathematical function $P(x) = x/15$ for $x = 0, 1, 2, 3, 4, 5$, write out the complete distribution defined by it and check the two properties to see whether or not this can be classified as a probability distribution. If so, the given function can be called a *probability function*.

8. Refer to exercise 7 to answer this question: Can the function $P(x) = x/9$ for $x = 1, 2, 3, 4$ be called a probability function?

 9. Find the mean and standard deviation of the probability distribution you found in exercise 1.

10. Find the mean and standard deviation of the probability distribution you found in exercise 6.

 11. For a family of 4 children, form the probability distribution using the number of girls as the random variable and then find the mean and standard deviation of this distribution.

 12. Repeat exercise 11 for a family of 5 children.

 13. From your answers for the mean in exercises 11 and 12, without doing the work, what do you think is the mean number of girls in a family of 6 children? A family of 7 children?

Section 3–4   The Binomial Distribution

In probability theory, there are many theoretical probability distributions that are best summarized with mathematical formulas. In this chapter, two such distributions are going to be considered. The first, the subject of this section, is the **binomial distribution.** The results of an experiment will follow a binomial distribution if three conditions are satisfied.

1. The experiment consists of a number (n) of repeated independent trials. This means that in each successive trial of the experiment, the outcome is independent of the outcomes in all other trials.
2. Each trial has exactly two possible outcomes of interest, one labeled "success," with a probability p, and one labeled "failure," with a probability q, where $q = 1 - p$.
3. The probabilities p and q remain stable over successive trials of the experiment.

EXAMPLE 3–33   If an experiment is to toss a coin three times, would the results follow a binomial distribution?

Here, a trial is a toss of the coin.

Solution   Every toss of a coin is independent of any other toss, so condition 1 is satisfied with $n = 3$. Each time the coin is tossed, there are two possible outcomes, heads and tails, so condition 2 is satisfied, with $p = \frac{1}{2}$ and $q = \frac{1}{2}$. Therefore, the answer to the question is: Yes, the results of this experiment do follow a binomial experiment.   ●

EXAMPLE 3–34   A student has not studied for a psychology exam and decides to answer all the multiple-choice questions by guessing. If there are 10 questions, each with 5 choices, only 1 of which is correct, would the student's results follow a binomial distribution?

Here, a trial is an attempt at a question.

Solution   Since the student is guessing, it can be assumed that every attempt to answer the question is independent of any other, so condition 1 is satisfied with $n = 10$.

You might think that condition 2 is violated here, since there are 5 possible outcomes for each question. However, there are only 2 outcomes *of interest:* the right answer, which he has 1 chance in 5 of obtaining, and the wrong answer, which he has 4 chances in 5 of obtaining. Therefore, condition 2 is also satisfied with $p = \frac{1}{5}$ and $q = \frac{4}{5}$. Again, the answer to the question is: Yes, the results of this experiment will follow a binomial distribution.   ●

To show how a binomial distribution can be constructed, first, take a look at two examples from scratch and then examine a couple of shortcuts.

EXAMPLE 3–35   Construct the binomial distribution of example 3–33.

Solution   The possibilities are 0 heads, 1 head, 2 heads, or all 3 heads.

$$P(0 \text{ heads}) = P(TTT) = \frac{1}{2} \times \frac{1}{2} \times \frac{1}{2} = \frac{1}{8}$$

$$P(1 \text{ head}) = 3 \times \frac{1}{2} \times \frac{1}{2} \times \frac{1}{2} = \frac{3}{8}$$

This second result is true because there are 3 ways of getting 1 head: *HTT, THT, TTH*, and each of these has a probability of $\frac{1}{2} \times \frac{1}{2} \times \frac{1}{2}$ of occurring. For a similar reason,

$$P(2 \text{ heads}) = 3 \times \frac{1}{2} \times \frac{1}{2} \times \frac{1}{2} = \frac{3}{8}$$

$$P(3 \text{ heads}) = P(HHH) = \frac{1}{2} \times \frac{1}{2} \times \frac{1}{2} = \frac{1}{8}.$$

This is the solution you should have found for exercise 1 of exercise set 3–3.

Thus, if x is the number of heads, the distribution becomes

x	$P(x)$
0	$\frac{1}{8}$
1	$\frac{3}{8}$
2	$\frac{3}{8}$
3	$\frac{1}{8}$.

Remember that because the values of x exhaust all possible outcomes, the second property of probability tells you that the sum of the probabilities in the distribution must equal 1. This is a convenient fact to use in checking your calculations.

138

EXAMPLE 3–36   Consider the experiment of rolling one die four times and defining a success as a 5 or 6. Construct the resulting binomial distribution.

Solution   Here, $n = 4$, $p = \frac{2}{6} = \frac{1}{3}$, $q = \frac{4}{6} = \frac{2}{3}$, and x can be 0, 1, 2, 3, or 4, representing no successes, 1 success, etc. There is just one way of getting no successes, i.e., all failures. There are 4 ways of getting 1 success, the 1 success being on first, second, third, or fourth roll. Similarly, check the following figures:

x	Number of ways of getting x
0	1
1	4
2	6
3	4
4	1.

The probability of getting no successes is given by

$$\frac{2}{3} \times \frac{2}{3} \times \frac{2}{3} \times \frac{2}{3} = \left(\frac{2}{3}\right)^4.$$

The probability of getting 1 success in any one of the 4 ways is given by

$$\frac{1}{3} \times \frac{2}{3} \times \frac{2}{3} \times \frac{2}{3} = \frac{1}{3} \times \left(\frac{2}{3}\right)^3.$$

Thus, $P(1 \text{ success}) = 4 \times \frac{1}{3} \times \left(\frac{2}{3}\right)^3.$

Continuing in this way, the following distribution is obtained:

x	$P(x)$
0	$\left(\frac{2}{3}\right)^4 \approx 0.198$
1	$4 \cdot \left(\frac{1}{3}\right) \cdot \left(\frac{2}{3}\right)^3 \approx 0.395$
2	$6 \cdot \left(\frac{1}{3}\right)^2 \cdot \left(\frac{2}{3}\right)^2 \approx 0.296$
3	$4 \cdot \left(\frac{1}{3}\right)^3 \cdot \left(\frac{2}{3}\right) \approx 0.099$
4	$\left(\frac{1}{3}\right)^4 \approx 0.012.$

Table 3–5

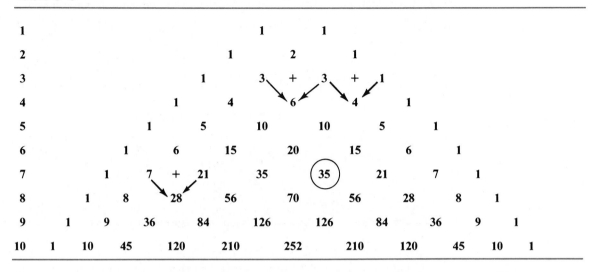

If you study the final probabilities in the solution to example 3–36 before they were converted to decimals, a pattern emerges. The required formula is

In mathematics, $y^0 = 1$ for any nonzero number y.

$$P(x \text{ successes in } n \text{ trials}) = w \cdot p^x \cdot q^{n-x}, \qquad (20)$$

where w is the number of ways of getting x successes in n trials, and p and q are as defined at the beginning of this section. Before going on, check to see if formula (20) gives the right answers to examples 3–35 and 3–36.

There is one problem remaining: what is an easy way to find w? One way is to use **Pascal's triangle,** 10 lines of which are reproduced in table 3–5. This table is generated by starting with the two 1s in line 1 and then adding pairs of numbers diagonally downwards, as in the few examples illustrated. *Note:* The first and last number in each line is 1.

You may recognize lines 3 and 4, since they were used in the solutions to examples 3–35 and 3–36, respectively. To use the triangle, use the line number that is equal to n. Then, the numbers in that line give w for no successes, one success, and so on, up to all n successes.

EXAMPLE 3–37   Find the number of ways of getting 3 successes in 7 trials.

Solution   Here, $n = 7$, $x = 3$. Referring to the 7th line of Pascal's triangle and counting: 0, 1, 2, 3 (*Note:* Always start at *zero,* not 1, and count until you reach x.), find the number 35 (circled in table 3–5). This, then, is w. Thus, there are 35 ways of getting 3 successes in 7 trials.   ●

What if n is larger than 10? One way would be to construct more lines of Pascal's triangle as described earlier. However, when n is quite large, it becomes cumbersome to use Pascal's triangle to determine the number of ways to obtain a particular value of x in a binomial experiment. Fortunately, a short-cut method is available. Before introducing this method, however, you need to learn about another mathematical concept: **n factorial.** Using factorial notation, the process of multiplying a number by all smaller integers can be summarized very conveniently.

---

n factorial, written $n!$, is defined by

$$n! = n \cdot (n - 1) \cdot (n - 2) \cdot \ldots \cdot 3 \cdot 2 \cdot 1,$$

when $n > 0$ and $0! = 1$.

Thus, $5! = 5 \cdot 4 \cdot 3 \cdot 2 \cdot 1 = 120$ and $3! = 3 \cdot 2 \cdot 1 = 6$.

Using this concept of factorial, w can be calculated by using the following formula:

$$w = \frac{n!}{x! (n - x)!}$$

and thus formula (20) can be rewritten as follows:

$$P(x \text{ successes in } n \text{ trials}) = \frac{n!}{x! (n - x)!} p^x q^{n - x}.$$

EXAMPLE 3–38   Check the answer to example 3–37 using the factorial method.

Solution   Again, $n = 7$ and $x = 3$. Thus,

$$w = \frac{7!}{3! (7 - 3)!} = \frac{7 \cdot 6 \cdot 5 \cdot 4 \cdot 3 \cdot 2 \cdot 1}{3 \cdot 2 \cdot 1 \cdot 4 \cdot 3 \cdot 2 \cdot 1} = 35, \text{ as before.}$$

EXAMPLE 3–39   Find the number of ways of getting 8 successes in 12 trials using the factorial method.

Solution   Here, $n = 12$, $x = 8$, and w is to be calculated. Thus, since $12! = 12 \cdot 11! = 12 \cdot 11 \cdot 10! = 12 \cdot 11 \cdot 10 \cdot 9!$, and so on,

$$w = \frac{12!}{8! (12 - 8)!} = \frac{12 \cdot 11 \cdot 10 \cdot 9 \cdot 8!}{8! \, 4!} = \frac{12 \cdot 11 \cdot 10 \cdot 9}{4 \cdot 3 \cdot 2 \cdot 1} = 495.$$

For large n, the factorial notation is much more convenient than using Pascal's triangle. However, as n continues to increase, the factorial method can result in very large numbers. Later in the text, you will see an even faster way to find binomial probabilities when n is large, without having to calculate w.

EXAMPLE 3-40   In example 3-34, find the probability that the student gets exactly 8 questions correct.

Solution   Here, $n = 10$, $p = \frac{1}{5}$, $q = \frac{4}{5}$, $x = 8$, and Pascal's triangle gives $w = 45$. Using the formula for w presented above, the identical result can be obtained without relying on the more cumbersome Pascal's triangle.

$$w = \frac{n!}{x!(n-x)!} = \frac{10!}{8!\ 2!} = \frac{10 \cdot 9 \cdot 8!}{8!\ 2!} = \frac{10 \cdot 9}{2 \cdot 1} = 45$$

Thus, using formula (20):

$$P(x = 8) = 45 \cdot \left(\frac{1}{5}\right)^8 \cdot \left(\frac{4}{5}\right)^2 \approx 0.00007.$$

EXAMPLE 3-41   In example 3-34, find the probability that the student gets more than 2 questions correct.

Solution   More than 2 questions correct is the complementary event of 2 or less questions correct, which is 0, 1, or 2 correct. So,

$$P(x \text{ is more than } 2) = 1 - P(x = 0, 1, \text{ or } 2)$$
$$= 1 - [P(x = 0) + P(x = 1) + P(x = 2)].$$

Use formula (20) to check that

$$P(x = 0) \approx 0.107, P(x = 1) \approx 0.268, P(x = 2) \approx 0.302, \text{ and then:}$$
$$P(x \text{ is more than } 2) = 1 - (0.107 + 0.268 + 0.302)$$
$$= 1 - 0.677$$
$$= 0.323.$$

The Mean and Standard Deviation of the Binomial Distribution

There are two simple formulas for finding the mean and standard deviation of a binomial distribution. They are

$$\mu = np \tag{21}$$

$$\sigma = \sqrt{npq}\ . \tag{22}$$

Using an example, it can be shown that for a binomial distribution these formulas give the same results as the more general formulas for mean and standard deviation of a probability distribution, i.e., formulas (18) and (19).

EXAMPLE 3–42 Find the mean and standard deviation of the distribution of example 3–35, first by using formulas (21) and (22), and then by using formulas (18) and (19).

Solution For tossing a coin 3 times, $n = 3$, $p = \frac{1}{2}$, $q = \frac{1}{2}$. Thus,

$$\mu = 3 \cdot \left(\frac{1}{2}\right) = 1.5$$

$$\sigma = \sqrt{3 \cdot \left(\frac{1}{2}\right) \cdot \left(\frac{1}{2}\right)} \approx 0.87,$$

according to formulas (21) and (22).

Now, using formulas (18) and (19):

x	$P(x)$	$x \cdot P(x)$	$x^2 \cdot P(x)$
0	$\frac{1}{8}$	0	0
1	$\frac{3}{8}$	$\frac{3}{8}$	$\frac{3}{8}$
2	$\frac{3}{8}$	$\frac{6}{8}$	$\frac{12}{8}$
3	$\frac{1}{8}$	$\frac{3}{8}$	$\frac{9}{8}$
		$\Sigma x \cdot P(x) = \frac{12}{8}$	$\Sigma x^2 \cdot P(x) = \frac{24}{8}$

and

$$\mu = \frac{12}{8} = 1.5$$

$$\sigma = \sqrt{\frac{24}{8} - (1.5)^2} = \sqrt{3 - 2.25} = \sqrt{0.75} \approx 0.87.$$

Notice that both pairs of answers agree.

Exercise Set 3–4

1. Construct the binomial distribution for tossing a coin 4 times.

2. A multiple-choice exam has 5 questions with 3 choices for each question, with only 1 choice being correct for each question. Construct the binomial distribution.

3. Find the mean and standard deviation of the distribution in exercise 1.

4. Find the mean and standard deviation of the distribution in exercise 2.

5. An airplane has 4 engines working independently of each other. The probability that any one fails is 4%. The plane will crash if 3 of the 4 or all 4 engines fail. What is the probability that the plane will crash?

6. You take a True/False quiz with 9 questions. You will get a D or an F if you get 6 or less questions correct. If you are guessing all the answers, what is the probability that you get a D or an F?

7. A baseball player has a 0.300 average. This means that he has a 30% chance of getting a hit each time he is at bat. What is the probability that he gets at least two hits in his next 5 times at bat, assuming all times at bat are independent of each other?

8. A basketball player makes 3 out of every 4 foul shots she attempts. What is the probability that she will make at least 10 of her next 12 foul shots, assuming all attempts are independent of each other?

9. An instructor gives his class 20 multiple-choice questions with 4 choices, only 1 of which is right, for each question. If the class as a whole is guessing, what would you expect the mean and standard deviation to be?

10. A roulette wheel is spun 100 times and you count the number of times red shows. If this experiment is repeated a large number of times, what would you expect the mean and standard deviation of the number of reds per 100 spins to be?

Section 3–5 The Normal Distribution, an Introduction

While the binomial distribution is useful in considering the outcomes of a wide assortment of experiments, there is another probability distribution that is even more important in statistics: the **normal distribution.** In fact, it is the most important probability distribution you will encounter in elementary statistics.

The normal distribution is especially important because many populations that are studied have distributions that fit the normal probability distribution quite closely. In addition, even when populations are not normally distributed, the statistics that are estimated using these populations may be normally distributed. As you will see in chapter 6, this turns out to be useful in testing the quality of your statistical work and in testing hypotheses about the population.

What is the normal distribution? Its mathematical equation is

$$y = \frac{1}{\sigma \sqrt{2\pi}} e^{-(x - \mu)^2/2\sigma^2}.$$

You see in this the symbols μ and σ, but unless you have studied calculus, this equation will mean nothing to you. Fortunately, it is not necessary to understand the equation of the normal distribution to be able to work with it. Rather than trying to fathom the equation, take a look at some of the shapes the graph of this equation can assume. You will see some examples in figure 3–3 on page 144.

Figure 3–3

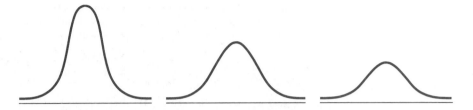

Since it is a continuous distribution, the normal distribution assumes the shape of a curve, rather than a polygon. This curve is called, appropriately enough, the **normal curve.** Although there are many different normal curves, they all have the distinctive bell shape. For this reason, the normal curve is often referred to as a **bell-shaped curve.**

The questions arise as to what determines which particular bell shape you are dealing with and where on the horizontal axis it is located. If you glance back at the equation of the normal curve, you will notice that the Greek letters μ and σ are involved. Recall that these are precisely the letters used for mean and standard deviation of a population. This has not changed now that you have reached this chapter. The normal distribution, being a probability distribution, like the binomial, also has a mean and standard deviation. It is the mean, μ, that determines where, along the horizontal axis, the normal curve is located; and it is the standard deviation, σ, that determines the particular bell shape you are dealing with. This will become clearer as you study the properties of the normal distribution.

Properties of the Normal Distribution

Normal Property 3–1   Symmetry
A typical normal curve is shown in figure 3–4.

Figure 3–4

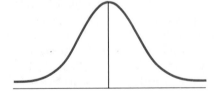

Notice that a central line has been drawn. Such a central line can be drawn for every normal curve and every normal curve has the property that it is symmetric with respect to this line. This means that the left half of the normal curve is an exact mirror reflection of the right half across this central line; or to look at it another way, if the normal curve were cut out of the page and the shape folded along the central line, the left half of the curve would fall exactly on the right half.

Normal Property 3–2   Center

Look again at the central line. It comes directly where the highest point on the normal curve is located. The highest point on the curve comes above the value on the horizontal axis that has the highest probability. The value with the highest probability is the value that would have the highest relative frequency were the experiment to be performed. But this is just the mode. The central line, then, comes at the mode of the normal distribution.

Because of the symmetry of the normal distribution about the central line, exactly half of the distribution falls below it and half above it. The value at the exact center of a distribution is called the median. Thus, the central line comes at the median of the normal distribution.

Again, because of the symmetry of the normal distribution, not only is it true that half the distribution falls below and half above the central line, but the two halves are identical. If the normal curve were a physical mass, would it not balance there? The balance point of a distribution is at the mean. Thus, the central line comes at the mean of the normal distribution.

In summary, the mean, median, and mode of any normal distribution are all equal, and the value that they are equal to is that value on the horizontal axis directly below the high point of the curve. Since all three of these measures are equal, it is only necessary to refer to one of them, and the mean is chosen to fulfill this role. Now you should be able to see why it is the mean which determines where, along the horizontal axis, the normal curve is located.

Normal Property 3–3   Tails

Take a look at figures 3–3 and 3–4. Notice that the curves are drawn approaching the horizontal axis without touching it. This is done to illustrate another property of the normal curve. The normal curve extends forever in both directions, getting closer and closer to the horizontal axis, the further out you go, without ever touching it. When a curve approaches a straight line like this, the straight line is referred to as an **asymptote** to the curve. The normal curve is **asymptotic** to the horizontal axis. No matter how far out you go, left or right, there is always space between the curve and the axis. Of course, after a while, the space becomes so minuscule that a pencil with a fine enough lead could not be found to show that the space is there. These parts of the normal curve to the left and right are referred to as the **tails** of the normal distribution.

Normal Property 3–4   Area

As with all probability distributions, the total area under the normal curve and above the horizontal axis is equal to 1 or 100%. You may wonder how this is possible when you were just informed, in normal property 3–3, that this area extends infinitely in both directions. How is it possible that an area that extends infinitely can have a finite value of 1? To illustrate that this is indeed possible, consider figure 3–5 on page 146.

Figure 3–5

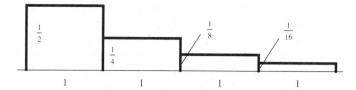

Here there is a series of rectangles, each with a base of length 1. The largest has a height of $\frac{1}{2}$, the next a height of $\frac{1}{4}$, the next $\frac{1}{8}$, and so on. Can you see that it is possible always to draw, in theory, another rectangle without reaching the horizontal axis? In other words, an infinity of rectangles can be drawn in this manner. The total area covered by these rectangles is called A.

Since the area of an individual rectangle is given by base $\times$ height, $A = \frac{1}{2} + \frac{1}{4} + \frac{1}{8} + \frac{1}{16} + \frac{1}{32} + \cdots$. Combining these fractions two at a time:

$$A = \tfrac{3}{4} + \tfrac{1}{8} + \tfrac{1}{16} + \tfrac{1}{32} + \cdots$$
$$A = \tfrac{7}{8} + \tfrac{1}{16} + \tfrac{1}{32} + \cdots$$
$$A = \tfrac{15}{16} + \tfrac{1}{32} + \cdots$$
$$A = \tfrac{31}{32} + \cdots.$$

Can you see what is happening? The more rectangles that are added in, the closer A gets to 1, while remaining smaller than 1. The study of limits reveals that when an infinity of rectangles are included, A becomes equal to 1. So, here is an example of an area extending infinitely, while its value is finite and equal to 1. The normal distribution has a similar property.

It follows from the symmetry of the normal distribution that the area under the normal curve to the left of the central line is 50% or 0.5, and the area to the right of the central line is also 50% or 0.5.

Normal Property 3–5   Shape

Examine the three normal curves in figure 3–6. The places on the curves indicated by the arrows are called **points of inflection.** To explain this concept, consider any particular part of a curve. It either has a curvature that opens upwards, in which case it is said to be **concave up** there, or it has a curvature that opens downwards, in which case it is said to be **concave down** there. The point at which a curve changes its concavity, i.e., from concave up to concave down, or vice versa, is called a point of inflection. All normal curves are concave down in the center and concave up in the tails. Thus, all normal curves have two points of inflection, and these are the points indicated by the arrows in figure 3–6. These points of inflection are exactly one standard deviation, σ, on either side of the mean, μ.

Figure 3–6

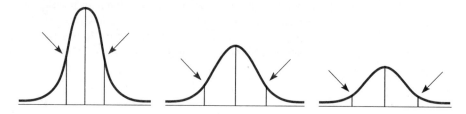

The three lines drawn on each of the three normal curves in figure 3–6 are, in order from left to right, at $\mu - \sigma$, μ, and $\mu + \sigma$. The distance between $\mu - \sigma$ and μ is σ, and the distance between μ and $\mu + \sigma$ is also σ. So, the two lengths between the three lines in each of the drawings are each of length σ. As you move left to right from one curve to the next, notice that these lengths ($= \sigma$) get larger.

To see why this is, recall that σ measures the spread of a distribution, and as you move from one curve in figure 3–6 to the next, they become more spread out. If one normal curve is less spread out than a second, then the value of the former's standard deviation is less than that of the latter's. The more spread out a normal distribution is, the larger the value of its standard deviation, and the further the points of inflection are from the center.

Notice, also, that the more spread out a normal distribution is, the lower its high point is. This is compensation, since it must always be true that the total *area* under the normal curve be 1 or 100%. What is of additional interest is the fact that, for *all* normal curves, the area under the normal curve between $\mu - \sigma$ and $\mu + \sigma$ is about 0.68 or 68%. By symmetry, the area between $\mu - \sigma$ and μ is 34%, and between μ and $\mu + \sigma$ also 34%. See figure 3–7.

Where this figure of 68% comes from is impossible to show without the use of calculus. So, once again, you are asked to accept this on trust.

Figure 3–7

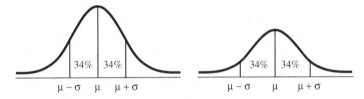

The Standard Normal Distribution

From the above discussion of the properties of the normal distribution, it should be clear to you that there are an infinite variety of normal distributions. Any particular normal distribution under consideration depends on the values of μ and σ. Now you are going to examine a particular normal distribution. This one is the one in which $\mu = 0$ and $\sigma = 1$. This is referred to as the **standard normal distribution.**

Notice that in figure 3–8 the central line is at 0 on the horizontal axis, or z-axis, and the points of inflection are above -1 and 1 on the axis. As is pointed

Figure 3–8

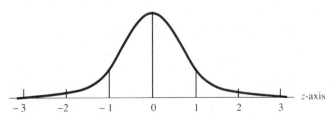

out above, the area under this normal curve between -1 and 1 is 68%. Frequently, other areas under the normal curve will be of interest. You might be interested in the area under the normal curve between 0 and 0.5, between -1 and 0.5, between 1.3 and 2, beyond 1.5, and so on. How are these areas computed? These calculations can be done from scratch, using methods from calculus. However, even with this knowledge, the computations are lengthy, dull, and routine. Fortunately, these computations have already been done and the results put in the form of a table. Look now at table 1 in appendix B. This is an example of such a table.

The letter z refers to the values along the horizontal axis. For instance, look down to 1.0 under z. The number immediately to the right of this, in the body of the table, is 0.3413. This is the 34% mentioned previously. The values in the body of the table give us the area between the central line and any particular value of z. There are ten columns in the body of the table so that you can find areas for values of z up to two decimal places. The letter A will be used to denote the area under the normal curve between the central line and a particular value of z. Suppose you wanted to find A for $z = 1.67$. Go down the column headed by z until you reach 1.6, then go across to the column headed by 0.07 and read off $A = 0.4525$. See figure 3–9.

Figure 3–9

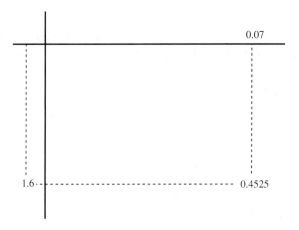

The interpretation of the result is that 45.25% of the area under the standard normal curve lies between the central line (where $z = 0$) and $z = 1.67$.

z may be positive or negative, but A, being an area, is always positive.

Notice that all values of z in the table are positive. This does not mean that you cannot use the table for negative values of z. The fact that the normal curve is symmetric allows you to do just that. For instance, the above result could be interpreted by saying that 45.25% of the area under the normal curve lies between $z = -1.67$ and the central line. It could further be deduced that the percentage of the total area under the normal curve between $z = -1.67$ and $z = 1.67$ is 90.5%, by addition. Now to look at some more examples that demonstrate the versatility of the table.

EXAMPLE 3–43   Find the area under the normal curve between z_1 = -0.63 and z_2 = 1.24.

Solution   A picture of the problem, with the required area shaded, is shown in figure 3–10.

Figure 3–10

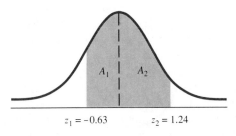

$z_1 = -0.63$ $z_2 = 1.24$

Now, for z_1 = -0.63, A_1 = 0.2357 and for z_2 = 1.24, A_2 = 0.3925. A is the area between the particular value of z and the central line. Since, in this example, the central line comes between the two values of z, to get the required area it is necessary to add, $A_1 + A_2$ = 0.6282 = 62.82%, the answer.

EXAMPLE 3–44   Find the probability that a particular value chosen at random from the standard normal distribution lies between z_1 = 1.3 and z_2 = 2.49.

Solution   As in all probability distributions, of which the normal is one, a probability can be calculated by finding the appropriate area. The situation is illustrated in figure 3–11.

Figure 3–11

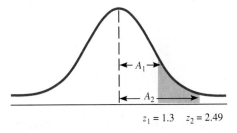

$z_1 = 1.3$ $z_2 = 2.49$

For z_1 = 1.3, A_1 = 0.4032 and for z_2 = 2.49, A_2 = 0.4936. This time the central line does not lie between the two values of z, so to get the required area it is necessary to subtract $A_2 - A_1$ = 0.0904. Thus the required probability is 0.0904 or 9%.

EXAMPLE 3–45   Find the area under the normal curve to the right of z = 2.56.

Solution   See figure 3–12.

Figure 3–12

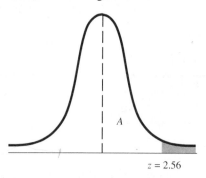

$z = 2.56$

For $z = 2.56$, $A = 0.4948$. But remember A gives the area between the central line and z. To get the shaded area, it is necessary to subtract this value of A from the area extending forever to the right of the central line, which, as pointed out earlier, is 0.5. Thus, $0.5 - A = 0.0052 = 0.52$ or 1%, the answer. ●

EXAMPLE 3–46   Find the probability that a particular value chosen at random from the standard normal distribution is greater than $z = -1.42$.

Solution   See figure 3–13.

Figure 3–13

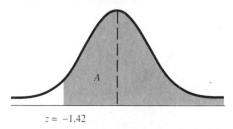

$z = -1.42$

If you were looking for the area to the left of $z = -1.42$, you could solve this problem as in example 3–45 by subtracting. But can you see that, in this example, *adding* 0.5 gives the necessary area? For $z = -1.42$, $A = 0.4222$ and $0.5 + A = 0.9222$ or about 92%, the required probability. ●

EXAMPLE 3–47   Between which two values of z does a centrally placed area of 80% of the standard normal curve lie?

Solution   This example is a little different from the previous ones since it requires using the table in reverse. The situation is shown in figure 3–14.

Figure 3–14

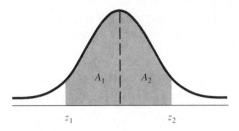

Since the area is to be centrally placed, $A_1 = A_2$ and z_1 must be the negative of z_2. From the statement of the example, $A_1 + A_2 = 80\%$, and, since A_1 and A_2 are equal, $A_1 = A_2 = 40\% = 0.4$. Looking through the body of the table, 0.3997 followed by 0.4015 can be found. The value of 0.3997 is closer to 0.4, since it is 0.0003 from 0.4 (as opposed to 0.0015). $A = 0.3997$ corresponds to $z = 1.28$. Therefore, $z_1 = -1.28$ and $z_2 = 1.28$, the two required values of z.   ●

Exercise Set 3–5

In exercises 1–6, find the area under the standard normal curve between the two indicated values of z.

1. $z_1 = 0$ and $z_2 = 1.8$
2. $z_1 = -2.05$ and $z_2 = 0$
3. $z_1 = -0.74$ and $z_2 = 1.36$
4. $z_1 = -2.92$ and $z_2 = 0.41$
5. $z_1 = 1.27$ and $z_2 = 2.13$
6. $z_1 = -2.99$ and $z_2 = -0.68$

In exercises 7 and 8, find the indicated area under the standard normal curve.

7. To the right of $z = 1.82$
8. To the left of $z = -0.50$
9. Find the probability that a value chosen at random from a standard normal distribution is smaller than $z = 0.57$.
10. Find the probability that a value chosen at random from a standard normal distribution is larger than $z = -1.27$.
11. Between which two values of z does a centrally placed area of 90% of the standard normal curve lie?
12. Between which two values of z does a centrally placed area of 75% of the standard normal curve lie?

Section 3–6   Nonstandardized Normal Distributions

Table 1 would not be very helpful if it could only be used for the standard normal distribution. Suppose you were dealing with a normal distribution with $\mu = 5$ and $\sigma = 3$, and you wanted to know the area under it between μ and $x = 11$. Now 11 is 6 units above $\mu = 5$, and 6 units is 2 standard deviations, since $\sigma = 3$. This problem, therefore, is equivalent to the problem of finding the area under the *standard* normal curve between the central line and $z = 2$. The answer is 0.4772 or 47.72%.

What the above example illustrates is a method of standardizing any normal curve. The connection between x and z is given by formula (12) in chapter 2 for standard scores:

$$z = \frac{x - \mu}{\sigma}.$$

Using the above figures,

$$z = \frac{11 - 5}{3} = \frac{6}{3} = 2.$$

The value of z resulting from formula (12) tells you how many standard deviations x is above or below the mean. Although there exists a whole family of normal distributions, respective areas are the same. Thus the area under *any* normal curve between its mean and, say, 2.37 standard deviations above the mean is exactly the same as the area under the *standard* normal curve between the central line and $z = 2.37$. Table 1, therefore, can be applied to any normal distribution.

Suppose a population of male heights follows a normal distribution with mean 69.6 inches and standard deviation 2.8 inches. Consider the following examples.

EXAMPLE 3–48   Find the area under this normal curve between 68 inches and 71 inches.

EXAMPLE 3–49   What percentage of this population is shorter than 67 inches?

EXAMPLE 3–50   If a man is chosen at random from this population, what is the probability that he is between 5 feet 10 inches and 6 feet?

EXAMPLE 3–51   Which height are the tallest 20% taller than?

As with the other examples in this chapter, it is always helpful to sketch a normal curve illustrating the situation.

Solution 3–48   See figure 3–15.

Figure 3–15

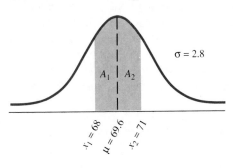

For $x_1 = 68$, $z_1 = \dfrac{68 - 69.6}{2.8} \approx -0.57$ and $A_1 = 0.2157$.

For $x_2 = 71$, $z_2 = \dfrac{71 - 69.6}{2.8} = 0.50$ and $A_2 = 0.1915$.

The answer is given by $A_1 + A_2 = 0.4072$ or about 41%.

Solution 3–49   See figure 3–16.

Figure 3–16

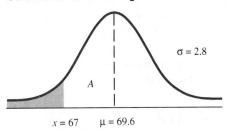

For $x = 67$, $z = \dfrac{67 - 69.6}{2.8} \approx -0.93$ and $A = 0.3238$. The answer is given by $0.5 - A = 0.1762$ or about 18%. Thus, 18% of this population is shorter than 67 inches.

Solution 3–50   5 feet 10 inches = 70 inches and 6 feet = 72 inches. See figure 3–17.

Figure 3–17

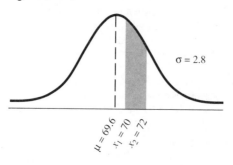

For $x_1 = 70$, $z_1 = \dfrac{70 - 69.6}{2.8} = 0.14$ and $A_1 = 0.0557$.

For $x_2 = 72$, $z_2 = \dfrac{72 - 69.6}{2.8} = 0.86$ and $A_2 = 0.3051$.

The answer is given by $A_2 - A_1 = 0.2494$ or about 25%. There is a 25% probability that a man, chosen at random from this population, is between 5 feet 10 inches and 6 feet.   ●

Solution 3–51   See figure 3–18.

Figure 3–18

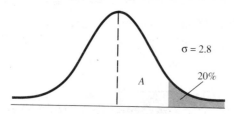

Since the shaded area = 20% = 0.20, $A = 0.5 - 0.20 = 0.30$. Using table 1 in reverse, find $A = 0.2995$ is closest to 0.30 and then $z = 0.84$. This means that the x that is needed is 0.84 of a standard deviation above the mean.

Solving $z = \dfrac{x - \mu}{\sigma}$ for x gives $x = \mu + z\sigma$. Thus,

$$x = \mu + 0.84\sigma$$

$$= 69.6 + 0.84(2.8)$$

$$\approx 71.95, \text{ very close to 6 ft.}$$

The tallest 20% are taller than 6 feet.   ●

EXAMPLE 3–52   A professor at a university gives her (large) class an examination and finds that the scores follow very closely to a normal curve with a mean of 71 and a standard deviation of 9. If she wishes no more than 7% of the class to fail, below which score is a fail? (When students ask an instructor, "Do you grade on a curve?", they assume (hope?) that the curve is normal, although they may not always know that that is what they have in mind!)

Solution   See figure 3-19.

Figure 3-19

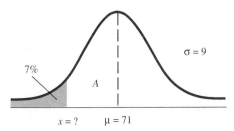

If the shaded area is 7%, $A = 43\% = 0.43$, which is closest to 0.4306 in the table, and $z = 1.48$. Thus x must be 1.48 standard deviations *below* the mean.

$$x = 71 - 1.48(9)$$
$$= 57.68$$

A student who has scored 57 or less has failed the exam.   ●

The Normal Approximation to the Binomial Distribution

As was promised in section 3-4, you will now examine a method of dealing with the binomial distribution when n is fairly large. The result that is needed is as follows:

> When n is large enough to make *both* np and nq larger than 5, the binomial distribution can be approximated very well by the normal distribution.

EXAMPLE 3-53   A teacher gives his class a multiple-choice test with 28 questions, where each question has 4 choices, only 1 of which is correct. What is the probability that a student, by guessing, gets between 8 and 10 questions (inclusive) correct?

Solution   As you have seen, the number of right answers found by guessing on a multiple-choice test follows a binomial distribution. In this case, $n = 28$, $p = \frac{1}{4}$, $q = \frac{3}{4}$. Thus, the mean and standard deviation of the binomial distribution are given by

$$\mu = 28 \cdot \tfrac{1}{4} = 7$$
$$\sigma = \sqrt{28 \cdot \tfrac{1}{4} \cdot \tfrac{3}{4}} \approx 2.29.$$

Since $np = 7$, $nq = 21$, which are both larger than 5, it is possible to approximate this binomial distribution by a normal distribution with the same mean and standard deviation. The question asks for the probability of getting between 8 and 10 (inclusive) questions correct.

The number of questions correct is a discrete variable, whereas the normal distribution is a continuous distribution. To allow for this, "between 8 and 10 (inclusive)" is transformed to "between 7.5 and 10.5," since these are the true class limits of the interval 8–10. This transformation is called a **continuity correction.**

Whenever the binomial distribution is approximated by the normal distribution, this continuity correction should be applied. The class limits of the intervals of values being examined should always be replaced by their *true* class limits. The examples in table 3–6 illustrate the procedure.

Table 3–6

Required Interval	Interval Used When Approximating with the Normal
1. 18–24	17.5–24.5
2. 18.00–19.50	17.75–19.75
3. 18 or more	17.5 to maximum possible
4. more than 18	18.5 to maximum possible
5. 18 or less	minimum possible to 18.5
6. less than 18	minimum possible to 17.5

For more on this, see chapter 6.

In examples 1 and 2 of table 3–6, it is being assumed that the intervals following those given are 25–31 and 20.00–21.50, respectively. Other possibilities might occur to you, in which case the intervals would be corrected differently for continuity. If, say, the interval following 18.00–19.50 had been 19.60–21.10, then it would be corrected for continuity by 17.95–19.55. In example 4, "more than 18" is the same interval as "19 or more," assuming only whole numbers are being considered, which is why it is corrected for continuity as shown.

Returning now to the solution, a picture of the situation is shown in figure 3–20.

Figure 3–20

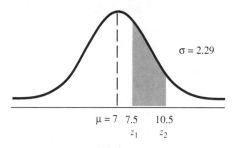

$\sigma = 2.29$

$\mu = 7$ 7.5   10.5

z_1 z_2

The problem is then completed as in example 3–50.

$$z_1 = \frac{7.5 - 7}{2.29} \approx 0.22 \text{ and } A_1 = 0.0871.$$

$$z_2 = \frac{10.5 - 7}{2.29} \approx 1.53 \text{ and } A_2 = 0.4370.$$

$$\text{Then, } A_2 - A_1 = 0.4370 - 0.0871$$

$$= 0.3499$$

$$\approx 0.350.$$

Thus, there is about a 35% probability that the student gets between 8 and 10 questions (inclusive) correct. ●

Although the calculations are not pretty, the solution to example 3–53 was worked out using only the binomial distribution formulas. To three decimal places, it worked out to 0.332, which is fairly close to the 0.350 found by using the normal approximation. The approximation gets better as np and nq get larger.

EXAMPLE 3–54   Find the probability that you get at least 12 sevens in 50 rolls of a pair of dice.

Solution   Since "seven" and "not seven" are the outcomes of interest, $p = \frac{1}{6}$, $q = \frac{5}{6}$ (see example 3–23), and $n = 50$. By formulas (21) and (22),

$$\mu = 8.33$$
$$\sigma = 2.64.$$

See figure 3–21.

Figure 3–21

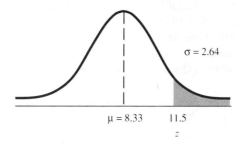

$\sigma = 2.64$

$\mu = 8.33$ $\quad$ 11.5

z

Since "at least 12" means "12 or more," this translates into "11.5 or more" for the continuous normal distribution.

Using the normal approximation:

$$z = \frac{11.5 - 8.33}{2.64} \approx 1.20.$$

$A = 0.3849$ and $0.5 - A = 0.1151$. There is about an 12% probability that you get 12 or more sevens in 50 rolls of a pair of dice. ●

The exact answer to example 3–48 is 0.117 to three decimal places. Good agreement, wouldn't you say?

In Retrospect

In chapter 2, you studied ranks, percentile ranks, and *z*-scores, which are three different ways of converting raw data—three separate measures of location. The question arises as to whether there is any connection between them. For instance, if the percentile rank of a particular raw score is known, can its *z*-score be computed, or vice versa? In general, the answer to this question is no. The methods of computing these two types of converted scores are completely different. However, if you are dealing with a normal distribution, table 1 enables you to answer the above question in the affirmative.

Consider examples 3–48 to 3–51. These concerned a normal distribution of male heights with mean 69.6 inches and standard deviation 2.8 inches. It is possible to answer the questions in examples 3–55 and 3–56.

EXAMPLE 3–55   What is the percentile rank of a height of 6 feet?

EXAMPLE 3–56   Which height is P_{25}?

Solution 3–55   Since 6 feet = 72 inches, to get its percentile rank, it is necessary to find the percentage of the distribution that falls below it. This is indicated by the shaded region in figure 3–22.

Figure 3–22

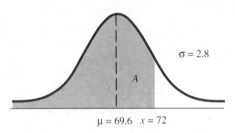

$\sigma = 2.8$

A

$\mu = 69.6 \quad x = 72$

Now,

$$z = \frac{72 - 69.6}{2.8} = 0.86,$$

and from table 1, $A = 0.3051$. Recall that this is the area under the curve between μ and x. Thus, area of shaded region = $0.5 + A = 0.8051$ or 80.51%. So, a height of 6 feet is at the 80th percentile, or 6 ft = P_{80}.   ●

Alternatively, it might be said that in the normal distribution, a *z*-score of 0.86 is at the 80th percentile. For nonmathematical distributions, no such connection exists between *z*-scores and percentile ranks.

Solution 3–56 Since x is to be at the 25th percentile, the area under the normal curve to the left of x must be 25%, as is shown in figure 3–23.

Figure 3–23

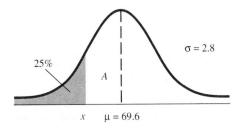

Since the area to the left of μ is 50%, the area between x and μ is $A = 25\%$ $= 0.25$. Using table 1, for $A = 0.25$, $z = 0.67$ (0.2486 is the closest value in the table to 0.25). However, from the figure, x can be seen to be *below* μ, which means it is necessary to take $z = -0.67$,

$$x = 69.6 - 0.67(2.8) \text{ inches}$$

$$\approx 67.7 \text{ inches.}$$

This, then, is the height at the 25th percentile,

$$P_{25} = 67.7 \text{ inches.} \qquad \bullet$$

Again, it might be said that in the normal distribution the 25th percentile has a z-score of -0.67.

In chapter 2, you were presented with an empirical rule stating that about 68% of a distribution falls between one standard deviation below the mean and one standard deviation above, while about 95% of a distribution falls between two standard deviations below the mean and two standard deviations above. Table 1 shows where these figures of 68% and 95% come from: $z = 1$ corresponds to $A = 0.3413$, doubled is about 68% and $z = 2$ corresponds to $A = 0.4772$, doubled is about 95%. The empirical rule works well when the distribution with which you are dealing is close to normal. It is, of course, less than satisfactory for distributions that deviate substantially from the normal.

The picture at the beginning of the chapter is of a machine in the Museum of Science and Industry in Los Angeles. The balls fall from a tube directly above the center of the normal distribution, painted on the glass. As they fall, they hit a series of 12 lines of obstacles, and at each they have a probability of $\frac{1}{2}$ of falling either way. So, where the balls fall follows a binomial distribution with $n = 12$, $p = \frac{1}{2}$, $q = \frac{1}{2}$. That this is approximated very well by a normal distribution is illustrated by the fact that after all the balls have fallen, they fit the normal distribution, painted on the glass, very well. This experiment is repeated over and over again every day to show that any one repetition is not a fluke.

HAVE NOT
COVERED

DO
THESE

Exercise Set 3–6

1. A normal distribution has $\mu = 34$ and $\sigma = 5$. Find the area under the curve of this distribution between 27 and 43.

2. A normal distribution has $\mu = 109$ and $\sigma = 21$. Find the area under the curve of this distribution between 70 and 90.

 3. A population of female heights is found to follow a normal distribution with mean 64.3 inches and standard deviation 2.6 inches. What percentage of this population is taller than 68 inches?

 4. If a woman is chosen at random from the population described in exercise 3, what is the probability that she is between 5 feet and 5 feet 2 inches tall?

 5. A population of light bulbs is found to have a mean life-time of 980 hours with a standard deviation of 115 hours. If these life-times are normally distributed, find the cut-off point for the longest-lasting 10% of the bulbs.

 6. A machine dispensing coffee into containers that are supposed to hold 16 oz is found to dispense a mean of 15.9 oz with a standard deviation of 0.2 oz. What percentage of the containers will have less than 16 oz in them if the weights dispensed are normally distributed?

 7. If a large university class has a set of examination scores that are normally distributed with a mean of 72 and a standard deviation of 8 and a B is awarded to those who score in the 80s, what percentage of the class earned a B? (*Note:* Since the normal distribution is continuous, a score in the 80s is a score between 79.5 and 89.5.)

 8. If the top 9% of the class in exercise 7 are to earn an A, what is the cut-off score for an A?

 9. Quarter-inch nuts are produced with their diameters following a normal distribution with mean 0.25 inches and standard deviation 0.01 inches. If the smallest 5% and the top 8% are deemed unacceptable, find the range of diameters of acceptable nuts.

 10. A psychologist finds that a large number of rats placed in a maze commit a mean of 20 errors with a standard deviation of 2 errors. If the number of errors follows a normal distribution, what percentage of the rats commit less than 15 errors?

11. A machine produces an important part for a video game. Each part produced is independent of other parts produced, and each has a probability of 5% of being defective. What is the probability that a batch of 200 of these parts has between 5 and 15, inclusive, that are defective?

12. An organization to help smokers give up smoking claims a 70% success rate (i.e., any smoker has a probability of 70% of being "cured"). Find the probability that less than 60 of the next 100 smokers are "cured" of smoking. Assume independence between smokers and note that "less than 60" is "59 or less."

13. In a standard normal distribution, find the percentile rank of $z = -1.35$.

14. In a standard normal distribution, find the percentile rank of $z = -2.16$.

15. In a standard normal distribution, find z corresponding to P_{85}.

16. In a standard normal distribution, find z corresponding to P_{68}.

17. In a normal distribution with mean 13.3 and standard deviation 2.6, find the percentile rank of 17.

18. In a normal distribution with mean 111.1 and standard deviation 11.1, find the percentile rank of 125.

19. In a normal distribution with mean 65 and standard deviation 3, which value corresponds to P_{30}?

20. In a normal distribution with mean 2.2 and standard deviation 0.16, which value corresponds to P_{45}?

Section 3-7   Summary of Terms

addition rule of probability 120
asymptote 145
asymptotic 145
bell-shaped curve 144
binomial distribution 136
chance 116
complement 119
concave down 146
concave up 146
conditional probability 122
continuity correction 156
evens or even odds 117
event 112
failure 112
independent 124
multiplication rule of probability 122

mutually exclusive 121
n factorial 140
normal curve 144
normal distribution 143
odds 116
outcome 112
Pascal's triangle 139
points of inflection 146
probability 111
probability distribution 129
random variable 128
sample space 112
simple event 112
standard normal distribution 147
success 112
symmetry 144
tails 145

Section 3-8   Summary of Symbols

Symbols	Meaning
$P(A)$	The probability of an event A
A'	The complement of an event A
$P(B \mid A)$	The probability of B, given A
n	The number of trials in a binomial distribution
p	The probability of a success in a binomial distribution
q	The probability of a failure in a binomial distribution
w	The number of ways of getting x successes in n trials of a binomial distribution
x	The number of successes in a binomial distribution
z	The number of standard deviations that a value in a normal distribution is above or below the mean

Section 3–9 Basic Properties of Probability

1. $0 \le P(E) \le 1$, where E is any event.

2. $\Sigma P(e) = 1$, where the summation is taken over all simple events e.

Section 3–10 Summary of Formulas

Formula	Use
$(14)\ P(E) = \dfrac{\text{number of simple events in } E}{\text{number of simple events in } S}$	To find the probability (theoretical or empirical) of an event E
$(15)\ P(A') = 1 - P(A)$	To find the probability of the complement of an event; this is sometimes easier than finding the probability of an event directly
$(16)\ P(A \text{ or } B) = P(A) + P(B) - P(A \text{ and } B)$ $(17)\ P(A \text{ and } B) = P(A)P(B \mid A)$	Compound probability rules
$(18)\ \mu = \Sigma x \cdot P(x)$	To find the mean of a probability distribution
$(19)\ \sigma = \sqrt{\Sigma x^2 \cdot P(x) - [\Sigma x \cdot P(x)]^2}$	To find the standard deviation of a probability distribution
$(20)\ P(x) = w \cdot p^x \cdot q^{n-x}$ where w is found from Pascal's triangle or using the formula: $w = \dfrac{n!}{x!\,(n-x)!}$	To find the probability of x successes in n trials of a binomial distribution
$(21)\ \mu = np$	To find the mean of a binomial distribution
$(22)\ \sigma = \sqrt{npq}$	To find the standard deviation of a binomial distribution

Chapter 3 Review Exercises

1. The integers $1, 2, \ldots, 10$ are each written on identical pieces of paper and placed in a hat. One piece of paper is drawn at random. Write the sample space for this experiment.

2. Suppose, in the experiment in exercise 1, two pieces of paper are drawn, one after the other, the second being drawn without replacing the first. How many simple events would there be in the sample space of this experiment?

3. If two jokers are added to an ordinary deck of cards and the cards shuffled, what is the probability that a card drawn at random is an ace?

4. What is the probability that a card drawn at random from the deck described in exercise 3 is a heart?

5. (Calculator problem) One of the games that French gamblers like the Chevalier de Méré (see text) played was to bet that in four rolls of a die, at least one ace would show. What is the probability of winning this bet?

6. (Calculator problem) In another French gambling game, they would bet that in twenty-four rolls of two dice, snake-eyes would show at least once. What is the probability of winning this bet?

Exercises 7 and 8 refer to this frequency polygon.

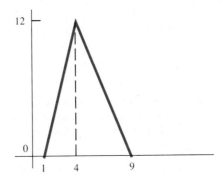

7. What is the probability that a value chosen at random from this distribution lies between 1 and 4 inclusive?

8. What is the probability that a value chosen at random from this distribution lies between 3 and 5 inclusive?

9. In the game of blackjack, a blackjack consists of an ace together with a ten, jack, queen, or king. What are the odds of being dealt a blackjack from a full deck of cards?

10. Suppose three jacks, one king, one ace, and five cards below a ten have already been dealt and you are dealt two cards from the remaining deck. What are the odds of you being dealt blackjack?

11. A die has been modified to have one face with two dots, two faces with four dots, and three faces with six dots. If x is the number of dots when the die is rolled, construct a probability distribution for this experiment.

12. If two dice, modified as in exercise 11, are rolled and x is the total of the dots on the two uppermost faces, construct a probability distribution for this experiment.

13. A coin is biased to favor heads, so that $P(H) = \frac{2}{3}$ and $P(T) = \frac{1}{3}$. If this coin is tossed eight times, what is the probability of getting at least seven heads?

14. Bruce likes to speed in his Porsche. He is about to take a trip that will take him through six radar speed traps. There is a 20% probability that he is caught by any one speed trap and these are independent. If Bruce plans to speed the whole way, what is the probability that he is caught in at least one speed trap?

DO
CIRCLED

15. Find the area under the standard normal distribution between $z_1 = -1.48$ and $z_2 = -0.48$.

16. Find the area under the standard normal distribution to the right of $z = 1.05$.

17. A normal distribution has $\mu = 0.08$ and $\sigma = 0.08$. Find the area under the curve of this distribution between 0 and 1.25.

18. A normal distribution has $\mu = 4,874$ and $\sigma = 636$. Find the area under the curve of this distribution to the right of 4,000.

19. A mail-order service receives a number of mail-orders per day and, over a period of time, this number is found to follow a normal distribution with mean 450 and standard deviation 25. What is the probability that on any day this service will receive more than 500 orders?

20. The ages of all faculty members at a large community college are found to be normally distributed with a mean of 43 and a standard deviation of 7. If there are exactly 300 on the faculty, how many of them are in their thirties?

21. The rents on apartments in a low-rent district are found to be normally distributed with a mean of $180 and a standard deviation of $30. What range of rents do the middle 50% of renters pay?

22. The time taken for a sprinter to run 100 yards is found to be normally distributed with a mean of 11.2 seconds and a standard deviation of 0.3 seconds. Find the probability that on a random run the sprinter will take between 10.5 and 10.9 seconds to run 100 yards.

23. A public relations firm has found that its mass-mailings have produced a 5% response rate. If it plans to mail 5,000 pieces in its next mailing, what is the probability that fewer than 200 respond? (*Hint:* A 5% response rate can be interpreted as a 5% probability that a person receiving a piece of the mailing will respond.)

24. In exercise 23, find the probability that *exactly* 250 respond. (Careful! Remember that the normal distribution is continuous.)

25. A population of schoolchildren's scores on a Raven test, a psychological test of intelligence, were found to be normally distributed with a mean of 31.0 and a standard deviation of 10.0. Find the score that is at the third quartile.

26. For the population described in exercise 25, find the percentile rank of a score on the Raven test of 20.

27. An agricultural researcher finds that a random sample of dairy cows produced yearly amounts of butterfat that were normally distributed with a mean of 350 lb and a standard deviation of 75 lb. Find the percentile rank of a cow that produces 400 lb of butterfat.

28. For the population described in exercise 27, find the amount of butterfat produced by the cow at the 20th percentile.

29. Find the mean of a normal distribution with standard deviation 7.5 if a value in the normal distribution $21 = P_{90}$.

30. Find the mean of a normal distribution with standard deviation 21 if a value in the normal distribution $7.5 = P_{20}$.

Hands-on Class/Student Project

The following are some experiments that you can perform yourself to test the ideas of probability theory in practice. If you worked with other students in the class for the projects at the end of chapters 1 and 2, work with these groups again.

1. Roll a pair of dice 360 times, recording the total of the uppermost faces each time. (If you are working in a group, you could share the load in doing this.) Check to see if your results conform to the theoretical result of example 3–28.

2. Combine your results with the results of another group or groups in the class to produce a larger sample of rolls. Does this (larger) sample give better agreement with example 3–28 than your sample alone?

3. Get a dime, a nickel, and a penny and toss all three into the air and note down how they land: *HHH, HHT, HTH,* and so on, where the first letter (*H* or *T*) in each group of three corresponds to the result for the dime, the second for the nickel, and the third for the penny. Repeat this 80 times.

4. Count how many times the dime came up heads, then how many times the nickel did, and finally how many times the penny did. Theory says that in 80 tosses each of the coins should show heads approximately 40 times. How well did this work for you in practice?

5. Combine your coin-tossing results with other students' results for the same experiment in the class. With these larger samples, do you get better agreement with the theory?

6. Form your own (empirical) probability distribution by taking the random variable x as the number of heads in each toss of the dime, nickel, and penny and calculating the percentage in your sample that $x = 0, 1, 2,$ or 3. How well do your results compare with the answer to example 3–35, when the (theoretical) fractions, there, are also converted to percentages? You can also combine your results with others' in the class to see (again) if the agreement improves.

7. In the chapter 1 project, you collected a sample of heights as variable 1. Heights are generally normally distributed (if males and females are considered separately). In chapter 2, you calculated the mean and standard deviation of this sample. (If you didn't do this for females and males separately, do so now.) Pretend, for a moment, that your sample (of the same gender as you) is a population, with mean and standard deviation that you calculated. Using the ideas of section 3–6, calculate what percentage of this "population" would have a height at your height or taller.

8. Calculate the actual percentage in your sample of heights of students having the same gender as you who are as tall or taller than you. Compare your answer to your answer to exercise 7. Again, better agreement might be attained if you combine your sample of heights with corresponding samples from others in the class.

9. The answer to exercise 10 of exercise set 3–2 is that there is a better than even chance that two people have the same birthday (day and month) in any group of 23 people or more. The data set for variable 2 (collected in chapter 1) is comprised of the birth dates of your sample of students. If you have more than 30 in your sample, you very likely do have a match if you ignore the *year* of birth. Check to see if you do! (In fact, if there are more than 23 students in your class, it is more likely than not that two of you have the same birthday.)

10. Combine your sample of birth dates (ignoring the year of birth) with as many other samples in the class as you can. Arbitrarily, divide this big sample into as many samples of exactly 23 as possible. Count how many of these samples of 23 birthdays do have at least one match. Since, in 23 birthdays, there is about a 50% probability that there is at least one match, approximately half of your samples of 23 birthdays should have a match, and of course half won't. This will work better the more samples of 23 birthdays you are able to divide your big sample into.

CHAPTER

4

The Central Limit Theorem

....

"Of the general character of those outlaws an
estimate may be formed from a few samples."

....

MACAULAY

The Care and Feeding of Statistics

Al Hibbs is a physicist at the Jet Propulsion Laboratory who speaks in the familiar Voice of the Voyager missions. Some years ago he and a friend used their knowledge of statistical analysis to beat the wheel at the Pioneer Club in Reno.

What they did sounds very simple. They took turns around the clock for several days, watching and recording the spins of a particular roulette wheel. If the wheel had been perfect, every number would have turned up in the record about the same number of times. But almost all man-made devices, including roulette wheels, are imperfect. The imperfections in the wheel Al Hibbs and his friend were watching made some numbers turn up more frequently than others. When their statistical analysis of the record showed them just which these numbers were, they placed their bets and ran an original stake of $125 up to $6,000 before the management changed the wheel.

Unfortunately no one can duplicate their feat any longer. The roulette wheels in Reno and Las Vegas are now regularly disassembled every day, and the parts of the different wheels are interchanged. All the wheels are still imperfect, but the imperfections vary from day to day. Al Hibbs and his friend had to record the results of 10,000 spins of the wheel to be sure the result of their statistical analysis would be accurate and reliable. With a record for only one day they could have lost their whole stake.

Of course there's more to statistical analysis than playing roulette. It is a useful and powerful tool for understanding critical relationships in complicated situations. Particularly in the sciences that deal with human beings where laboratory experiments are all but impossible—the biomedical sciences and the social sciences—statistical analysis can uncover hidden connections and help to establish explanations.

Putting together the necessary large set of data on human beings can be very expensive. The especially high cost in medical research explains why studies there have sometimes misused statistical analysis by applying it to sets of observations too small to give reliable and accurate results.

Another problem in using statistics on human beings is often overlooked. Suppose someone wanted to estimate the average age of gamblers in Reno and made a survey of players on a particular day. If a reunion of veterans from World War I just happened to fill the hotels in town that day, the answer wouldn't be "right" no matter how carefully the statistical analysis had been carried out or how many people had been surveyed. More precisely, the estimate obtained for the average age of gamblers would be correct for just that day, but wrong for all other days.

It is easy to see what is wrong in this example, but the same kind of error can appear in research also. A conclusion about the treatment of a human disorder based on observations of patients at just one hospital can be flawed in exactly the same way as the conclusion from the sample of gamblers in Reno. Such a sample is described as "biased." In everyday speech we would say it is "not a fair sample." The Reno sample was obviously biased with too many people in their 80s, but a sample of patients from a single hospital or a single doctor's practice may be biased in ways that cannot be discovered so easily.

Perhaps it is the flaws in clinical studies that used small samples or samples with a hidden bias that have made people somehow suspicious of the results of clinical trials obtained with statistical analysis. Whatever the reason, patients and even some doctors may remain unconvinced by the results of elaborate clinical trials in which many patients were observed over long periods and the data were subjected to careful statistical analysis.

The drug Laetrile is an example of that kind of unfounded skepticism. Clinical trials of Laetrile showed that the drug had no effect on cancer and indeed had unwanted side effects. Still people continue to stream across the U.S. border to get Laetrile in Mexican clinics.

So where does that leave us? Well, statistical analysis is like any other human activity. It must be carried out correctly, with due understanding and care, but human beings make mistakes and not every use is correct. On the other hand when it is done right— whether by Al Hibbs or a medical research team— believing the answer and acting on it can yield real benefits.

"The Care and Feeding of Statistics" by Mildred Goldberger, in *Scientific View*, April 1982; *Los Angeles Times*.

......

Chapter Overview
When Is a Sample Representative?

You are strongly urged to read the article at the beginning of this chapter, "The Care and Feeding of Statistics." It will give you an idea of what this chapter is about. In this chapter, you will learn why statisticians sample, the nature of random sampling, other possible sampling schemes, and the most important mathematical result used in elementary statistics: the central limit theorem.

......

Section 4–1   Sampling Methods

Why Sampling?

Researchers in many fields often wish to learn about a population. It may, however, be impossible, or at least impractical (given the amount of effort or expense involved), to obtain the data on a whole population.

If a manufacturer producing light bulbs claims that the bulbs last a certain length of time, and you want to test this, it would be impossible to test the whole population, since no bulbs would then be left for ordinary use.

If you wanted to examine some characteristic of outlaws, you would be hard pressed to contact *every* outlaw. Thus, gathering population data would be a practical impossibility.

To find out anything about *every* person living in California (by the 1980 census, a state of over 23 million people) would not be impossible. Given the effort and expense involved, however, it would be a formidable task indeed. Even the Census Bureau only *attempts* this every ten years and is never successful in obtaining a response from everyone.

With the correct statistical techniques, however, it is possible to make creditable inferences about a population from a relatively small sample. You will be studying these techniques in part 3 of this book. To use any of these techniques effectively, it is important to make sure that the sample being used accurately represents, or reflects, the population of which it is a part. In other words, you must ensure, somehow, that your sample is a **representative sample** of its parent population. If, for any reason, it is unrepresentative, it matters little how sophisticated the statistical techniques applied to it are. The results are going to be useless as far as telling you anything reliable about the population. To take an extreme example, consider a study of male undergraduate heights at a certain university. If you choose as your sample the university's basketball team, do you think this would give you accurate information about the population of all the university's undergraduate heights? If you want to know about typical gambler ages and you draw your sample primarily from World War I veterans, as in the article at the beginning of this chapter, are you likely to get a representative sample of ages in the population? In both cases, the answer is certainly not. You can never be 100% sure that the sample is sufficiently representative of the entire underlying population. As long as you are dealing with sample data, rather than population data, there is *always* room for error. Using statistical methods, however, you can determine the size of the possible error quite accurately.

Random Sampling

The techniques described in this introductory text are based on the concept of **random sampling.** Statisticians have a very definite idea of what this means. Suppose you wanted a sample of size $n = 1$. (Of course, you would never try to make inferences about a population based on a sample this small; but the concept is easier to grasp if you start small and work up.) What is a *random* sample with only one value in it? To draw one value, at random, from a population, statisticians make the following requirement:

> Any member of the population must have an equal probability of being chosen as any other member.

There are several ways that this can be achieved. But take a look first at a method that would not qualify as "drawing at random." If a "volunteer" is to be chosen for a dangerous mission, a method often employed is that of drawing straws. Straws of various lengths are prepared, one for each member of the population, and then drawn, the lengths being concealed by the hand holding them. The person drawing the shortest straw becomes the "volunteer." The statistician would argue that this is not a random process. How are the straws held? Once the shortest straw has been drawn, the remaining drawers have *no* chance of being selected. Therefore, not all members of the population have an equal probability of being selected.

Another method of choosing the "volunteer" would be to give each member of the population a number. Then write the numbers on identical pieces of paper, put the papers in a hat, shuffle them well, and draw one out. Each member of the population would necessarily have an equal probability of being selected. This would be an acceptable method of drawing a sample of one, at random. If the population is relatively small, this would even be a practical way of proceeding. But, what if the population is large? Statisticians do not spend their time putting thousands of identical pieces of paper into a hat. Rather, they use a **random number table.** One such table is table 7 in appendix B.

If, for example, you wish to draw one at random from a population of size $N = 1,000$, you could label the members of the population $000, 001, 002, \ldots,$ $998, 999$. Then, arbitrarily choosing a place in the random number table, you can find a three-digit number to give you the one member of the population you are looking for. Since in a true random number table all combinations and permutations of three digits occur with equal frequency, this method qualifies as a random process, according to the definition of what it means to draw one at random: every member has an equal probability of being selected.

Now, suppose you wish to draw a random sample larger than $n = 1$. Besides the requirement that every member of the population have an equal probability of being included in the sample, a second condition is required. Suppose you want a sample of size n. Then, the second condition states that every sample of size n that can possibly be drawn from the population must have an equal probability of being selected as *your* sample.

Therefore, the two conditions for a **simple random sample** are:

1. Every member of the population must have an equal probability of being included in the sample.
2. Every sample of the required size n that can be drawn from the population must have an equal probability of being the sample selected.

How could these conditions be violated? Here is an example: Suppose you wanted a random sample of people from Los Angeles County. There are 77 cities in Los Angeles County. Suppose you decide to take a sample of size $n = 77$ by taking one person, at random, from each of these cities. This would not qualify as a simple random sample for two reasons. First, not every member of the population has an equal probability of being selected. A person living in Rolling Hills (pop. 2,039) has probability $\frac{1}{2,039}$ of being selected, whereas a person living in Los Angeles City (pop. 2,950,010) only has probability $\frac{1}{2,950,010}$ of being selected. Second, even if the subgroups of the population (in this case the 77 cities of Los Angeles County) are of equal size, condition two would still be violated by the sample procedure. Not every sample of size $n = 77$ has an equal probability of being *the* random sample. A sample with $n = 77$ and having 2 people from Los Angeles City and none from Rolling Hills, for example, has a probability of 0 of being *the* sample.

The use of the random number table can be extended to samples larger than $n = 1$. In a true random number table, not only do the digits 0, 1, 2, 3, 4, 5, 6, 7, 8, 9 occur with equal frequency, but every combination and permutation of two digits, three digits, four digits, etc., occurs with equal frequency as every other. Since the early days of the study of statistics, much research has been undertaken in the construction of random number tables. At first, statisticians thought they might use any tables, such as mathematical tables or telephone books, but these proved totally unsatisfactory. In Britain, for example, they found that telephone numbers had a marked underrepresentation of 5s and 9s. Why? This was in the early days of the telephone, with no automatic dialing, you simply picked up the receiver and told the operator the number you wanted. The telephone company deliberately avoided 5s and 9s in their assignment of numbers to subscribers, because over the telephone line these two digits were often confused.

Don't choose numbers "at random" from a random number table. The numbers are already arranged at random. If you use a random process, you might undo the randomness already there.

To return to the population of size $N = 1,000$: Suppose you wanted a sample of size $n = 38$. After arbitrarily choosing a starting place, as was discussed earlier, move down (or across) the random number table in a systematic manner until you have 38 different three-digit numbers. You might end up with these: 875, 228, 026, 200, 205, and so on. Then, you could return to the population and choose the 875th, the 228th, the 26th members, and so on, and this would constitute your simple random sample of $n = 38$. It would satisfy both conditions of a simple random sample.

Suppose the size of the population is less than 1,000, say 750. You could still label the members of the population 000, 001, 002, . . . , and so on, but this would come to an end at 749. The random number table could be used in the same way; however, three-digit combinations giving numbers larger than 749 would have to be discarded (the 875 above, for example) and you would continue reading from the table until you had n numbers of the right size.

If the population size N is greater than 1,000, then four-digit, five-digit, or longer combinations would have to be used to label the members of the population. Thus, if N were larger than 1,000 but smaller than 10,000, then the labels 0000, 0001, 0002, . . . , and so on, would have to be used.

Some people, on first seeing this scheme, think that 0000 is far less likely to appear than, say, 2346 in the random number table. Not so in a true random number table, since every combination and permutation of digits occurs with equal frequency.

Nowadays, it is possible to generate random numbers with the push of a button—certain calculators and computers have this capability. In earlier days, other devices were used. One of the oldest random number generators is a die, or two or more dice. As long as dice are well-constructed as well as shaken and rolled well, they are perfectly good random number generators, and, what is more, they are still in use today! Their principal use is in board games and games of chance, however, not in statistical research!

Why Random Sampling?

You may be wondering, at this point, why make all this fuss over what does or does not constitute a simple random sample? After all, wouldn't the sampling scheme described above give a representative sample of Los Angeles County? It may very well. However, the scheme is not entirely objective, since human judgment was used to decide how to subdivide the population before proceeding. Again, this is not necessarily bad. There are perfectly respectable sampling schemes combining human judgment and objective methods that are in use today. You will examine a couple of these shortly. If these are used, however, modifications have to be made in the statistical techniques discussed later. Since this is a first course, the important statistical techniques *based on simple random sampling* are presented. The modifications necessary for sampling schemes other than simple random sampling will be left to more advanced texts.

Any sampling technique must have an element of randomness to it. It is not satisfactory to take *any* sample from the population. Bias can creep into your results if this is done. Here is a famous example of this. In 1936, the two main presidential candidates were Franklin D. Roosevelt and Alfred Landon. A magazine of the day, the *Literary Digest,* conducted an opinion poll before the election. The size of their sample was 2,000,000, gigantic by today's standards, in which opinion pollsters routinely make fairly accurate predictions based on results of samples of a few hundred. Based on the results of their opinion poll, the *Literary Digest* predicted Landon would win by a landslide. As it turned out, Roosevelt was the one who won by a landslide! The *Literary Digest* was biased in their method of selection. They polled subscribers to their magazine, owners of telephones, and registered car owners, principally. Now, in the depression-plagued United States of the 1930s, if you subscribed to the *Literary Digest* or owned a telephone or a car, you tended to be better off than most. Those who were better off tended to vote Republican rather than Democratic. Thus, despite its enormous size, the *Literary Digest*'s sample was biased—not just a little, but *extremely* biased. Predictably, the magazine did not last long after that.

If statisticians wish to make reliable inferences about a population from a sample, their samples must be unbiased. The *Literary Digest*'s errors have been examined ad nauseam. Their false results were glaring. The biases created by nonrandom sampling methods often are less spectacular and may go unnoticed unless the researcher is alert to this potential problem. To avoid *any* bias, to the best of their abilities, statisticians take great care in the selection of their samples.

Other Sampling Schemes

As in the Los Angeles County example, statisticians sometimes want to ensure that certain subdivisions of the population are represented in the sample.

Then, a simple random sample is taken from each subdivision (or stratum) of interest in the population, and these samples are combined to make the whole sample. This is known as **stratified random sampling.**

Again, as in the above example, statisticians may want to ensure that subdivisions of the population are represented, but in proportion to their size. Taking an equal-sized sample from Rolling Hills as from Los Angeles, a city 1,000 times as large, is out of proportion, they reason. This is taken into account, and larger samples are drawn from larger subdivisions and smaller samples from smaller subdivisions. This is known as **quota sampling.**

Sometimes the number of subdivisions in the population is so large that it would be impossible to obtain a reasonably sized sample if a random sample were taken from *every* subdivision. A process sometimes used is first to draw a random sample of the subdivisions and then to sample from those subdivisions chosen.

This is known as **cluster sampling.** If the subdivisions are small enough, *every* member of the population in the subdivisions selected may be included in the sample.

A population may, for all practical purposes, be infinite. A manufacturing process producing light bulbs results in such a population. In theory, there will never be an infinite number of light bulbs produced, but, in practice, no end to the process can be seen. Therefore, you can never assign a number to *every* member of the population for use with random numbers. In such a case, if a sample is desired for testing, usually every tenth bulb (or every twentieth or some other number) coming off the assembly line is selected until the sample is filled. This is known as **systematic sampling.** This is generally as good as random sampling of light bulbs (or whatever) produced at *that* time. The reason is that the tenth, twentieth, and so on, should be no different from any other. However, the sample may be totally unrepresentative of the product produced years earlier or later, since the machinery can change or become worn.

All the above schemes are valid and, to a greater or lesser extent, have been used. The reader who is interested in learning more about them and others is encouraged to refer to books on sampling or experimental design. The discussion in this book will be confined to simple random sampling.

Exercise Set 4–1

1. Using the random number table, list 10 random numbers that could be used to sample from a population of size $N = 8,375$.

2. Using the random number table, list 15 random numbers that could be used to sample from a population of size $N = 100$.

3. The following is a list of random numbers. State what the smallest and largest sizes of a population might be so that these could be used to sample from it.
   591   509   601   395   526   637   006   574   429   694   132

4. Do the same as exercise 3 for these random numbers.
   4239   9207   1394   1395   0257   4599   8350   0647   4303

In exercises 5–10, a sampling scheme is described. Identify it as one of the sampling schemes listed in the text.

5. Data for a population is listed on 100 pages. A sample of size $n = 100$ is drawn by selecting one at random from each of the 100 pages.

 6. A state is divided into precincts for voting purposes. A polling organization selects 50 of these precincts at random and polls a random sample of 10 voters from each of these precincts.

 7. To obtain a sample of people from a certain city, a number is assigned to each person in the city and a table of random numbers is used to draw a sample of 100.

 8. For a sample of people from a city, researchers wish to include 20 whites, 20 blacks, 20 Chicano/Hispanics, 20 Asians, and 20 native Americans. A table of random numbers is used to sample from each of these groups.

9. It is known that in a certain city 50% are white, 30% are Chicano/Hispanic, 10% are black, 6% are Asian, and 4% are native American. A table of random numbers is used to obtain a sample of 100 whites, 60 Chicano/Hispanics, 20 blacks, 12 Asians, and 8 native Americans.

 10. To obtain a sample of tires from a tire factory, every seventh tire produced on a certain day is included in the sample until 20 tires are obtained.

Section 4-2   The Central Limit Theorem

Sampling for What?

Statisticians use **sampling** when taking a census of the entire population is impractical. Data from the sample is used to calculate statistics, which are estimates of the corresponding population parameters. A sample might be drawn, its range calculated, and this value used as an estimate for the population range, or $\bar{x}$ calculated to estimate μ, or s calculated to estimate σ. How do you estimate a population parameter from its corresponding sample statistic and how do you determine the accuracy of your estimate? These will be the first problems that are tackled in part 3 of this book. To begin this process, you will first be introduced to a concept that initially might seem strange to you.

The Frequency Distribution of a Statistic

None of the words in the term **frequency distribution of a statistic** are new to you at this point in the text. What do they mean, however, when put together like this?

Recall that a statistic is defined as a single number that tells you something about a sample, such as range, mean, proportion, and so on. You could draw a sample of size n from a population and calculate the statistic you were interested in, such as the range. You could do this again with a second sample of the same size, n. And with a third sample, size n. And a fourth. And a fifth. And so on, until you had drawn *all possible* samples of size n from the population. Suppose this were done and the range calculated each time. Given even a fair-sized population, there would be a huge number of different samples of the given size, n. For each of the samples, the range is calculated.

Now, you have a huge number of numbers, each being the value of the range of a different sample from the given population. Back to chapter 1! What do you do with all these numbers? Why, organize them into a frequency distribution, of course, so that you can examine what is going on. The range is an example of a statistic. Thus, what you end up with is a frequency distribution of a statistic—in this particular case, the frequency distribution of the sample range. If you repeated the above steps replacing sample range by sample mean ($\bar{x}$), you would wind up with a frequency distribution of $\bar{x}$. This last frequency distribution, the frequency distribution of $\bar{x}$, is of particular interest because a mathematical result about it is a key that unlocks much of what is to follow. Before going into it, however, take a look at a simple, if unrealistic, example of the concept "frequency distribution of a statistic."

EXAMPLE 4–1   Suppose you had a very small population: 2, 3, 5, 6, 9, 11. For this population, $\mu = 6$, $\sigma = 3.16$, and $R = 9$. Take *every* sample of size $n = 3$ from this population, calculate $\bar{x}$ and R, and form their frequency distributions.

Solution   First, verify for yourself the values of μ, σ, and R given. All the samples with the corresponding $\bar{x}$ and R are shown in table 4–1.

Table 4–1

Sample	$\bar{x}$	R
2, 3, 5	3.3	3
2, 3, 6	3.7	4
2, 3, 9	4.7	7
2, 3, 11	5.3	9
2, 5, 6	4.3	4
2, 5, 9	5.3	7
2, 5, 11	6.0	9
2, 6, 9	5.7	7
2, 6, 11	6.3	9
2, 9, 11	7.3	9
3, 5, 6	4.7	3
3, 5, 9	5.7	6
3, 5, 11	6.3	8
3, 6, 9	6.0	6
3, 6, 11	6.7	8
3, 9, 11	7.7	8
5, 6, 9	6.7	4
5, 6, 11	7.3	6
5, 9, 11	8.3	6
6, 9, 11	8.7	5

The frequency distribution of R is shown in table 4–2.

Mean = 6.4
Standard deviation
= 1.98

Table 4–2

Class	f
3	2
4	3
5	1
6	4
7	3
8	3
9	4

The frequency distribution of $\bar{x}$ is shown in table 4-3.

Mean = 5.98
Standard deviation
= 1.42

Table 4-3

Class	f
3.0–3.9	2
4.0–4.9	3
5.0–5.9	4
6.0–6.9	6
7.0–7.9	3
8.0–8.9	2

And the histograms are shown in figures 4-1 and 4-2.

Figure 4-1

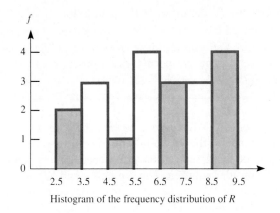

Histogram of the frequency distribution of R

Figure 4-2

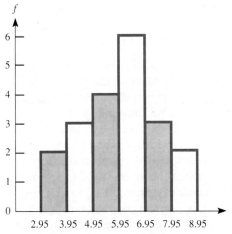

Histogram of the frequency distribution of $\bar{x}$

Notice that, as with all frequency distributions, the frequency distribution of a statistic has (1) a mean, (2) a standard deviation, and (3) a shape. These three properties of the frequency distribution of the *sample mean* $\bar{x}$ are the subject of the mathematical result mentioned above, and stated now.

The Central Limit Theorem

If all possible samples of size n are drawn from a population with mean $= \mu$ and standard deviation $= \sigma$, and $\bar{x}$, the sample mean, is calculated for each sample, then the frequency distribution of $\bar{x}$, thus obtained, has the following three properties:

1. Its mean $= \mu$, the mean of the population.
2. Its standard deviation $= \sigma/\sqrt{n}$, the standard deviation of the population divided by the square root of the size of each of the samples.
3. It will tend to have a normal distribution, regardless of the shape of the population.

Notice that the central limit theorem (CLT) is only concerned with the frequency distribution of $\bar{x}$ and not with the frequency distribution of any other statistic.

The proof of this theorem is beyond the scope of this book. However, it is of vital importance to the whole of statistical inference, so its meaning and implications will be discussed in this and succeeding chapters.

First, consider notation. The symbols μ and SD, with suitable subscripts, will be used to denote the mean and standard deviation of frequency distributions of statistics. Thus, $\mu_{\bar{x}}$ and $SD_{\bar{x}}$ will be used when referring to the mean and standard deviation (respectively) of the frequency distribution of $\bar{x}$. Similarly, μ_R and SD_R will have the same functions for the frequency distribution of the sample range; and so on for other statistics.

Take a look now at each of the three parts of the CLT in turn.

1. This states that $\mu_{\bar{x}} = \mu$, or, in words, that the mean of all the sample means is equal to the mean of the population. If you think about this for a while, this makes common sense. Some samples will have means larger than μ, while others will have means that are smaller. It seems logical that the mean of all the means will be μ. In example 4–1, $\mu = 6$, while $\mu_{\bar{x}} = 5.98$. The discrepancy is due to round-off error. Notice that $\mu_R = 6.4$, nowhere near the population range of 9. Again, common sense should tell you that the mean of the sample ranges has no reason to be close to the population range. It is possible, in any particular sample, to have a range equal to the population range, if both the lowest and highest values in the population are present in the sample. It is also possible to have a sample with a range *smaller* than the population range, but impossible to have one larger. Therefore, the average of the sample ranges could not possibly equal the population range. All that can be said is the μ_R will necessarily be smaller than the population range. It is different with the sample means.

2. This states that

$$SD_{\bar{x}} = \sigma/\sqrt{n}. \tag{23}$$

What this implies is that the frequency distribution of $\bar{x}$ is less dispersed than the population and its dispersion diminishes as n increases. If $n = 4$, $SD_{\bar{x}}$ is $\frac{1}{2}$ the value of σ, since $\sqrt{4} = 2$. If $n = 49$, $SD_{\bar{x}}$ is $\frac{1}{7}$ the value of σ, since $\sqrt{49} = 7$. Believe it or not, this makes common sense, too. Suppose, for example, that the population is a population of adult male heights with $\mu = 5$ feet 10 inches. While it is not that rare to find individual men as tall as 6 feet 6 inches or as short as 5 feet 2 inches, how likely is it to find 49 men whose *mean* height is 6 feet 6 inches or 5 feet 2 inches? Not that probable, wouldn't you agree? It would be even less probable if you were dealing with a sample of 100 men. Thus, while individual men's heights may spread over a wide range, means of samples of these men's heights are unlikely to be spread over such a wide range. Furthermore, the more heights there are in the samples, the less spread out the means are likely to be. This is the essence of this, the second result of the CLT.

The formula $SD_{\bar{x}} = \sigma/\sqrt{n}$ is technically only true for infinitely large populations. For finite populations, there exists what is called a **finite population correction,** which modifies the formula thus: For a finite population of size N,

$$SD_{\bar{x}} = \frac{\sigma}{\sqrt{n}} \sqrt{\frac{N-n}{N-1}}. \tag{24}$$

If $\sqrt{\frac{N-n}{N-1}}$ is close to 1, formula (24) will give almost the identical value as formula (23).

If N is very large in comparison to n, then the correction factor, $\sqrt{\frac{N-n}{N-1}}$, is very close to 1, and so formula (24) is reliable. For example, with $N = 1,000,000$ and $n = 100$,

$$\sqrt{\frac{N-n}{N-1}} = 0.9999505.$$

In example 4–1, $\sigma = 3.16$, $N = 6$, $n = 3$, and N is *not* very large in comparison to n. Using the finite population correction, $\frac{\sigma}{\sqrt{n}} \sqrt{\frac{N-n}{N-1}}$ = 1.41. It was found that $SD_{\bar{x}} = 1.42$. Notice, there is very good agreement between the two estimates.

3. To illustrate this result, take a look at figure 4–3. The shape of the population is said to be irrelevant. So, various shapes are illustrated, bearing in mind results 1 and 2 above.

Figure 4–3

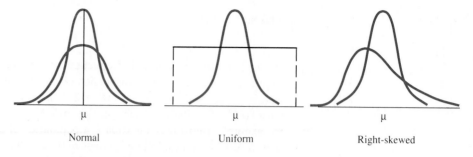

                μ μ μ

         Normal   Uniform   Right-skewed

In each case, the graph of the frequency distribution of $\bar{x}$, which is the thin normal distribution, has been superimposed on the graph of the population. What this result of the CLT states is that, no matter what the shape of the population, the shape of the frequency distribution of $\bar{x}$ tends to be a normal distribution. Of the three results of the CLT, this is the one that most often defies intuition. Many people expect the shape of the frequency distribution of $\bar{x}$ to mirror the shape of the population. However, there will be many more samples, especially if n is fairly large, whose means are close to μ than there are samples whose means are far away. Hence, the pileup in the middle, so characteristic of the bell-shaped curve.

A final word on the use of the word "tends" in the statement of the central limit theorem: The implication of the use of this word is that the frequency distribution of $\bar{x}$ will not follow the exact lines of a normal curve; rather, it will come close. This is true, but how close is "close"? This depends on the shape of the distribution and the size of n. With symmetric populations, such as the normal and uniform, the approximation to the normal distribution is extremely good, even with small values of n; and for larger values of n, even better. With skewed distributions, the approximation to the normal distribution is not so good for small values of n. As n becomes larger, the approximation becomes better and better. For a heavily skewed distribution, there may be traces of this skewness in the frequency distribution of $\bar{x}$, even for n as large as 40. Notice in example 4–1 that the frequency distribution of $\bar{x}$ (figure 4–2) does have a shape approximating a normal distribution, whereas the frequency distribution of R (figure 4–1) does not.

Now, take a look at another, more realistic example illustrating the central limit theorem.

EXAMPLE 4–2 Consider a population of 151 female students' weights. This is small for a population, but they are actual weights and will serve to illustrate the CLT. For this population, $\mu = 117.83$ lb, $\sigma = 14.53$ lb, and its histogram is shown in figure 4–4. Discuss the CLT for this population.

Figure 4-4

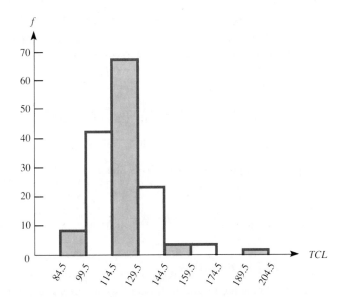

Solution   As you can see from figure 4-4, the population is right-skewed.

In order to illustrate the CLT, using a computer, 100 samples, each of size $n = 36$, were drawn from this population and $\bar{x}$ calculated for each. These 100 $\bar{x}$'s form an approximation for the frequency distribution of $\bar{x}$. The reason that they do not form the complete frequency distribution of $\bar{x}$ is that there are literally an astronomical number of possible samples of size 36 that could be drawn from the population. The computer was asked to draw only 100 of these. But, since these 100 samples were all simple random samples, together they should provide a good approximation for the complete frequency distribution of $\bar{x}$. The histogram of this frequency distribution is shown in figure 4-5.

Figure 4-5

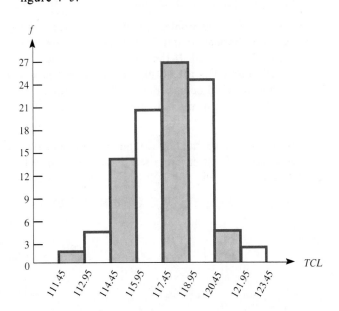

Notice, despite the right-skewedness in the population, this frequency distribution of $\bar{x}$ is close to normal. With larger samples, this approximation to the normal curve would have been even better.

The computer reveals that the mean of this frequency distribution of $\bar{x}$ is 117.79, while the standard deviation is 2.19. Thus, $\mu_{\bar{x}} = 117.79$ and $SD_{\bar{x}} \approx 2.19$.

According to the central limit theorem, $\mu_{\bar{x}} = \mu = 117.83$ and $SD_{\bar{x}} = \dfrac{\sigma}{\sqrt{n}} = \dfrac{14.53}{\sqrt{36}} \approx 2.42$. Notice that there is good agreement between these two distinct estimates of $\mu_{\bar{x}}$ and $SD_{\bar{x}}$. The agreement is even closer if the finite population correction factor is used. Then,

$$SD_{\bar{x}} = \frac{\sigma}{\sqrt{n}} \sqrt{\frac{N - n}{N - 1}} = 2.12.$$

●

The CLT in Practice

It would now be helpful to take a look at several examples using the CLT. Consider a large population of female weights with $\mu = 117.83$ lb and $\sigma = 14.53$ lb.

EXAMPLE 4–3   What is the probability that a female student drawn at random from this population will weigh between 115 and 120 lb?

EXAMPLE 4–4   If a random sample of size 36 is drawn from this population, what is the probability that its mean is between 115 and 120 lb?

EXAMPLE 4–5   If a random sample of size 81 is drawn from this population, what is the probability that its mean is larger than 120 lb?

Solution 4–3   This problem cannot be solved using what you know about probabilities under the normal distribution, since a population of weights is typically right-skewed. If the population were normally distributed, a solution could be found using the normal table. Later you will see that a solution to this problem would be possible if another frequency distribution could be found such that it fit the data and for which a mathematical formula could be written. Here, however, there is no apparent candidate for such a distribution. ●

Solution 4–4   The mean of a random sample of size 36 will fall somewhere in the normal distribution promised by part 3 of the CLT. Parts 1 and 2 of the CLT give

$$\mu_{\bar{x}} = 117.83 \text{ lb}$$

and

$$SD_{\bar{x}} = \frac{\sigma}{\sqrt{n}} = \frac{14.53}{\sqrt{36}} = 2.42 \text{ lb.}$$

Thus, to answer the question, it is necessary to find the area of the shaded region in figure 4–6.

Figure 4–6

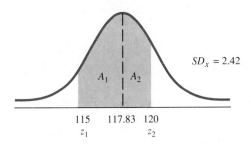

$SD_x = 2.42$

A_1 A_2

115   117.83 120
z_1 z_2

Now, $z_1 = \dfrac{115 - 117.83}{2.42} = -1.17$ and $A_1 = 0.3790$

$z_2 = \dfrac{120 - 117.83}{2.42} = 0.90$ and $A_2 = 0.3159$

$A_1 + A_2 = 0.6949.$

So, there is a probability of about 69% that the mean of a random sample of size 36 drawn from the population falls between 115 and 120 lb. ●

Solution 4–5   According to the CLT, the frequency distribution of $\bar{x}$ ($n = 81$) will be a normal distribution with $\mu_{\bar{x}} = 117.83$ lb and $SD_{\bar{x}} = \dfrac{14.53}{\sqrt{81}} = 1.61$ lb. To answer the question, it is necessary to find the area of the shaded region shown in figure 4–7.

Figure 4–7

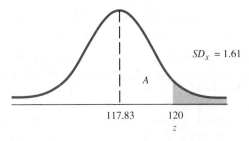

$SD_x = 1.61$

A

117.83   120
             z

Now, $z = \dfrac{120 - 117.83}{1.61} = 1.35$ and $A = 0.4115$

$0.5 - A = 0.0885.$

So, there is a probability of 8.85% that the mean of a random sample of size 81 drawn from the population is larger than 120 lb. ●

Consider a large population of male heights with $\mu = 70$ inches and $\sigma = 2.8$ inches.

EXAMPLE 4–6   What is the probability that a man, chosen at random, is taller than 67 inches?

EXAMPLE 4–7   If a random sample of size 50 is drawn from the population, what is the probability that its mean is within 1 inch of the population mean?

Solution 4–6   Unlike weights, heights do generally follow closely to a normal distribution. Thus, this is a question similar to the questions studied in chapter 3.

Figure 4–8

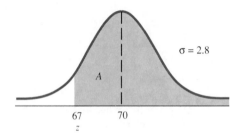

$$\text{Now, } z = \frac{67 - 70}{2.8} = -1.07 \quad \text{and} \quad A = 0.3577$$
$$0.5 + A = 0.8577 \approx 0.86 \text{ (refer to figure 4–8)}.$$

There is about an 86% probability that a man, chosen at random from the population, will be taller than 67 inches.   ●

Solution 4–7   According to the CLT, the frequency distribution of $\bar{x}$ ($n = 50$) will be a normal distribution with $\mu_{\bar{x}} = 70$ inches and $SD_{\bar{x}} = \dfrac{2.8}{\sqrt{50}} \approx 0.40$ inches.

To be within 1 inch of $\mu = 70$ inches, the sample mean must be between 1 inch less than μ (69 in.) and 1 inch above μ (71 in.). See figure 4–9.

Figure 4–9

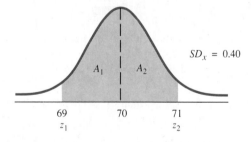

$$\text{Now, } z_1 = \frac{69 - 70}{0.4} = -2.5$$

$$z_2 = \frac{71 - 70}{0.4} = 2.5$$

$$A_1 = A_2 = 0.4938$$

$$A_1 + A_2 = 0.9876.$$

There is almost a 99% probability that the mean of a random sample of size 50 will be within 1 inch of the population mean. ●

Exercise Set 4-2

You should find that there are 21 different samples of size 5 that you can construct from the given population.

1. List all samples of size 5 from this fictitious population:

2   5   6   8   10   12   13.

Calculate $\bar{x}$ for each sample, form a frequency distribution of $\bar{x}$, calculate the mean and standard deviation of this frequency distribution and compare them to the mean and standard deviation of the population.

2. Repeat exercise 1, replacing $\bar{x}$ by $\tilde{x}$.

Exercises 3–6 refer to a population of boys' heights, which may be assumed to have a normal distribution with mean 54 inches and standard deviation 2.5 inches.

3. What percentage of the population would have heights between 53 inches and 56 inches?

4. If a boy is chosen at random from this population, what is the probability that he is taller than 52 inches?

5. If all possible samples of size 25 are drawn from this population, what percentage of them would have means between 53 inches and 55 inches?

6. If a random sample of size 81 is drawn from this population, what is the probability that the mean of this sample is larger than 53.5 inches?

7. A population of families in a certain city have a mean income of $13,000 and a standard deviation of $7,000. What is the probability that a random sample of 100 of these families have a mean income between $13,500 and $14,000?

8. A manufacturer of light bulbs produces bulbs that last a mean of 950 hours with a standard deviation of 120 hours. What is the probability that the mean life-time of a random sample of 10 of these bulbs is less than 900 hours?

9. A particular brand of cigarettes has a mean nicotine content of 15.2 mg with a standard deviation of 1.6 mg. What is the probability that a random sample of 20 of these cigarettes has a mean nicotine content of more than 16 mg?

10. If the time spent on commercials during half-hour TV programs has a mean of 5 minutes with a standard deviation of 1 minute, what is the probability that a random sample of 36 half-hour TV programs have commercials that last within $\frac{1}{4}$ minute of the mean?

Are you now beginning to see how the CLT can be used to answer questions about sample means? Unfortunately, the above examples assumed complete knowledge of the population: mean, standard deviation, and shape. In practice, researchers usually are ignorant of populations and draw random samples to find out about them. The next chapter will begin to explain how the CLT is used for statistical inference. Before going on, however, take a look at one more statistical truth that helps explain something people often wonder about. This statistical truth is called the law of large numbers.

Section 4–3   The Law of Large Numbers

The **law of large numbers** can be stated as follows:

> The larger the random sample drawn from a population, the more representative it is of the population.

What this means is that a random sample of size 100 will better reflect the population than one of size 50. And one of size 200 will be even better. This may not come as any surprise to you and you might wonder why a law is needed at all.

What is significant about the law of large numbers is that the size of the *population* is not mentioned in it at all. In fact, the size of the population is irrelevant to the conclusion of the law of large numbers. A random sample of size 50 is not as good as a random sample of size 100, whether the population is of size 1,000 or 1,000,000.

This is why public opinion polls predicting outcomes of elections are often so accurate even though the sample may only include a few hundred voters. This is true whether the election is for mayor of a city, governor of a state, or president of the country. How can they be so accurate? After all, the *Literary Digest* had 2 million in their sample and they blew it. The key difference is in the randomness of the sample. If you have a biased sample, it matters little how large it is, your results are likely to be biased. However, if your sample is truly random, 100 is large, 50 is large. In fact, to statisticians, anything over 30 is usually considered large.

Here is an analogy to help you understand this idea. Suppose you are making a homemade soup in a pot on the stove. If you stir the soup well, wouldn't a teaspoon of it give you a very good idea of what the whole soup tastes like? Now, ask yourself this: would it matter if you had a small pot making soup only for yourself or if you had a gigantic tureen with enough soup for dozens of people? If the soup is stirred, a teaspoonful gives you an idea of the taste of the whole pot, large or small.

Stirring the soup is analogous to making sure the sample is random. A teaspoonful is a sample and the pot is the population. The size of the pot or of the population is irrelevant. A relatively small sample can give you valuable information about the population, provided that the sample is *random*.

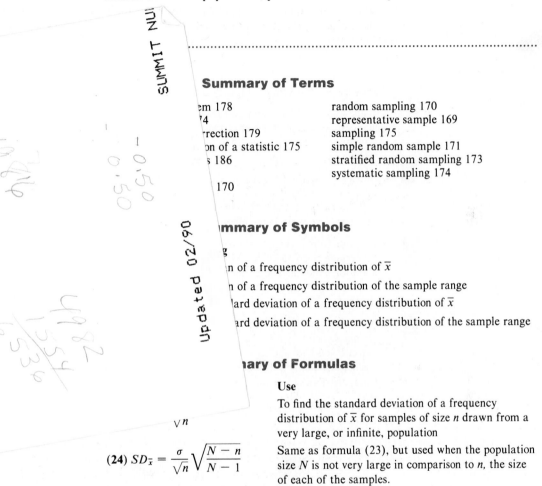

Summary of Terms

em 178   random sampling 170
'4   representative sample 169
rrection 179   sampling 175
on of a statistic 175   simple random sample 171
s 186   stratified random sampling 173
                                          systematic sampling 174
170

mmary of Symbols

g

n of a frequency distribution of $\bar{x}$

n of a frequency distribution of the sample range

lard deviation of a frequency distribution of $\bar{x}$

ard deviation of a frequency distribution of the sample range

ary of Formulas

Use

To find the standard deviation of a frequency distribution of $\bar{x}$ for samples of size n drawn from a very large, or infinite, population

$$\text{(24)} \quad SD_{\bar{x}} = \frac{\sigma}{\sqrt{n}} \sqrt{\frac{N - n}{N - 1}}$$

Same as formula (23), but used when the population size N is not very large in comparison to n, the size of each of the samples.

Chapter 4 Review Exercises

In exercises 1–4, identify the sampling technique used.

1. A telephone company obtains a sample of its subscribers by including every fiftieth subscriber on its list.

2. A random number table is used to obtain a sample of students from a college.

3. The government of a country wishes to obtain a sample of its inhabitants. To do this, it obtains a random sample of the towns and cities within its borders and then includes a random sample of inhabitants of those cities chosen.

4. A sociologist wishes to sample the employees of a company so that there are equal numbers of men and women in the sample. To this end, equal-sized random samples from the male and female populations of the employees are drawn.

 5. A random sample of 64 families is selected from the population of a certain city, which has mean family income of $11,000 with standard deviation $4,000. What is the probability that the mean income in the sample is between $10,500 and $11,500?

 6. The mean weight of pears grown by a certain orchard is 6 oz with a standard deviation of 0.5 oz. If a random sample of 10 pears is selected, how likely is it that the mean of this sample is less than 5.5 oz?

 7. If the heights of City College's female students are normally distributed with a standard deviation of 3 inches, and if 8% of the mean heights of random samples of 25 female students are less than 63.3 inches, what is the mean height of all students on the campus?

 8. Suppose 6-packs of a brand of soda pop can be considered to be random samples of all cans produced. If it is found that 10% of these 6-packs have mean content more than 12.1 oz and it can be assumed that the standard deviation of the content of all cans is 0.2 oz, what is the mean content of all cans produced by this company?

 9. Students from a large freshman psychology class participate in an experiment concerning sleeping habits. They are asked to record how long they sleep each night for a month. These times are divided up randomly into samples of 16 and it is found that 15% of these samples have means of less than 7 hours. If the mean of all the times is 7 hours 40 minutes, what is the standard deviation of all the times?

 10. It can be argued that your final adult height is a mean of a sample of many variables such as your parents' heights, your diet while growing up, the amount of exercise you had, and so on. It is for this reason that heights generally follow a normal distribution, they being a distribution of "sample means." On the other hand, distributions of weights are generally right-skewed. Why do you suppose that is?

Hands-on Class/Student Project

If you completed the project at the end of chapter 1, get together with the students with whom you participated then, retrieve the data that you collected at that time, and perform the following exercises.

1. Discuss with your group the method by which you selected the sample when you did this project in chapter 1. Did your method come close to any of the sampling schemes described in section 4–1?

2. Discuss with your group how you would go about selecting your sample if you wanted to make more certain that it was representative of the population of students at your college. In other words, which of the sampling schemes discussed in section 4–1 would you favor?

3. Share your sample of variable 1, the students' heights, with all the other groups in the class. Depending on the number of groups in the class and how many were included in each group's sample, this should give each group several hundred heights to work with. When the heights were collected, care was probably not taken with sampling methods, so even though there are several hundred heights, it is not necessarily a representative sample of the population of all heights at your college. Instead, let us consider these several hundred heights as a population in its own right, say population H.

4. Apply descriptive statistics' methods to population H; i.e., classify it into a frequency distribution, draw a histogram, and calculate its mean and standard deviation.

5. By using a computer or random number table, select 20 different random samples, each of size 15, from population H and calculate $\overline{x}$, the sample mean of each.

6. Share your sample of 20 sample means with each of the other groups in the class. Each group will now have the same fairly large sample of means of samples of size 15 drawn from population H.

7. Classify this sample of sample means into a frequency distribution, draw its histogram, and calculate its mean and standard deviation.

8. Compare your answers for exercise 7 to the corresponding answers found by other groups in the class. All groups should have exactly the same answers at this point.

9. Check the three parts of the central limit theorem:
   a. Does the mean of the sample means ($\mu_{\overline{x}}$), which you found in exercise 7, equal the mean of population H (μ) found in exercise 4? Is it at least close?
   b. Does the standard deviation of population H (σ) found in exercise 4 divided by $\sqrt{15}$ come close to the standard deviation ($SD_{\overline{x}}$) found in exercise 7? (*Note:* You may get better agreement if you use the finite population correction.)
   c. Does the histogram you drew in exercise 7 look close to being normally distributed?

   (*Note:* The fairly large sample of sample means that you are working with is *not* the frequency distribution of means of samples of size 15 from population H, since you and your classmates did not collect *all possible* samples of size 15 from population H. However, since you gathered random samples, it should be representative of this frequency distribution.)

10. The above exercises could be repeated for any of the variables 3, 4, 5, 6, or 11 collected in chapter 1, if any of those are of more interest to the class than heights of students.

3

Inferential Statistics

.....

"You haven't told me yet," said Lady Nuttal, "what it is your fiancé does for a living."

"He's a statistician," replied Lamia, with an annoying sense of being on the defensive.

Lady Nuttal was obviously taken aback. It had not occurred to her that statisticians entered into normal social relationships. The species, she would have surmised, was perpetuated in some collateral manner, like mules.

"But Aunt Sara, it's a very interesting profession," said Lamia warmly.

"I don't doubt it," said her aunt, who obviously doubted it very much. "To express anything important in mere figures is so plainly impossible that there must be endless scope for well-paid advice on how to do it. But don't you think that life with a statistician would be rather, shall we say, humdrum?"

Lamia was silent. She felt reluctant to discuss the surprising depth of emotional possibility which she had discovered below Edward's numerical veneer.

"It's not the figures themselves," she said finally, "it's what you do with them that matters."

.....

The Undoing of Lamia Gurdleneck

K. A. C. MANDEVILLE

CHAPTER
5

Estimation

....

". . . they may be expected with confidence,
and must be borne with philosophy."

....

CHARLES DICKENS

Political Polls—Much More Than Meets the Eye

By KENNETH REICH, *Times Staff Writer*

When the latest California Field Poll on the race for the GOP gubernatorial nomination was released April 2, it appeared that Lt. Gov. Mike Curb had made a dramatic resurgence against Atty. Gen. George Deukmejian. Curb led 50% to 40%, while he had trailed in January 32% to 43%.

But the margin of error in a poll with a sample of 229 voters, as this most recent poll had, is six percentage points in either direction, and there are slight chances the error could be even greater.

Statistically, Field reported this week, there is one chance in 128 that Deukmejian actually led Curb among the state's Republicans at the time of the survey by two percentage points, and there is even one chance in 2,940 that Deukmejian led Curb 50% to 40%, rather than the other way around.

"These are the kind of things that should be talked about," Field said in an interview. "Scrutiny will make all polls less impressive. . . . They are actually fragile measures (of opinion)."

Two other pollsters who do surveys publicly in California—I. A. Lewis of The Times and Steve Teichner of Teichner Associates—express some agreement.

'Never Any Certainty'

"There is simply no time that things are 100% certain," remarked Lewis. "There is never any certainty in probability. . . .

"The public has a peculiar attitude about pollsters. They think of us as scientists when we're artists. A poll is not a gum machine where you put in a nickel and get out your gum. It takes experience and skill and a great deal of careful attention to get the right answers."

Lewis' remarks refer to the fact that both he and Field do not simply compute and release raw survey results. They probe to find out as best they can who among those surveyed is likely to vote, and, when people don't initially express a preference between candidates, they press to find their leanings. The results, when released, often are weighted according to these probes. Unlikely voters are often left out. . . .

Field, in his most recent survey, first asked who an individual intended to vote for in a particular race without giving any names of candidates. In the Republican gubernatorial primary, 16% named Curb and 11% Deukmejian without prodding.

In a second question to the 73% who named no one, Field then listed Curb and Deukmejian and asked for their preference. Many more answered this and the cumulative results showed 45% for Curb and 37% for Deukmejian. Eighteen percent said they were still undecided.

Field then "probed" with a third question: "Well, just suppose you had to make up your mind today, in the governor's race, who would you vote for?"

This elicited 5% more support for Curb, bringing him to 50% and 3% more for Deukmejian, bringing him to 40%. The other 10% still said they were undecided.

Lewis also probes. In his most recent poll, he sought to elicit leanings by asking: "Well, as of today, do you lean more toward Curb or Deukmejian or someone else?"

But he also probed among those who expressed a preference in another way, asking: "Are you certain that you're going to vote for your candidate, or is it possible that something might happen between now and the primary election that might change your mind?"

This is intended to get some indication of depth of commitment. Lewis said recently that while Curb might be leading at the moment, he saw some indication that Curb's support was less solid than Deukmejian's.

Teichner up to now has done no probing, although he said in an interview that he will do some in his last survey before the election.

"One reason we do not push the undecided is that it's my feeling you don't get a true base for your results when you do so at early stages," he said.

···

Chapter Objectives
What Do You Infer?

Suppose you interviewed a random sample of students at your college and found that the mean age in this sample was 22. Does this mean that the mean age of *all* students attending your school is 22? From the discussion in the previous chapter, you must answer this question, "not necessarily." There, you saw that although the mean of a population of weights was 117.8 lb, 100 random samples, each of size 36, had sample means varying between 112 lb and 122 lb. Some samples had means less than μ = 117.8 lb, while some had means larger than μ. In practice, researchers will draw *one* sample to make an inference about the population from which it was drawn. This chapter discusses the ideas that statisticians use to make meaningful inferences about a population from a single random sample.

···

Section 5–1 Confidence Interval Estimates with Known σ

Point Estimation

"Almost always" is used here because, for example, as you have seen, the sample range is often a poor estimate of the population range.

The process of drawing a random sample from a population, calculating $\bar{x}$, and using this value as an **estimate** for μ is an example of **point estimation.** More generally, any statistic can be calculated from a random sample and used to estimate the corresponding parameter in the population. Almost always, the best *point* estimate for a population parameter is the corresponding sample statistic, provided that the sample is random. Thus:

$\bar{x}$ is the best point estimate for μ

$\tilde{x}$ is the best point estimate for $\tilde{\mu}$

s is the best point estimate for σ.

After analyzing your sample results, you can state, "we estimate that the mean age of all students attending our school is 22 years." From the above discussion, can you see that this estimate may be way off, very close, or dead on? It might be an overestimate or an underestimate. Needless to say, statisticians and others find this situation less than satisfactory. An alternative method of analyzing the sample results is called for.

Interval Estimation

A preferable way of presenting such a result as the age example is to say that the mean age of all students attending the school falls somewhere between 21 and 23 years, or 20 and 24 years, and so on. Such a way of stating a result is called an **interval estimate.** What it is saying, in effect, is that the true value of the population parameter falls in some interval, rather than trying to pinpoint it.

If you are presenting a range of values, rather than a single value, you should be more confident of your estimate. Now, anyone can be 100% confident that the mean age of students at your school is between 0 and 100 years; or, if they wanted to be more daring, between 15 and 50 years. But neither of these intervals is of any real practical value as far as giving useful information about the population. Even if the interval were as narrow as between 20 and 30 years, it would be of limited use. However, suppose you were able to say that you are 80% confident that the true mean lies between 21.5 and 22.5 years; or that you are 95% confident that μ lies between 21.2 and 22.8 years. Would these not be pieces of information of far greater practical significance?

The two statements toward the end of the previous paragraph are examples of **confidence interval estimates.** Since this is somewhat of a mouthful, in this book, these will be abbreviated CIE's.

> A **confidence interval estimate (CIE)** is an interval of values calculated to estimate a population parameter from a random sample with a stated degree of confidence.

It is more informative to say that μ lies between 21.5 and 22.5 than to say it lies between 21.2 and 22.8, since the latter interval is wider and, therefore, allows for more possible values for μ. However, as you will discover, you do not have complete freedom to choose as narrow an interval as you please. The narrower your interval estimate is, the less confidence you will have that μ is contained in it. It is possible to have 100% confidence that the true mean age lies between 0 and 100, and practically 0% confidence that μ is *exactly* 22, the point estimate. Intervals between these two extremes will give a confidence level somewhere between 0% and 100% that they include the true value of μ. The problem of how to choose an appropriate interval and, correspondingly, how much confidence this allows is taken up now.

The CLT and CIE's

Suppose a random sample of size n is drawn from a population with mean, μ, and standard deviation, σ. The mean of this sample, $\bar{x}$, is calculated. Without actually performing the task, it can be theorized that this $\bar{x}$ is one of the many, many $\bar{x}$'s that would occur had *all* possible samples of size n from this population been drawn. This (theoretical) collection of $\bar{x}$'s would form a frequency distribution of $\bar{x}$. The central limit theorem (CLT), discussed in the previous chapter, promises that this frequency distribution of $\bar{x}$: (1) has mean $\mu_{\bar{x}} = \mu$, (2) has standard deviation $SD_{\bar{x}} = \sigma/\sqrt{n}$, and (3) tends to have a normal distribution. The one $\bar{x}$, that has been calculated, lies somewhere in this distribution. Figure 5-1 is a picture of the situation.

Figure 5-1

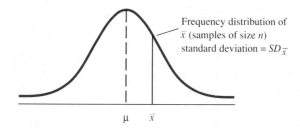

Frequency distribution of $\bar{x}$ (samples of size n) standard deviation $= SD_{\bar{x}}$

μ $\bar{x}$

The calculated $\bar{x}$ has been placed on the normal distribution that represents the frequency distribution of $\bar{x}$. Of course, in reality, you have no idea where this $\bar{x}$ lies in relation to $\mu_{\bar{x}}$, or μ, since μ is not known; $\bar{x}$ might be anywhere in this normal distribution of sample means. Consider two possibilities: (a) $\bar{x}$ lies within one standard deviation of μ, (b) $\bar{x}$ does not lie within one standard deviation of μ. The two situations are shown in figure 5-2.

Figure 5-2

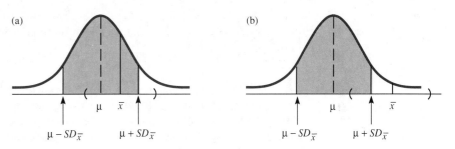

The shaded region in both figures is the area under the normal curve between one standard deviation below the mean and one standard deviation above.

Since μ is unknown (the reason for this whole exercise is to estimate μ), it is also unknown which of the two situations, (a) or (b), has occurred. First, concentrate on situation (a). The parentheses marked on the horizontal axis indicate the points $\bar{x} - SD_{\bar{x}}$ and $\bar{x} + SD_{\bar{x}}$. Notice that μ does lie between these two points. In fact, you should satisfy yourself that if $\bar{x}$ falls anywhere in the shaded region, then μ will lie between $\bar{x} - SD_{\bar{x}}$ and $\bar{x} + SD_{\bar{x}}$. Another way of stating this is to say that if $\bar{x}$ lies within one standard deviation of μ, then μ must lie within one standard deviation of $\bar{x}$. This is just like saying that if you are standing within 3 feet of a friend, then your friend must be within 3 feet of you.

In situation (b), the reverse is true. Again, the parentheses on the horizontal axis indicate the points $\bar{x} - SD_{\bar{x}}$ and $\bar{x} + SD_{\bar{x}}$. But notice this time, μ does *not* lie between these two points. In fact, is it not true that if $\bar{x}$ falls anywhere in the *nonshaded* regions, μ will not lie between $\bar{x} - SD_{\bar{x}}$ and $\bar{x} + SD_{\bar{x}}$? Consider now this question: What is the *probability* that the sample mean conforms with situation (a) rather than situation (b)? Can you see that the answer to this question is the same as the answer to the question: What is the probability that $\bar{x}$ falls in the shaded region? This can be answered by reference to the discussion in chapter 3 on the use of the normal table. Since the area of the shaded region is 68% (corresponding to $z = \pm 1$), the probability that $\bar{x}$ falls within it is 68%. Thus, *whenever* a random sample is drawn, there is a 68% probability that situation (a) has occurred. It can therefore be said that 68% of the intervals $\bar{x} - SD_{\bar{x}}$ to $\bar{x} + SD_{\bar{x}}$ contain μ.

This conclusion that has now been reached can be stated in several ways:

68% of the intervals $\bar{x} - SD_{\bar{x}}$ to $\bar{x} + SD_{\bar{x}}$ contain μ.
There is a 68% chance that μ lies between $\bar{x} - SD_{\bar{x}}$ and $\bar{x} + SD_{\bar{x}}$.
There is a 68% probability that μ lies in the interval $\bar{x} \pm SD_{\bar{x}}$ (read: $\bar{x}$ plus or minus $SD_{\bar{x}}$).
We are 68% confident that μ lies in the interval $\bar{x} \pm SD_{\bar{x}}$.
A 68% confidence interval estimate for μ is $\bar{x} \pm SD_{\bar{x}}$.
A 68% CIE for μ is $\bar{x} \pm SD_{\bar{x}}$.

The last sentence, being the shortest and also fitting in with the previous discussion on CIE's, will be the preferred method of stating the result. Take a look at an example.

EXAMPLE 5–1 A random sample of 50 students is chosen from a certain college and their mean age is found to be 22.3 years. It is known that the standard deviation of *all* the ages of the students at this college is 4.2 years. Find a 68% CIE for the mean age of all the students.

Solution Since 50 is the size of the sample, $n = 50$. The mean age of *these* 50 students is 22.3 years. Thus, $\bar{x} = 22.3$. The standard deviation of *all* students is 4.2 years. Since *all* students are included, this is the *population* standard deviation. Thus, $\sigma = 4.2$.

Now, by part 2 of CLT, $SD_{\bar{x}} = \dfrac{\sigma}{\sqrt{n}} = \dfrac{4.2}{\sqrt{50}} \approx 0.59.$

Thus, a 68% CIE for μ is 22.3 ± 0.59.

Since the mean age of all students is a population mean, μ, this is the required 68% CIE.

To see that 22.3 ± 0.59 is indeed an interval, subtract and add 0.59 to 22.3, in turn.

$$22.3 \pm 0.59 = 21.71 \text{ to } 22.89$$

The interpretation of the result is that the mean age of all students at this college is somewhere between 21.71 years and 22.89 years, and you can be 68% confident of this result. ●

Confidence Expanding

To be 68% confident of a result is not satisfactory for many purposes. After all, this level of confidence means that there is a 32% probability that such an interval does not contain μ. The confidence level can, however, be increased. The above discussion could be repeated, but, this time, replace the two possibilities above by these: (a) $\bar{x}$ lies within *two* standard deviations of μ, (b) $\bar{x}$ does not lie within *two* standard deviations of μ. The two situations are illustrated in figure 5-3.

Figure 5-3

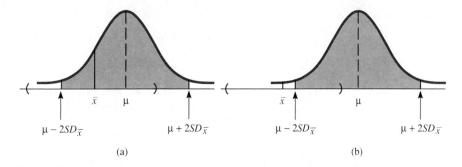

(a)   (b)

The parentheses in these two figures represent the points $\bar{x} - 2SD_{\bar{x}}$ and $\bar{x} + 2SD_{\bar{x}}$. The same argument applies. If $\bar{x}$ falls in the shaded region, μ is included in the interval $\bar{x} \pm 2SD_{\bar{x}}$, and if $\bar{x}$ falls outside the shaded region, μ will not lie in this interval. This time, however, the shaded region comprises 95% of the area under the normal curve. Hence, the interval $\bar{x} \pm 2SD_{\bar{x}}$ constitutes a 95% CIE for μ. In other words, you can construct an interval for which you have 95% confidence that μ is contained in it.

In making this change, there is a gain and a loss. The gain is that you have substantially increased the amount of confidence you have in your interval estimate for μ. The loss is that the interval you can present is necessarily wider. By adding and subtracting *two* standard deviations, instead of one, the resulting interval must be twice as wide.

Thus, in example 5-1:

A 95% CIE for μ is $22.3 \pm 2(0.59)$

$$= 22.3 \pm 1.18$$

$$= 21.12 \text{ to } 23.48.$$

Despite the width of this interval, it is not nearly so wide as the intervals with which you could be 100% confident: 0 to 100, 15 to 50, or even 20 to 30. And 95% confidence is not much less than 100% confidence.

The two confidence levels discussed thus far, 68% and 95%, are not the only ones available to you. *Any* desired level of confidence can be attained by making a slight modification. To achieve a desired level of confidence, choose this value as the area of the shaded regions in the above figures. Of course, doing this will change the number of standard deviations added to and subtracted from $\bar{x}$. As was done in the past, call this number of standard deviations z. Using the normal table, you should satisfy yourself that the following values for z are appropriate:

For 68% confidence, use $z = \pm 1$.
For 75% confidence, use $z = \pm 1.15$.
For 95% confidence, use $z = \pm 1.96$ (often, ± 2 is used here).
For 99% confidence, use $z = \pm 2.58$.

Notice that for a higher level of confidence, larger values of z are required, causing wider interval estimates. If a narrow interval is required, while maintaining a high level of confidence, this can be achieved by starting with a larger sample size. This is a fact because, by part 2 of the CLT, $SD_{\bar{x}} = \dfrac{\sigma}{\sqrt{n}}$ and, thus, increasing n will *reduce* the value of $SD_{\bar{x}}$.

EXAMPLE 5-2   Suppose you performed the experiment described in example 5-1 but used a random sample of $n = 100$ and found $\bar{x} = 22.5$. Find (a) a 95% CIE for μ, (b) a 99% CIE for μ.

Solution   In this new example, $SD_{\bar{x}} = \dfrac{4.2}{\sqrt{100}} = 0.42$.

a. A 95% CIE for μ is $22.5 \pm 2(0.42)$

$$= 21.66 \text{ to } 23.34.$$

(Notice that this is a narrower interval than the 95% CIE for μ found using a sample of only 50.)

b. For 99% confidence, $z = \pm 2.58$.
A 99% CIE for μ is $22.5 \pm 2.58(0.42)$

$$= 22.5 \pm 1.08$$

$$= 21.42 \text{ to } 23.58.$$

In summary:

> A CIE for μ is $\bar{x} \pm zSD_{\bar{x}}$, where $SD_{\bar{x}} = \dfrac{\sigma}{\sqrt{n}}$ and the value of z depends on the level of confidence required and the normal table.

Since this is an important result, it will be designated as formula (25).

$$\text{A CIE for } \mu \text{ is } \bar{x} \pm zSD_{\bar{x}}. \tag{25}$$

Exercise Set 5-1

1. If $\sigma = 8.4$ for a population and a random sample of size 49, drawn from this population, is found to have mean 73.2, find a 68% CIE for μ, the mean of the population.

2. If $\sigma = 1.6$ for a population and a random sample of size 25, drawn from this population, is found to have mean 19.6, find a 95% CIE for μ, the mean of the population.

3. A random sample of 36 students attending a community college is found to have mean age 26.7. Find a 90% CIE for the mean age of all students attending this college, if the standard deviation of all the ages can be assumed to be 3.6.

4. The lengths of 100 fish caught in a California lake were found to have a mean length of 16.7 in. If the population standard deviation is known to be 2.8 in., find a 99% CIE for the mean length of all fish in the lake.

5. A random sample of 50 newborn baby girls is found to have a mean weight of 7.2 lb. If the standard deviation of all baby girls' weights is 1.1 lb, find an 80% CIE for the mean weight of all newborn girls.

6. The Environmental Protection Agency (EPA) tests a random sample of 12 cars of a new make and finds that their mean highway mileage is 32.3 mpg. Assuming that σ is 3.5 mpg, find a 98% CIE for the mileage of all cars of this make.

7. A supervisor of a large office observes a random sample of 20 clerical workers and records the time required for each worker to learn a certain task. It is found that the mean of these times is 43.2 min. If the supervisor knows from previous studies that the population standard deviation is about 5 minutes, find a 95% CIE for the mean time necessary for all clerical workers to learn the task.

8. State agencies provide foster families with assistance for the foster children they take care of. It is found that a random sample of 40 such families selected from a certain community received a mean of $6.50 per child per day. If the population standard deviation is $1.50, find a 90% CIE for the mean aid received by all families in this community.

9. A random sample of teenagers received the following weekly allowances in dollars:

   10   12   5   7   4   10   8   12   5   15   8   10   20   16   8   14
   12   10   5   8   10   6   5   10.

If the population of teenagers' allowances has a standard deviation of $2, find a 95% CIE for the mean allowance of this population.

10. In a test of memory, it was found that the numbers of items recalled (out of 20) by a random sample of students were as follows:

   10   12   20   6   13   15   10   18   8   7   9   14   12.

If the population variance is known to be 4.6, find a 99% CIE for the mean number of items recalled by all students in the population.

Section 5–2   CIE's with Unknown σ

Formula (25) is used to estimate μ, the *unknown* population mean. When μ is unknown, it is unlikely (though possible) that σ would be known. More often than not, *both* μ and σ are unknown. In these cases, formula (25) is useless, since, if σ is unknown, $SD_{\bar{x}}$ cannot be calculated. Occasionally, historical values can be used. For instance, researchers who have worked extensively with adult human heights know that the standard deviation of these is usually about 3 inches. Therefore, if you were working with a sample of adult human heights and you used $\sigma = 3$, you should obtain fairly accurate results. However, suppose you were working with data for which you had no idea about the size of σ. What can be done then?

   If the size of the sample is not too small (the usual guideline is n larger than 30), the sample standard deviation s is a good estimate for σ. Thus, with a large sample, $SD_{\bar{x}}$ can be estimated by $s/\sqrt{n}$. This quantity, $s/\sqrt{n}$, is referred to as the **standard error** and will be labeled $SE_{\bar{x}}$, to distinguish it from the actual standard deviation of the frequency distribution of $\bar{x}$, $SD_{\bar{x}}$. The standard error is given by,

$$SE_{\bar{x}} = \frac{s}{\sqrt{n}}. \tag{26}$$

   Even if σ is unknown, a CIE can still be produced, since it is always possible to calculate s from the sample results in front of you, using the methods of chapter 2. Thus:

> For $n > 30$, a CIE for μ is $\bar{x} \pm zSE_{\bar{x}}$, where $SE_{\bar{x}}$ is given by formula (26), and, as before, the value of z depends on the level of confidence required.

EXAMPLE 5–3   A random sample of male students attending a certain college were asked how many siblings (brothers and sisters) they had. Here are the results.

$$
\begin{array}{ccccccccccccccccc}
0 & 2 & 2 & 2 & 3 & 0 & 1 & 5 & 2 & 6 & 4 & 2 & 2 & 0 & 0 & 2 & 3 & 4 \\
4 & 3 & 3 & 2 & 1 & 1 & 2 & 1 & 3 & 1 & 1 & 1 & 1 & 1 & 3 & 1 & 0
\end{array}
$$

Find a 95% CIE for the mean number of siblings that all the students at this college have.

Solution   Check the following:

$$n = 35$$

$$\bar{x} = 1.97$$

$$s = 1.44$$

$$SE_{\bar{x}} = 0.24.$$

Thus, a 95% CIE for μ is $1.97 \pm 2(0.24)$

$$= 1.49 \text{ to } 2.45.$$

You can conclude, with 95% confidence, that the true mean number of siblings of all students on this campus is somewhere between 1.49 and 2.45.   ●

Exercise Set 5–2

1. If, for a random sample, it is found that $n = 49$, $\bar{x} = 6.3$, $s = 3.5$, find a 68% CIE for μ.

2. If, for a random sample, it is found that $n = 50$, $\bar{x} = 16.3$, $s = 2.5$, find an 85% CIE for μ.

 3. In a foreign country, a sample of 35 students was found to have a mean number of siblings of 3.6 with a standard deviation of 1.20. Find a 95% CIE for the mean number of siblings of all students in this country.

 4. A sociologist wishes to determine the mean age at which regular marijuana users in high school first tried it. It is found that a random sample of 75 of these reported a mean of 15.4 years old with a standard deviation of 2.4. Find a 98% CIE for the mean age of first trying marijuana in the population.

 5. In a study of traffic, the traffic police found that a random sample of 100 cars using a section of freeway at peak times held a mean number of 1.38 occupants with a standard deviation of 0.25 people. Find a 90% CIE for the mean number of occupants of all cars using this section of freeway at peak times.

 6. The mean length of dating before marriage for a random sample of 54 married couples selected from a community turns out to be 164 days with a standard deviation of 36 days. Find a 95% CIE for the mean length of dating of all married couples in this community.

7. A random sample of smokers drawn from office workers in a large city produced the following numbers of cigarettes smoked daily:

20 25 10 15 18 24 20 35 40 32 12 8 6 21 25
19 28 20 45 18 30 20 10 25 14.

Find, if possible, a 68% CIE for the mean number of cigarettes smoked daily by all office workers in this city.

 8. Mr. Spock of the USS Enterprise is investigating life on the newly discovered planet Sigma-Alpha 4. He finds that a random sample of these beings weigh the following numbers of kilograms:

15 21 32 27 14 19 22 21 36 27 30 23 25 18
22 35 21 19 16 30 18 20 15 24 25 28 31 28
25 19 20 18 30 31 30 20.

Find, if possible, a 95% CIE for the mean weight of all inhabitants of this planet.

 9. In a study of the common housefly, a random sample of 64 of these wonderful creatures produced this frequency distribution.

Length (mm)	f
4.9–5.1	2
5.2–5.4	11
5.5–5.7	15
5.8–6.0	21
6.1–6.3	11
6.4–6.6	3
6.7–6.9	1

Find a 99% CIE for the mean length of all houseflies.

10. In 1980, a study was conducted in Los Angeles on the prices of goods in convenience stores. This frequency distribution gives the prices for a gallon of milk in a random sample of these stores.

Price (¢)	f
190–194	2
195–199	8
200–204	11
205–209	18
210–214	13
215–219	4

Find a 98% CIE for the mean price of a gallon of milk in all these stores.

Section 5–3 Determining Sample Size When Estimating μ

It should be clear from the previous examples and discussion that the more confidence that is required, the wider the resulting interval estimate will be. However, you can maintain a high level of confidence *and* obtain a narrower interval by increasing the sample size to begin with. Just how large should n, the sample size, be? The temptation is to answer this question, "as large as possible." After all, the law of large numbers promises more representative results, the larger the sample is. However, it is possible that a very large sample produces an interval estimate for μ that is narrower than you absolutely require. Then you have wasted time, effort, and, possibly, money in drawing a sample larger than you need. Fortunately, there is a method for deciding the size of the sample that is optimal for your needs. Formula (25) states that a CIE for μ is $\bar{x} \pm zSD_{\bar{x}}$. The quantity that is added to and subtracted from $\bar{x}$ is referred to as the **error in the estimate** and is labeled with the letter E. Thus,

$$E = zSD_{\bar{x}}.$$

To **determine sample size** required, the *desired* size of E must be known in advance. Here are some examples:

The 68% CIE for mean age was calculated earlier as 21.71 to 22.89. This has a width of 1.18 yr. Suppose you wanted it to be 1 year or less. Then, instead of 0.59 being added to and subtracted from $\bar{x}$, you would require that 0.5 or less be that amount. Thus, you would desire $E = 0.5$.

The 95% CIE for number of siblings was found to be 1.97 ± 0.48. This may be interpreted by saying that you have estimated the mean number of siblings to *within* 0.48. You may *wish* to make the estimate to within 0.35. Then, the error E should be taken to be 0.35.

Suppose you wish to estimate mean income in a community to within $100.00. Then, take $E = 100$.

Since $SD_{\bar{x}} = \dfrac{\sigma}{\sqrt{n}}$, the necessary steps to determine the sample size follow:

$$E = zSD_{\bar{x}}$$

$$E = z\frac{\sigma}{\sqrt{n}}$$

$$E\sqrt{n} = z\sigma \qquad \text{by multiplying both sides of the equation by } \sqrt{n}$$

$$\sqrt{n} = \frac{z\sigma}{E} \qquad \text{by dividing both sides of the equation by } E$$

$$n = \left(\frac{z\sigma}{E}\right)^2 \qquad \text{by squaring both sides}$$

To determine the size of a random sample necessary to estimate μ to within a certain quantity, E, with a prescribed level of confidence, use the formula,

$$n = \left(\frac{z\sigma}{E}\right)^2. \tag{27}$$

EXAMPLE 5–4   You wish to estimate the mean height of a population of adult males to within $\frac{1}{4}$ inch with 95% confidence. How large a random sample should you draw?

Solution   For 95% confidence, z is about 2. $E = \frac{1}{4} = 0.25$ from the statement of the problem. σ can be taken to be 3 in. for adult heights.

$$\text{Thus, } n = \left(\frac{2 \cdot 3}{0.25}\right)^2 = (24)^2 = 576.$$

To satisfy the requirements of the problem, a random sample of at least 576 is required.   ●

Notice that in example 5–4 you had to have an idea of the size of σ in order to apply formula (27). What if you have no idea what σ is? The usual procedure, followed in this case, is first to take a random sample to calculate s, which is used to estimate σ. (Since σ is being estimated, the size of this preliminary sample should be at least 30.) Then, the estimated value of σ is used in formula (27) to obtain the size of the sample actually needed. If necessary, you return to the population to draw sufficient additional sample values that are combined with the original sample to form one sample large enough to fit the requirements. Since, then, the sampling has been accomplished in two stages, this is called **two-stage sampling.**

EXAMPLE 5–5   Suppose you wish to estimate the mean family income in a certain large city. You wish to make the estimate to within $500.00 with 90% confidence.

Since σ is unknown, a preliminary random sample of 50 is drawn and s is calculated to be $6,500.00. How many additional incomes are required?

Solution   Using the figure of 6,500 as an estimate for σ, noting that $E = 500$, and determining from the normal table that $z = 1.65$ for 90% confidence, find that

$$n = \left(\frac{1.65 \cdot 6,500}{500}\right)^2 = 460.1025.$$

Since n is the size of the sample, it must necessarily be a whole number. Thus, here, n should be 460 or 461. Based on the argument that it is better to include slightly too many in the sample than too few, the value for n, calculated from formula (27), is usually rounded *up*. Therefore, you need a random sample of 461 incomes. But you have already drawn a random sample of 50. Thus, you need an additional 411 incomes.   ●

Suppose these additional 411 incomes are gathered, combined with the original 50, and the following are calculated from this sample of 461:

$$\bar{x} = \$15,300.00, \; s = \$6,470.00.$$

$$\text{Then, } SE_{\bar{x}} = \frac{s}{\sqrt{n}} = \frac{6,470}{\sqrt{461}} \approx \$301.34.$$

Thus, the 90% CIE for μ is $\$15,300 \pm 1.65(301.34)$

$$= \$15,300 \pm 497.21$$

$$= \$14,802.79 \text{ to } \$15,797.21.$$

Notice that the error in the estimate is $497.21, which is indeed less than the required $500.00.

This completes the discussion of estimation of a population mean. The next section explains how these ideas can be adapted to estimate other population parameters.

Exercise Set 5–3 $(1-9)$

1. Suppose you require 90% confidence in estimating a mean and $E = 0.8$, $\sigma = 3.7$. Find n, the required sample size.

2. Suppose you require 98% confidence in estimating a mean and $E = 350$, $\sigma = 1,700$. Find n, the required sample size.

3. How large a sample must be drawn to estimate a population mean with an error in the estimate no greater than 0.5 with 68% confidence if the population standard deviation can be assumed to be 4.5?

4. How large a sample must be drawn to estimate a population mean to within 0.9 units with 95% confidence if the population standard deviation is 3?

5. Suppose the lifetimes of bulbs made by a certain process are known to have a standard deviation of 110 hours. How large a sample must be taken to be 95% confident that the estimate for the mean will not differ from the true mean by more than 25 hours?

6. The standard deviation of IQ scores is usually taken to be 15. How large a sample of a population of college students' IQ scores must be used to be 99% confident that you have estimated the true mean IQ score to within 4 points?

7. A zoologist wants to determine the gestation period (length of pregnancy) of a previously unstudied mammal. After observing 30 of these mammals, she discovers that the standard deviation of this sample is 3.5 days. How many more of these mammals must she study to be 95% confident that she can estimate the true gestation period to within one day?

8. A consumers' group wishes to determine the true mean net weight of crackers packed in boxes marked "Net Wt 32 oz." A preliminary sample of 50 of these boxes reveals a standard deviation of 0.8 oz. How many more boxes must the group weigh to be 98% confident that they have estimated the true mean net weight to within 0.2 oz?

9. It is known that 95% of all vehicles traveling a certain section of freeway (speed limit 55 mph) travel between 51 and 59 mph, and, furthermore, that these speeds follow a normal distribution. How large a sample of vehicles must be monitored to be 90% confident that the true mean speed has been estimated to within 0.5 mph?

10. If it is known that 90% of the boxes of crackers in exercise 8 weigh between 31 and 33 oz, with the weights following a normal distribution, how large a sample must be tested to estimate the true mean net weight to within 0.2 oz with 99% confidence?

Section 5–4 Proportion

When dealing with qualitative data, it is impossible to calculate a mean. However, you may instead wish to know what **proportion** of a sample or population possesses a certain quality. For instance, if in a sample of 50 women, 30 have blue eyes, you can say that the proportion of blue-eyed women in the sample is $\frac{30}{50}$ or $\frac{3}{5}$ or 0.6 or 60%. Alternatively, it can be said that, in this sample, the proportion of non-blue-eyed women is 0.4 or 40%.

The word proportion was introduced briefly in chapter 1 under the discussion on relative frequency. The calculation of a proportion is very similar to the calculation of an empirical probability. As with probability, an occurrence of the quality of interest is labeled a success, while an occurrence not possessing the quality of interest is labeled a failure. In this way, a sample or population is **dichotomized** (i.e., cut in two). In chapter 10, data that cannot be dichotomized in this fashion will be discussed. In a sample, the proportion of successes is calculated by counting the number of successes occurring and dividing by the sample size. This will be designated by p. Analogously, the proportion of failures, designated by q, is calculated by counting the number of failures and dividing by the sample size. Since the number of successes added to the number of failures must equal the sample size, it must always be true that $p + q = 1$ or, equivalently, $q = 1 - p$. (*Note:* In the blue-eyed women example, $0.6 + 0.4 = 1$.) To get the proportion of successes for a population, divide the number of successes by the population size; and, do similarly for failures. Population proportions will be designated by the capital letters P and Q. As with a sample, $P + Q = 1$ or $Q = 1 - P$.

Estimating Proportion

If p is a sample statistic, then P will be the corresponding population parameter. Thus, for a random sample drawn from a population, p will be the best *point estimate* for P. But, as was done for means, can a CIE for P be found? If so, it would be of the form $p \pm zSD_p$ or, if necessary, $p \pm zSE_p$, where SD_p represents the standard deviation of the frequency distribution of sample proportions and SE_p represents the corresponding standard error. For means, it was possible to develop formulas for $SD_{\bar{x}}$ and $SE_{\bar{x}}$ by appealing to the central limit theorem. But the CLT is concerned only with the frequency distribution of sample means, not with proportions. However, there is a little trick that allows the results of the CLT to be used.

The Indicator Variable

When you are dealing with data that has been dichotomized into successes and failures, a variable, x, can be assigned to "indicate" (numerically) in which category each observation falls. The variable x only assumes two values: 1, if a value in the data set is a success, and 0, if a failure. Use of an **indicator variable** such as this changes the qualitative data into ones and zeros, which are quantitative. Any set of quantitative data has a mean and standard deviation, therefore the CLT can be used. For instance, in the example about eye-color of women, a 1 would be assigned if a woman has blue eyes and a 0 if she has any other color. After collecting the data, you would end up with 30 ones and 20 zeroes. How many ones and zeros would you have in a sample of size n if the proportion of successes is p? The proportion p was calculated as follows:

$$p = \frac{\text{number of successes in sample}}{n}$$

This was how $p = \frac{30}{50} = 0.6$ for blue-eyed women was calculated. Now, multiply both sides of this equation by n to get

$$np = \text{number of successes in sample.}$$

Since a 1 is assigned for each success, np must be the number of ones. [For blue-eyed women, $n = 50$, $p = 0.6$, and $(50)(0.6) = 30$.] Analogously, nq is the number of zeros in the sample. A frequency distribution can now be formed, and from it, the mean and standard deviation calculated. Since np is the number of ones in the sample, it is also the *frequency* of ones. Thus:

x	f	fx	fx^2
1	np	np	np
0	nq	0	0
		$\Sigma fx = np$	$\Sigma fx^2 = np.$

Notes: 1. Since the number of successes + the number of failures is the sample size, $np + nq = n$, and so $\Sigma f = n$, as it should be for a frequency distribution.

2. There is no need to use coding since x is already expressed in very simple numbers.

To complete the calculations:

$$\bar{x} = \frac{\Sigma fx}{n} = \frac{np}{n} = p$$

$$s = \sqrt{\frac{\Sigma fx^2}{n} - \left(\frac{\Sigma fx}{n}\right)^2} = \sqrt{\frac{np}{n} - p^2}$$

$$= \sqrt{p - p^2} = \sqrt{p(1 - p)}, \text{ since } ab - ac = a(b - c)$$

$$= \sqrt{pq}, \text{ since } q = 1 - p.$$

By employing the indicator variable, it has been discovered that the data thus formed has mean $= p$ and standard deviation $= \sqrt{pq}$.

Exactly the same procedure could be applied to a population, with N, P, Q, replacing n, p, q, arriving at the following:

$$\mu = P \qquad \sigma = \sqrt{PQ}.$$

The CLT states that if all possible samples of size n are drawn from a population with mean $= \mu$, standard deviation $= \sigma$, and $\bar{x}$ calculated for each, then the frequency distribution of $\bar{x}$ (1) has mean $= \mu$, (2) has standard deviation $= \sigma/\sqrt{n}$, (3) tends to have a normal distribution. Because of "the little trick" of using an indicator variable, a similar result about the frequency distribution of p can be stated.

If all possible samples of size n are drawn from a population with a (true) proportion of successes $= P$, and p is calculated for each sample, then the frequency distribution of these sample proportions p (1) has mean $= P$, the true population proportion; (2) has standard deviation $= \sqrt{PQ}/\sqrt{n}$, the standard deviation of the sampling distribution of the proportions; and (3) tends to be (or tends toward) a normal distribution.

Here, the CLT has simply been rewritten replacing $\bar{x}$, μ, and σ by p, P, and $\sqrt{PQ}$, respectively, which are their calculated values when the indicator variable is used. Using analogous notation to that developed in the previous chapter,

$$SD_p = \frac{\sqrt{PQ}}{\sqrt{n}} = \sqrt{\frac{PQ}{n}} \text{ since, from algebra, } \frac{\sqrt{a}}{\sqrt{b}} = \sqrt{\frac{a}{b}}.$$

Thus,

$$SD_p = \sqrt{\frac{PQ}{n}}. \tag{28}$$

The central limit theorem assures that p is normally distributed with $\mu_p = P$. Therefore, by reasoning similar to that used in section 5–1, a CIE for P is given by the interval

$$p \pm SD_p,$$

where SD_p is given by formula (28).

This is the best way to estimate P that has been developed so far. However, notice that it involves SD_p, which, in turn, uses P in its calculation. If you are attempting to estimate P, it is presumably unknown. Thus, in practice, SD_p cannot be calculated. Just as you can estimate $SD_{\bar{x}}$ by $SE_{\bar{x}}$ when σ is unknown, you ought to be able to estimate SD_p by SE_p when P is unknown; but how? The best *point* estimate for P is p and for Q is q. Thus, the best estimate for SD_p is

$$SE_p = \sqrt{\frac{pq}{n}}.$$

This can always be calculated, since p, q, n are all calculated from the sample results in front of you. In practice, P is always estimated as follows:

$$\text{A CIE for } P \text{ is } p \pm zSE_p, \text{ where } SE_p = \sqrt{\frac{pq}{n}}. \qquad (29)$$

EXAMPLE 5-6 A survey was taken of a random sample of 302 students at a particular university. Of these, 161 stated that they had smoked marijuana (at least once). Find a 95% CIE for the proportion of all the students at this university who have smoked marijuana.

Solution The sample results reveal $p = \frac{161}{302} \approx 0.533$, $q = 1 - p \approx 1 - 0.533 = 0.467$, and $n = 302$. Therefore, $SE_p = \sqrt{\frac{(0.533)(0.467)}{302}} \approx 0.0287$, and for 95% confidence, take $z = 2$. Thus,

$$\text{a 95\% CIE for } P \text{ is} \quad 0.533 \pm 2(0.0287)$$
$$\approx 0.533 \pm 0.057$$
$$= 0.476 \text{ to } 0.590.$$

The sample proportion p could be changed to a percentage immediately: $p = 53.3\%$. If this is done, *be careful!* The standard error should then also be changed to a percentage: $SE_p = 2.87\%$. Don't change one and not the other. The CIE, $53.3 \pm 2(0.0287)$, would give the *wrong* answer.

This result could also be represented in percentage terms as 47.6% to 59.0%.

You can be 95% confident that between 47.6% and 59.0% of this university's students have smoked marijuana. ●

Exercise Set 5-4   (1-9)

1. In a random sample of size $n = 81$, p is found to be 0.1. Find a 68% CIE for P.

2. In a random sample of size $n = 100$, p is found to be 0.35. Find a 95% CIE for P.

3. In a random sample of size 64, there are found to be 32 successes. Find a 90% CIE for the population proportion of successes.

4. In a random sample of size 120, there are found to be 84 successes. Find a 98% CIE for the population proportion of successes.

5. A die is rolled 300 times and 6 shows up 60 times. What is a 99% CIE for the true proportion of times a 6 will show?

6. In a sample of 60 students attending a large university, 27 favored the continuation of certain physical education requirements. Construct an 80% CIE for the true proportion of students attending this university who favor the continuation of P.E. requirements.

7. A public health physician wants to know what percentage of patients with a certain lung disorder are smokers. It is found that of 350 such patients 200 are smokers. Assuming that these 350 form a random sample, find a 95% CIE for the true percentage of patients with this lung disorder who are smokers.

8. In a nationwide survey of college students, of a random sample of 1,635 students, 1,202 said they favored legalization of marijuana. Construct a 98% CIE for the true percentage of United States students who favor legalization.

 9. In a random sample of 150 males in the labor force of a moderately sized city, it is found that 16 are unemployed. Find a 90% CIE for the percentage of male unemployment in this city.

 10. A manufacturing process producing Rubik's cube does not turn out usable cubes 100% of the time. In an effort to determine what percentage of the cubes produced are defects, a random sample of 90 of them included 6 defective ones. Find a 99% CIE for the true percentage of defective cubes produced by this process.

Section 5–5 Determining Sample Size When Estimating *P*

Just as in section 5–3, when estimating a mean, it is possible to decide *in advance* what size sample should be drawn to estimate a proportion to within a certain error *E*. Here are the steps to take for proportion.

$$E = zSD_p$$

$$E = z\sqrt{\frac{PQ}{n}}$$

$$E = z\frac{\sqrt{PQ}}{\sqrt{n}} \quad \text{since} \quad \sqrt{\frac{a}{b}} = \frac{\sqrt{a}}{\sqrt{b}}$$

$$E\sqrt{n} = z\sqrt{PQ} \quad \text{by multiplying both sides by } \sqrt{n}$$

$$\sqrt{n} = \frac{z\sqrt{PQ}}{E} \quad \text{by dividing both sides by } E$$

$$n = \frac{z^2PQ}{E^2} \quad \text{squaring both sides}$$

Now, a similar problem as before has to be confronted. Since the final object of the exercise is to estimate *P* (and *Q*), they are unknown. You might consider replacing them by *p* and *q*. But, at this point, these are also unknown, since the sample has not been drawn yet. One idea is to use two-stage sampling, as was done in section 5–3. There is a better way, however. Recall that $P + Q = 1$. Consider some representative values to see what happens to the *product, PQ,* as *P* changes.

Table 5–1

P	Q	PQ	PQ as Percent
0.1	0.9	0.09	9
0.2	0.8	0.16	16
0.3	0.7	0.21	21
0.4	0.6	0.24	24
0.5	0.5	0.25	25
0.6	0.4	0.24	24
0.7	0.3	0.21	21
0.8	0.2	0.16	16
0.9	0.1	0.09	9

Of course, there are many values missing from table 5–1. (If $P = 0.25$, then $Q = 0.75$ and $PQ = 0.1875$. Notice that 0.25 would occur between 0.2 and 0.3 in the column headed P in the table, and then the value found for PQ, 0.1875, is a value between 0.16 and 0.21, the corresponding values of PQ.) Despite the missing values, examine the *trend* in the last column. The largest value that PQ can attain is 0.25 or 25%. If you are skeptical about this, try to choose values of P and Q that would make their product larger than 0.25, remembering that $P + Q$ *must* be equal to 1.

Now, return to the formula developed for n, so far,

$$n = \frac{z^2 PQ}{E^2}.$$

If PQ is replaced by the largest possible value that it can attain, this will guarantee that no sample size larger than the value of n calculated by doing this will be needed. Thus,

$$n = \frac{z^2 \cdot 0.25}{E^2}.$$

To remove the decimal, multiply the numerator and the denominator of the fraction by 4 to get

$$n = \frac{z^2}{4E^2}. \tag{30}$$

EXAMPLE 5–7   Suppose you wish to estimate a proportion with 95% confidence such that the maximum error in your estimate is 3%. How large a random sample would you need?

Solution   From the statement of the problem, $E = 3\% = 0.03$, and for 95% confidence, take $z = 2$. Then,

$$n = \frac{2^2}{4(0.03)^2}$$

$$n = \frac{4}{4(0.0009)}$$

$$n \approx 1{,}111.1.$$

You would need a random sample of 1,112. (As in section 5–3, round up, just to be sure.)   ●

Note: If it turns out that the proportion that you are estimating is very large or very small, formula (30) might give you a sample size much larger than you actually need. Therefore, if you have a strong prior sense that p is close to 0 or 1, a two-stage sampling process might be preferable.

EXAMPLE 5–8   Suppose that in example 5–7 you want to estimate the proportion of defective screwdrivers produced by a manufacturing process. You examine 1,112 randomly selected screwdrivers and find 48 are defective. Find a 95% CIE for P, the proportion of all screwdrivers that are defective.

Solution

$$p = \frac{48}{1{,}112} \approx 0.0432$$

$$q \approx 1 - 0.0432 = 0.9568$$

$$SE_p \approx \sqrt{\frac{(0.0432)(0.9568)}{1{,}112}} = 0.0061$$

and a 95% CIE for P is $0.0432 \pm 2(0.0061) = 0.0432 \pm 0.0122$.   ●

Notice that the error in the estimate is only 1.22%, much less than the 3% that was the requirement in example 5–7. This is because the product pq $= (0.0432)(0.09568) \approx 0.0413$, is much less than 0.25, the maximum it could be. This maximum is only approached when p and q are close to 0.5 or 50%. Here, 0.0413 is about $\frac{1}{6}$ of 0.25.

EXAMPLE 5–9   It is desired to estimate the proportion of defective screwdrivers produced by the manufacturing process in example 5–8. If it is known from past experience that no more than 5% defectives are ever produced, how many screwdrivers should be included in the random sample to be 95% confident that the error in the estimate is no more than 3%?

Solution   Instead of using formula (30), it is better to use the formula developed at the beginning of this section,

$$n = \frac{z^2 PQ}{E^2},$$

and estimate P by 0.05, since past experience tells you that P is never larger than 5%.

As in example 5–7, $z = 2$, $E = 0.03$, and so

$$n = \frac{2^2(0.05)(0.95)}{(0.03)^2}$$

$$n = 211.11$$

$$n \approx 212.$$

A sample of 212 is needed.   ●

Notice that examples 5–7 and 5–9 asked you to use the same amount of confidence and the same value for E, but that the final answers for n came out radically different.

Formula (30) should be used when you have no idea what P is. If you have a rough idea of what P might be and this value is far from 50%, it is worthwhile using this value (and the corresponding value of Q) rather than formula (30) to calculate n.

Exercise Set 5–5 $\left(1-9\right)$

1. If it is required that $E = 0.04$, find n for 68% confidence in estimating P.

2. If it is required that $E = 0.065$, find n for 95% confidence in estimating P.

3. To obtain data for her thesis, a psychology major plans to interview a random sample of people. She wants to be 90% confident that she can estimate the true proportion of people who react to a certain situation in a certain way to within 0.02. How many people should she interview?

4. A new drug for alleviating migraine headaches is proposed and its effectiveness is to be tested. If it is desired to be 99% confident of estimating the true proportion of migraine sufferers who are helped by the drug to within 0.05, how many are needed for the study?

$z^2 pq$

5. Smith is thinking of running for Congress. How many votes should he poll in his district to determine the support he has if he wants to be 95% confident that he has estimated his true support to within 4%?

6. A cryptographer wants to determine the relative frequencies of letters in a foreign language. How many letters should be included in a random sample of words from the language if the cryptographer would like to be 99% sure that the true relative frequencies have been estimated with an error in the estimate of less than 1 in 100?

7. A manufacturer would like to sample her product for defects. She knows from past experience that there are never more than 5% defects. How many should be included in a random sample if she would like to be 90% confident of estimating the true percentage of defects to within $2\frac{1}{2}$%?

8. If it is known that at least 85% of high school students who enter twelfth grade complete it, find the size of a random sample of twelfth-graders necessary to estimate the true percentage of students who enter twelfth grade and complete it to within 3.75% with 97.5% confidence.

9. Read the second paragraph of the article at the beginning of the chapter. With the size of the sample given there, what level of confidence are they assuming to get the given margin of error? (*Note:* Assume they used the maximum possible value for the product pq.)

10. Now read the third paragraph of the article at the beginning of the chapter. Explain how Field may have calculated the probabilities given there.

Section 5–6 Estimating Standard Deviation

When discussing the estimation of μ in section 5–1, it was stated that with a large enough sample, s is a good estimate for σ, and $SE_{\bar{x}} = s/\sqrt{n}$ therefore is a good estimate for $SD_{\bar{x}} = \sigma/\sqrt{n}$. In reality, how good an estimate is it? As for μ and P, this question would be answered if a standard error could be attached to the standard deviation itself and a CIE calculated in the form

$$s \pm zSE_s.$$

Early statisticians, believing that every population was normally distributed, developed a formula for SD_s, the standard deviation of the frequency distribution of s, based on the assumption of normality.

$$SD_s = \frac{\sigma}{\sqrt{2n}}$$

Since σ is unknown, SD_s is estimated by SE_s, using s,

$$SE_s = \frac{s}{\sqrt{2n}}. \tag{31}$$

Formula (31) is valid only under two conditions: the population follows a normal distribution (or close to it) and your sample size is larger than 30. If the population is substantially non-normal, formula (31) will underestimate the true value of SE_s. There does not exist a truly valid method for calculating SE_s when dealing with non-normal distributions. If you ever do use formula (31) to give yourself an idea about the accuracy of s as an estimate for σ, it would be wise to view your answer with healthy skepticism. It may well be an underestimate.

EXAMPLE 5–10   Example 5–3 concerned a sample of students with respect to the number of siblings each had. You were asked to check that $n = 35$, $\bar{x} = 1.97$, $s = 1.44$. Find a 99% CIE for σ.

Solution

$$SE_s = \frac{s}{\sqrt{2n}} = \frac{1.44}{\sqrt{70}} \approx 0.17$$

Thus, a 99% CIE for σ is $1.44 \pm 2.58(0.17)$

$$= 1.44 \pm 0.44$$

$$= 1.00 \text{ to } 1.88.$$ ●

If $\sigma = 1.00$,

$$SD_{\bar{x}} = \frac{1.00}{\sqrt{35}} \approx 0.17.$$

If $\sigma = 1.88$,

$$SD_{\bar{x}} = \frac{1.88}{\sqrt{35}} \approx 0.32.$$

When estimating μ for this example, it was found that $SE_{\bar{x}} = 0.24$. Using the result of example 5–10, it can be seen that $SD_{\bar{x}}$ may be as low as 0.17 or as high as 0.32, if you use the extremes of the above interval for σ. In fact, the histogram of this sample exhibits slight right-skewness. Therefore, the CIE for σ may be narrower than it ought to be.

In summary, use formula (31) to attach a standard error to your estimate for σ. However, don't place too much faith in it, especially if you suspect that the population you are dealing with is substantially non-normal.

Exercise Set 5–6

1. If a random sample from a normal population is found to have $n = 50$, $s = 5.7$, find a 68% CIE for σ.

2. If a random sample from a normal population is found to have $n = 100$, $s = 8.3$, find a 95% CIE for σ.

 3. A random sample of 64 heights is found to have mean 64.3 inches and standard deviation 2.8 inches. Use formula (31) to obtain a 90% CIE for the standard deviation of the population of heights from which the sample was drawn. How valid do you think your answer to this problem is?

4. The weights of the sample in exercise 3 were found to have a mean of 140 lb with a standard deviation of 28 lb. Use formula (31) to obtain a 98% CIE for the standard deviation of the population of weights from which the sample was drawn. How valid do you think your answer to this problem is?

 5. A random sample of 175 family incomes was found to have a standard deviation of $6,350. Use formula (31) to obtain a 95% CIE for the population standard deviation of incomes and state how valid you believe your answer to be.

 6. An educational psychologist wishes to test the eighth grade of a junior high school with an aptitude test. It is known that scores on this test invariably follow a normal distribution. A preliminary sample of 25 of these eighth-graders reveals a standard deviation of their scores of 12.6. Construct a 99% CIE for the standard deviation of all eighth-grade scores on this test and state how valid you believe your answer to be.

 7. The amount of butterfat produced annually by dairy cows is generally normally distributed. If a random sample of 200 of these cows reveals a standard deviation of 83 lb of butterfat, construct a 96% CIE for the true standard deviation.

8. Of those adults who reach 40 years of age, it is found that the age of death is normally distributed. If a random sample of 35 adults, 40 years and older is observed until they are all dead and the results show their mean ages of death to be 73 with a standard deviation of 7.5 years, construct a 95% CIE for the true standard deviation.

....

Section 5–7   Summary of Terms

confidence interval estimate (CIE) 196   interval estimate 196
determining sample size 205   point estimate 195
dichotomized 208   proportion 208
error in the estimate 205   standard error 202
estimate 195   two-stage sampling 206
indicator variable 209

Section 5–8   Summary of Symbols

Symbol	Meaning
E	The desired error in the estimate when determining sample size
p	The proportion of successes in a sample
P	The proportion of successes in a population
q	The proportion of failures in a sample
Q	The proportion of failures in a population
SD_p	The standard deviation of a frequency distribution of a sample proportion
SD_s	The standard deviation of a frequency distribution of a sample standard deviation

$SD_{\bar{x}}$	The standard deviation of a frequency distribution of $\bar{x}$ (see chapter 4)
SE_p	The standard error of a sample proportion
SE_s	The standard error of a sample standard deviation
$SE_{\bar{x}}$	The standard error of a sample mean

Section 5–9   Summary of Formulas

Formula

(25) $\bar{x} \pm zSD_{\bar{x}}$

(26) $SE_{\bar{x}} = \dfrac{s}{\sqrt{n}}$

(27) $n = \left(\dfrac{z\sigma}{E}\right)^2$

(28) $SD_p = \sqrt{\dfrac{PQ}{n}}$

(29) $p \pm zSE_p,$

  where $SE_p = \sqrt{\dfrac{pq}{n}}$

(30) $n = \dfrac{z^2}{4E^2}$

(31) $SE_s = \dfrac{s}{\sqrt{2n}}$

Use

A confidence interval estimate for μ

Formula for the standard error of the sample mean

Formula for determining the sample size when estimating a sample mean; generally rounded up to the next whole number

To find the standard deviation of the frequency distribution of p

A CIE for P, a population proportion

To determine a sample size when estimating a proportion and you have no idea of the rough size of P

To find the standard error when estimating a standard deviation of a population following a normal distribution

Chapter 5 Review Exercises

1. If, for a population, $\sigma = 8.3$, and a random sample drawn from this population is such that $n = 16$, $\bar{x} = 16.2$, find a 95% CIE for μ, the mean of the population.

2. If, for a population, $\sigma = 245$, and a random sample drawn from this population is such that $n = 25$, $\bar{x} = 2{,}120$, find a 98% CIE for μ, the mean of the population.

3. If a random sample drawn from a population is such that $n = 49$, $\bar{x} = 72.4$, $s = 3.6$, find a 90% CIE for the mean of the population.

4. If a random sample drawn from a population is such that $n = 60$, $\bar{x} = 0.32$, $s = 0.07$, find a 68% CIE for the mean of the population.

 5. A random sample of 81 lengths of pipe for plumbing is drawn from a large shipment of pipes. The sample of pipes is found to have a mean interior diameter of 1.04 in. with a standard deviation of 0.02 in. Find a 95% CIE for the mean interior diameter of the shipment.

6. A random sample of 40 sacks of rice, allegedly weighing 25 lb each, drawn from a large shipment is found to have a mean weight of 24.72 lb with a standard deviation of 0.63 lb. Find a 90% CIE for the true mean weight of the shipment.

7. A psychologist wishes to determine the mean score on a standardized test. He administers the test to a sample of 250 subjects and finds the mean to be 134.6. If the standard deviation of scores on this test has been found to be about 20 points for past users, find a 99% CIE for the probable mean, were everybody to take it.

8. A college claims that it has streamlined its enrollment procedures, and a student can complete it in 30 minutes or less. Some students dispute this and sample 50 students at random and find that they average 32 minutes. When pressed, the college administration admitted to a standard deviation of 10 minutes. Assuming this is correct, find a 95% CIE for the mean registration time of all students. Do you think the college has made a fair claim?

9. If a consumers' group allows a standard deviation of 2 mpg on gasoline consumption of new cars, how many runs must it test a new car to be 98% confident that it has estimated the true gasoline consumption to within 0.25 mpg?

10. Bats, being blind, use echolocation to detect objects in their paths. They emit pulses at certain intervals. If a special machine is used to measure the lengths of the intervals between pulses, how many should be measured to be 90% confident that you can estimate the mean length to within 5 milliseconds, if the standard deviation can be taken to be 25 milliseconds?

11. The staff of a newspaper is thinking of running a new feature. A random sample of 225 of its subscribers reveals that 108 would be interested. Find a 95% CIE for the true proportion of the subscribers who would be interested.

12. A random sample of 1,000 Americans was selected and asked if they know the name of the current president. Of these, 73% answered correctly. Construct a 98% CIE for the percentage of all Americans who know the name of the current president.

13. If you wanted to estimate the proportion of people in your city who earn more than $20,000, how many would you have to include in your sample to be 90% confident that the error in your estimate will be no more than 5%?

14. An opinion poll claims to have interviewed a random sample of 1,600 voters. After printing the results, they announce that they may be in error by 3 percentage points either way. If it is assumed that they are using 95% confidence, are these two pieces of information consistent? What is the maximum amount of confidence they could claim and still be consistent?

15. The standard deviation of a sample of 45 measurements made of a meter stick at different times produces a standard deviation of 2.5 mm. Find a 99% CIE for the true standard deviation of such measurements. What assumption must you make?

16. The standard deviation of a random sample of 60 flight times of the same run by an airline is found to be 6.2 minutes. If it can be assumed that these times follow a normal distribution, construct a 90% CIE for the true standard deviation of these times.

Hands-on Class/Student Project

Some of the following exercises are a continuation of the exercises for the class/student project at the end of chapter 4. So retrieve the data that you collected then and get together in the same groups as you did before.

1. In exercise 5 of the class/student project of chapter 4, you were asked to select 20 samples, each of size 15, from population H and to find $\bar{x}$, the sample mean of each. Use these means in exercise 2.

2. For each of the 20 samples, use its mean and the standard deviation of population H, which you calculated in exercise 4 of class/student project of chapter 4, to form a 95% confidence interval estimate (CIE) for the mean of population H.

3. Of your 20 CIE's, how many do include the mean of population H (μ)? (*Note:* For 95% confidence, there ought to be a 95% probability that any one of your CIE's does contain μ. So, if the theory worked perfectly in practice, you should find that 95% [or 19 out of 20] of your CIE's do include μ.)

4. Combine your results of exercise 3 with the corresponding results from all the other groups in the class. Do you get better agreement with the 95% figure when more samples are considered?

5. Since the size of each of your samples is only 15, your CIE's are probably quite wide intervals. Let us say that you wanted to estimate the mean height of a population of students at your college. Discuss in your group how narrow an interval you would like your CIE to be. Put another way, decide among yourselves how large the error in your estimate (E) should be. Now use this value of E and your desired value of confidence to calculate how large a sample from your chosen population you would need to measure. (*Note:* If population H is a similar population to the one that you want to estimate, you could use population H's standard deviation as an estimate for your population's standard deviation. Otherwise, a population of heights of the same gender usually has a standard deviation of about 3 inches.)

6. Repeat exercise 5 by changing E, or the level of confidence required, or both, to see what effect these changes have on the size of the sample you would need.

7. Look in the business section of your newspaper and find the stock results for the previous day's trading on the New York Stock Exchange. The last column usually gives the change in price, for example, $+1/4$, $-1/2$, or just blank (meaning no change). Devise a sampling scheme and sample 100 of these changes.

8. In the sample of exercise 7, count how many are $+1/2$ or higher and convert this to a proportion of your sample.

9. Use the sample proportion you found in exercise 8 to form a 90% CIE for the proportion of all changes on the NYSE that day that were $+1/2$ or higher.

10. Compare your results to others in the class. (If you are really ambitious, you could count all the stock changes for that day and find the true proportion that finished at $+1/2$ or higher, and then see if your CIE included this proportion.)

Hypothesis Testing

"The great tragedy of science-—the slaying of a
beautiful hypothesis by an ugly fact."

THOMAS HENRY HUXLEY

Reprinted with special
permission of NAS, Inc.

Reprinted with special
permission of NAS, Inc.

Sally Forth   **By Greg Howard**

CATHY © 1982 UNIVERSAL
PRESS SYNDICATE.
Reprinted with permission.

CATHY   **by Cathy Guisewite**

....

Chapter Overview
Do You Believe Everything You See on TV?

While watching your favorite TV program, a commercial is aired that shows a man claiming that a "revolutionary" new flashlight battery will stand at least 40 hours of continuous use. A skeptical consumer's group wishes to test this manufacturer's claim by burning out a random sample of these batteries while keeping a record of the time before burnout. Suppose they find that the mean time that these batteries lasted is 39 hours. Of course, if they found it to be 40 hours or more, they would have proved nothing against the advertising claim. However, have they proved anything if they find the mean to be 39 hours? Can they issue warnings against or sue the manufacturer? Or were they just unlucky with the particular sample they chose? You might be inclined to think that 39 is close to 40, so give the manufacturer the benefit of the doubt. If so, would you still think the same way if the mean of the sample turned out to be 38? Or 37? 36? 35? Where would you draw the line? How far below 40 would the mean have to be for you to conclude, beyond reasonable doubt, that the manufacturer is misrepresenting the product? Do not expect to be able to answer these questions right away. No one could without more knowledge of statistical procedures and more facts about the sample or population. The questions are raised to illustrate the type of issues involved in the hypothesis-testing procedure, the subject of this chapter.

....

Section 6-1 The Hypothesis-Testing Procedure

Typically, a **hypothesis** about a population is formed. This hypothesis may be in the form of a claim, such as the flashlight battery one, or simply a statement of "truth," such as the net weight of coffee in a jar for sale is 16 ounces. If, as is the case more often than not, the population cannot be tested completely, then a random sample is drawn from the population and the sample results are examined. After the sample results are compared to the hypothesis, a decision must be reached. The decision will be either to accept the hypothesis or to reject the hypothesis. Somewhere a line has to be drawn that will indicate how to reach this decision.

This, in a nutshell, is the hypothesis-testing procedure. Many different statistical tests have been developed to help you reach decisions. In this and succeeding chapters, you will learn about some of them. You will notice similarities to the ideas of the preceding chapter on estimation. However, there are fundamental differences as well. In estimation, little or no knowledge was assumed about the population; a random sample was drawn and analyzed to find out about the population. In hypothesis testing, you begin with a speculative piece of information about the population but have no idea whether or not it is true; a random sample is drawn to test the validity of this initial belief about the population.

Hypotheses

When somebody claims something to be true about a population, this is a **hypothesis**. To begin, consider hypotheses about the mean of a population μ. Typically, the claim is made that μ is equal to some number, which will be named μ_H (for **hypothesized mean**). This is called the **null hypothesis** and labeled H_0. Thus, the first line of a hypothesis-testing problem will be of the form

$$H_0 : \mu = \mu_H.$$

The colon indicates that the null hypothesis is the "claim" that μ is equal to the number μ_H. It is called the null hypothesis since it could be written

$$H_0 : \mu - \mu_H = 0$$

or, in words, that the difference between μ and μ_H is zero (or null). No matter how the hypothesis is stated, H_0 *will always include the equals sign*. For the example about the flashlight batteries, the null hypothesis would be stated

$$H_0 : \mu = 40.$$

You will have reached the conclusion of the procedure when you can come to a decision as to whether you can accept this as true or reject it. In the case of the batteries, when would you want to reject this hypothesis? You may want to reject it if you find that the mean life of the sample of batteries is "substantially" less than 40. Usually, then, before proceeding, an **alternative hypothesis**, labeled H_1, is set up to show under what conditions you wish to reject H_0. For batteries, it would be

$$H_1 : \mu < 40.$$

You will need more information before you can decide what "substantially" is.

The consumers' group would not be interested in rejecting the manufacturer's claim if the sample results indicated that $\mu > 40$. The alternative hypothesis, H_1, will always consist of an inequality indicating the *direction* of rejection. Thus it could be

$$H_1 : \mu < \mu_H$$

or

$$H_1 : \mu > \mu_H$$

or

$$H_1 : \mu \neq \mu_H.$$

The choice of alternative hypothesis depends on the statement of the null hypothesis. The TV commercial claimed that the batteries last *at least* 40 hours, i.e., 40 hours or more. The consumers' group wishes to test the truth of this: is H_0 true or do the batteries in fact last, on the average, *less* than 40 hours? That is why the symbol for "is less than" ($<$) is used in the above alternative hypothesis.

EXAMPLE 6-1 Suppose a friend of yours claims that the mean age of the students attending your school is no more than 23. How would you set up the hypotheses for testing?

Solution Since your friend says "no more than 23," he/she is claiming that it is 23 or less. You will prove your friend wrong if you can show it is more than 23. Thus the relevant hypotheses are

$$H_0 : \mu = 23$$

$$H_1 : \mu > 23.$$

Notice that, although the claim states that the mean is 23 or less, H_0 is *not* set up as : $\mu < 23$. The null hypothesis is *always* set up with the equals sign. There is a good reason for this, as you will see shortly.

EXAMPLE 6-2 A coffee manufacturer has a machine that dispenses instant coffee with jars labeled "Net Wt 16 oz." The manufacturer wishes to draw a random sample of these jars to test whether the correct amount is being dispensed into the jars. It will be a cause for concern if either too little *or* too much is dispensed by the machine. Too little could cause trouble with angry consumers and too much would cut into the profits. Set up the hypotheses.

Solution
$$H_0 : \mu = 16$$

$$H_1 : \mu \neq 16$$

Here, H_0 will be rejected if the sample results indicate that μ is substantially larger than *or* smaller than 16.

Exercise Set 6-1

In these exercises, state the null and alternative hypotheses to set up each problem.

1. It is claimed that the mean of a population is no more than 65.

2. It is claimed that the mean of a population is at least 43.

3. It is desired to test whether the mean of a population is 30 or not.

4. It is desired to test whether the mean of a population is no more than 8.

 5. A friend of yours thinks that the mean height of students attending your school is 65 inches. You believe this to be an underestimate.

 6. A manufacturing process is supposed to turn out $\frac{1}{4}''$ rivets. It would be a cause for concern if they were too large or too small.

 7. A psychologist believes it will take at least a half an hour for subjects to learn a certain task.

8. A sociologist believes that the mean annual income in a certain section of the city is no more then $10,000.

 9. Boxes of cereal are marked "Net Wt 12 oz."

 10. The administration of a community college claims that, on the average, the students travel no more than 5 miles to school.

Section 6–2 The z-score Test Using σ

There are a couple of other things that ought to be decided *before* the sample is drawn and the results examined. Before going into these, however, take a look at a simple test using the battery example.

For the flashlight batteries, so far, the following hypotheses have been made:

$$H_0 : \mu = 40$$

$$H_1 : \mu < 40.$$

Suppose you obtain a random sample of 45 of these batteries and let them run down while timing this process. You find that the mean life of the 45 batteries is 39.1 hours. In symbols, you have found

$$n = 45$$

$$\bar{x} = 39.1.$$

Now, thinking theoretically for a moment: your sample is just one of all possible samples of size 45 that could have been drawn from the population. Suppose that H_0 is true. If all possible $\bar{x}$'s are calculated, then the central limit theorem (CLT) guarantees the result shown in figure 6–1.

Figure 6–1

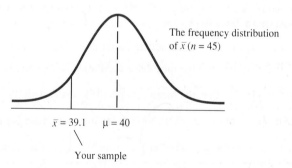

The frequency distribution of $\bar{x}$ ($n = 45$)

$\bar{x} = 39.1$ $\mu = 40$

Your sample

After example 6–1, it was stated that there is a good reason for putting = in H_0. It is so you know where to center the distribution.

Part 3 of the CLT guarantees the normal shape. Part 1 guarantees that it is centered at μ, which is assumed to be 40 under the null hypothesis. Part 2 guarantees that the standard deviation of this normal distribution is $SD_{\bar{x}} = \sigma/\sqrt{n}$.

Although $\bar{x} = 39.1$ has been drawn in on this normal distribution, it is not yet known where to place it, except that it is to the left of center. The closer $\bar{x}$ is to the center, the more inclined you will be to accept H_0, and the further away, the more inclined you will be to reject H_0. What determines how near $\bar{x}$ is to μ_H

is the spread of the normal distribution $SD_{\bar{x}}$. This in turn depends on σ, the standard deviation of the population. Suppose that you have worked extensively with various flashlight batteries and know that the standard deviation of these type of batteries is about 6 hours. In other words, assume that $\sigma = 6$. Then, $SD_{\bar{x}} = 6/\sqrt{45} \approx 0.89$.

Now convert $\bar{x}$ into a standard score.

$$z = \frac{39.1 - 40}{0.89}$$

$$z \approx -1.01$$

This tells you that the particular sample mean you have found is slightly more than one standard deviation below the hypothesized mean. From the normal table, $A = 0.3438$, which, in turn, tells you that the area under the normal curve to the left of $\bar{x} = 39.1$ is $0.5 - 0.3438 = 0.1562 \approx 15.6\%$. The above discussion was predicated on the assumption that H_0 is true. Thus, if H_0 is true, there is a probability of 15.6% that you obtain $\bar{x}$ as small as, or smaller than, 39.1. If you now decide to reject H_0, there will be a probability of 15.6% that this will be an incorrect decision. This probability will be called the **P-value,** or P-val for short.

With everything else unchanged, suppose you had found $\bar{x} = 38.3$. Then,

$$z = \frac{38.3 - 40}{0.89}$$

$$z \approx -1.91$$

$$A = 0.4719$$

$$\text{P-val} = 0.0281 \approx 2.8\%.$$

This time, if you decide to reject H_0, there will be a probability of only 2.8% that you are making an incorrect decision. Would you not feel more comfortable in rejecting H_0 in this case than in the above case when $\bar{x} = 39.1$?

Errors, Significance Level and *P*-value

The *reality* of any hypothesis-testing situation allows for two possibilities: H_0 is true or H_0 is false. There are also two possibilities when arriving at a *decision:* accept H_0 or reject H_0. This allows for four possible results as illustrated by table 6–1.

Table 6–1

		Decision		
		Accept H_0	Reject H_0	
Reality	H_0 is true	A	B	
	H_0 is false	C	D	

Two of the four possible results, boxes A and D, amount to correct decisions, the kind you would like to achieve. You *would* like to accept H_0 if, in reality, it is true, and you *would* like to reject H_0 if, in reality, it is false. But, since the reality is unknown to you, it is also possible that you blunder. Box B illustrates one such error: rejecting H_0 when, in reality, it is true. This is called an error of the first type, or, more simply, a **type I error.** Box C illustrates the other error that can be made, a **type II error:** accepting H_0 when, in reality, it is false.

The situation is not unlike the decision faced by a jury in the trial of a defendant. There are two possible correct decisions they can make: finding an innocent defendant innocent or a guilty defendant guilty. But juries can make two types of errors too: finding an innocent defendant guilty or a guilty defendant innocent. The jury must hear all the evidence, debate it among themselves, and then come to a decision in the hope that they have minimized all possibilities of making an error. Likewise, the statistician, in coming to a decision, would like to minimize the possibility of making an error. To do this, the statistician sets up the null hypothesis, which is assumed to be true. This is analogous to the court presuming that a defendant is innocent. Then, the statistician examines the sample results in preparation for coming to a decision. This is like the jury examining and debating the evidence preparatory to *their* decision.

Just as there can be serious consequences when a jury commits an error— an innocent person being thrown behind bars, or a killer going free—so there can be serious consequences if a statistician commits either type of error described above.

When the Salk vaccine was being tested in the 1950s as a possible prevention for polio, it was important not to accept it if it was useless, thereby raising false hopes for wiping out polio, not to mention possible harmful side effects. Nor was it desirable to reject it if it did the trick, depriving thousands of people, particularly children, of the chance of avoiding a crippling disease. Great care was taken with statistical procedures to avoid committing either type of error and the story had a happy ending: the Salk vaccine was found to be effective and polio was all but wiped out.

There is no certain way of avoiding the two types of errors completely. Rather, what the statistician does is to minimize the chances of committing the errors. A type I error is the error of rejecting a true hypothesis. While the statistician is performing the test, H_0, the null hypothesis, is continually assumed to be true. When the P-value is finally calculated, it tells the statistician the probability of finding a value of $\bar{x}$ so deviant from μ_H when H_0 is true. If the statistician then decides to reject H_0, the size of the P-value equals the probability of committing a type I error. To minimize the chances of committing a type I error, H_0 is rejected only when the P-value is suitably small. How small is "suitably small"? The answer depends primarily on the consequences of the decision, rather than on statistical theory. For instance, in the previous battery example, for $\bar{x} = 39.1$, P-val $= 15.6\%$ was calculated. The decision as to whether to reject the manufacturer's claim rests more with the consumers' group than with the statistician. Are they willing to take action against the manufacturer and thereby take a 15.6% chance of committing a type I error, i.e., in rejecting a claim that may be true?

In short, for any hypothesis test,

P-value = probability of committing a type I error.

One way to assure a small chance of committing a type I error is to pick an acceptably small, specific P-value even before the sample is drawn. Then, if the P-value for the sample is smaller than or equal to this preselected value, H_0 will be rejected; if the sample P-value is larger, H_0 will not be rejected. This preselected value is called the **significance level.** For instance, a very common choice for significance level is 0.05 or 5%. The letter that will be used for significance level is the lowercase Greek letter α (alpha). Notice that the significance level equals the P-value. Thus, the significance level is the probability of a type I error, i.e., the probability that H_0 will be rejected when H_0 is true. When $\alpha = 0.05$, the decision is made according to the following scheme:

if P-val ≤ 0.05, reject H_0
if P-val > 0.05, accept H_0.

Thus, in the battery example, when $\bar{x} = 39.1$, using a 5% significance level, you would be forced to accept H_0, since the corresponding P-value is 0.156. On the other hand, when $\bar{x} = 38.3$, P-val $= 0.028$. This *is* less than 0.05, and you could reject H_0 at the 5% significance level.

Another common choice for α is 0.01. This can be used if you want to be very sure that you are not making a type I error when rejecting H_0. What does it mean to reject a hypothesis at the 1% significance level? A hypothesis has been rejected and there is only 1 chance out of 100 (or even a smaller chance) that a type I error has been committed. In fact, both values, 0.05 and 0.01, are arbitrary. *Any* value could be chosen for α, the significance level. Choosing a significance level is a convenient way of indicating to the reader of the research how significantly the sample results are at variance with H_0. One can often find sentences at the end of research papers such as,

"The results were statistically significant at the 5% level."

This is just another way of indicating the probability of making a type I error. Newspapers and popular magazines will sometimes state that some result was "statistically significant," as if there were some universal law that enables you to decide when a result is significant or not. The phrase "statistically significant" is meaningless unless some (preferably prechosen) numerical level is attached.

Returning now to the battery example and working backwards, suppose you want $\alpha = 0.05$, i.e., you will take action against the manufacturer if there is a 5% (or less) chance that $\bar{x}$ is that far away from μ_H. The situation is pictured in figure 6–2.

Figure 6–2

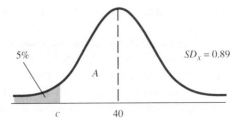

What value is c such that the area of the left tail is 5%? For 5% in the tail, $A = 0.45$ and, from table 1, $z = -1.65$ (negative because it is left of center), and

$$c = 40 - 1.65 \, (0.89)$$

$$c = 38.53.$$

This tells you that if $\bar{x}$ in your sample turns out to be 38.53 or less, it is in the tail of the sampling distribution, and thus is far enough away to be able to reject H_0 with less than 5% probability of committing a type I error. If $\bar{x}$, on the other hand, is greater than 38.53, you must either accept H_0 or take a bigger risk than 5% in committing a type I error. The value, 38.53, is called the "critical value." This value marks "the line" that was referred to earlier when this question was posed: "Where do you draw the line," so that you can make a decision? One possible answer is: "At 38.53." Of course, if $\alpha = 0.01$, or if some other significance level is chosen, the line would necessarily have to be drawn in a different place.

Minimizing the Type II Error

The above discussion allows you to arrange for the probability of committing a type I error to be as small as you please. What about minimizing the probability of committing a type II error, that of accepting a hypothesis that is, in reality, false? This is a much more difficult proposition. If, in the battery example, $\bar{x} = 39.1$, and you decide to accept that $\mu = 40$, couldn't you also have accepted that $\mu = 39.5$, $\mu = 39$, or, in fact, any value close to 39.1? Obviously, only *one* value is the true value of μ, and it could be argued that if that true value is 39.8, say, then accepting that $\mu = 40$ is committing a type II error. That H_0 hits on the true value *exactly* is an extremely unlikely proposition. It then becomes a question of what is important to the person conducting the research. For instance, would it be important to the consumers' group if μ were, in reality, 39.8 instead of 40? Since, with this sample, $SD_{\bar{x}}$ is 0.89, it would be impossible to detect such a small difference as that between 39.8 and 40. To do so would require a much larger sample size.

The most important decision to be made when attempting to minimize the type II error is the decision as to which alternative hypothesis (or hypotheses) are important. It probably would matter little to the consumers' group if μ were 39.8 and the manufacturer claimed 40. When it has been decided which alternative hypotheses are important to detect, the sample size must be decided in such a way that there is no danger of accepting those alternative hypotheses when accepting H_0. The discussion of these ideas is left to a more advanced statistics course.

There is another way of dealing with the problem, which many researchers use, but which may seem like a cop-out to you. If you never accept H_0, you can never commit a type II error. Often the phrase "fail to reject H_0" is used instead. Not rejecting H_0 is not the same as accepting it, it is simply saying that the sample evidence was not damning enough to be able to reject H_0 out of hand. Researchers will often use terminology such as "the sample results were not significant at the 5% level, and, therefore, we have not proved H_0 false." This is like a jury finding that there *is* reasonable doubt that the defendant is guilty. They would then find him/her "not guilty," which is *not* the same as declaring him/ her innocent.

Now, take a look at the hypothesis-testing procedure, comparing it to a trial in a court of law.

Hypothesis Test	**Trial**
1. State hypotheses: $H_0 : \mu = \mu_H$ $H_1 : \mu \ ? \ \mu_H$? could be $<$, $>$, or $\neq$.	**1.** Defendant faces trial on a charge and is presumed innocent.
2. α, the significance level, is chosen. Common choices are 0.05 and 0.01, but any values are permitted.	**2.** Proof of guilt must be beyond a reasonable doubt.
3. The type of test to be used is decided.	**3.** The type of trial, by jury, judge, and so on, is decided.
4. Sample data is collected.	**4.** Evidence is collected.
5. Mathematical computations are made until a *P*-value is calculated according to which test was chosen in step 3.	**5.** Evidence is presented to and debated by the jury.

Hypothesis Test	**Trial**
6. (a) If P-val ≤ the value chosen for α in step 2, H_0 is rejected.	**6.** (a) If there is no reasonable doubt, the jury or judge finds the defendant guilty.
(b) If P-val > the value chosen for α in step 2, then fail to reject H_0.	(b) If there is reasonable doubt, the jury or judge finds the defendant not guilty.
7. Interpret the decision in step 6 and act accordingly.	**7.** Sentence or free the defendant.

All hypothesis tests follow this skeletal outline. What changes from test to test is the choice of test (step 3) and, therefore, the computations involved (step 5). So far, you have only met one test. It is the **z-score test using** σ and uses the formula

$$z = \frac{\bar{x} - \mu_H}{\sigma/\sqrt{n}}. \qquad (32)$$

The denominator, $\sigma/\sqrt{n}$, is the formula for $SD_{\bar{x}}$ from part 2 of the CLT.

After calculating z from formula (32), the P-value is found, using the normal table.

EXAMPLE 6–3   Examine the battery example again, by presenting the test without all the attendant explanations.

Solution

1. $H_0 : \mu = 40$

   $H_1 : \mu < 40$
2. $\alpha = 0.05$
3. Use the z-score test using σ, and assume $\sigma = 6$.
4. Sample results:

$$n = 45$$

$$\bar{x} = 38.3$$

5. $$z = \frac{38.3 - 40}{6/\sqrt{45}} \approx -1.91$$

6. $A = 0.4719$ and $P\text{-val} = 0.0281 < 0.05 \, (= \alpha)$.

   Reject H_0.

7. The advertising claim by the manufacturer is exaggeration. Warn
   consumers, or take legal action, or both. ●

   The steps are numbered to correspond to the outline above.
   Sometimes researchers, rather than choosing a significance level and making
   a decision about H_0, will simply present the P-value and leave the decision up to
   the reader.

EXAMPLE 6–4   A video game enthusiast claims that he can average at least
30,000 points in Pac Man, with a standard deviation of 9,000 points. The
scores in his next 15 games average 27,500. Assuming that these 15 games
constitute a random sample of all his games, test this claim by giving the
resulting P-value together with a conclusion.

Solution

$$H_0 : \mu = 30{,}000$$

$$H_1 : \mu < 30{,}000$$

Use the z-score test using σ, with $\sigma = 9{,}000$. Sample results:

$$n = 15$$

$$\bar{x} = 27{,}500$$

$$z = \frac{27{,}500 - 30{,}000}{9{,}000 / \sqrt{15}}$$

$$\approx -1.08$$

$A = 0.3599$ and $P\text{-val} = 0.1401$.

   Notice that you could not reject the null hypothesis at a 5%, or even a
10%, level. There is not sufficient evidence to dispute the video game
enthusiast's claim at the 5% (or 10%) significance level. ●

Exercise Set 6–2

1. It is claimed that the mean of a population, μ, is 55 or more. A random sample
   from the population reveals that $n = 50$, $\bar{x} = 51$. If σ is assumed to be 20, and the
   z-score test using σ is applied, what is the resulting P-value?

2. It is believed that the mean of a population, μ, is no more than 83. A random
   sample from the population reveals that $n = 100$, $\bar{x} = 89$. If σ is assumed to be 30,
   and the z-score test using σ is applied, what is the resulting P-value?

3. A farmer samples, at random, 200 oranges from his crop. He only wishes to market them if they weigh, on the average, 4 oz or more. If he weighs the sample of 200 and they weigh a total of 760 oz, would he wish to market his crop if he chooses to use a 5% significance level? Assume a population standard deviation of 1.1 oz.

4. A new drug on the market is claimed by its manufacturers to reduce overweight women by at least 10 lb per month, with a standard deviation of 2 lb. Ten women, chosen at random, reported losing 6, 12, 7, 8, 15, 9, 10, 13, 4, and 5 lb within a month. Does this data support the claim at the 0.01 level?

5. Boxes of cereal are marked "Net Wt 12 oz." A random sample of this brand reveals a mean of only 11.5 oz. If 25 boxes were weighed and the standard deviation in the population is 1.5 oz, do you think the makers of this brand are selling us short? If somebody answered yes to the previous question, what is the probability that he is committing a type I error?

6. A sociologist believes that the mean income in a certain section of the city is no more than $10,000. To test this, another sociologist finds that the mean of a random sample of 40 families from this section is $11,775. If the second sociologist, using a standard deviation of $6,000, declares that the first is mistaken, what is the probability that the second is committing a type I error?

7. A light-bulb manufacturer regularly advertises that his bulbs last 900 hours with a standard deviation of 75 hours. A random sample is chosen before each campaign to make sure the claim is correct. If one such sample of 20 bulbs shows a mean of 925 hours, can the advertising claim be considered an underestimate at the 0.05 level?

8. A psychologist believes that it will take at least an hour for certain disturbed children to learn a task. A random sample of 30 of these children resulted in a mean of 50 minutes to learn the task. Should the psychologist modify her belief at the 0.01 level if the population standard deviation can be assumed to be 15 minutes?

9. An automobile manufacturer claims that his new model has a gas consumption of at least 30 mpg, with a variance of 24. If a random sample of 8 models produce the consumptions 25, 27, 31, 30, 28, 23, 24, 32, is the manufacturer's claim feasible at the 0.05 level?

10. The variance of the lengths of bass in California lakes is 8 sq in. One lake owner claims that the bass in his lake average at least 1 foot. If a random sample of 80 bass drawn from the lake have a mean length of 8 inches, how valid do you think the lake-owner's claim is?

Section 6–3 The *z*-score Test Using *s*

Another Test

The *z*-score test using σ is an excellent test and can be used to test a hypothesis about a population mean using a sample of any size. However, it does have one weakness. The *z*-score test using σ assumes knowledge of σ, the population standard deviation, and this is not always known. In the discussion of CIE's in chapter 5, *s,* the sample standard deviation, was shown to be a good estimate for σ, *provided* the sample is sufficiently large. The usual guideline in using *s* in place of σ is that *n* should be larger than 30. The test, with this modification, is then called the **z-score test using s,** and in place of formula (32), this formula is used:

$$z = \frac{\bar{x} - \mu_H}{s/\sqrt{n}}, \, n > 30. \tag{33}$$

Because *s* is assumed to be a particularly good estimate of σ for samples of $n \geq 30$, the *P*-value is found in the same way, using the normal table.

EXAMPLE 6–5 The manufacturers of the batteries could argue that $\sigma = 6$ is not valid for *their* batteries. Perform the test again if the sample of 45 batteries showed a mean of 38.3 and a standard deviation of 7.8.

Solution

$$H_0 : \mu = 40$$

$$H_1 : \mu < 40$$

$$\alpha = 0.05$$

Use the *z*-score test using *s*. Sample results:

$$n = 45 \text{ (which is larger than 30)}$$

$$\bar{x} = 38.3$$

$$s = 7.8$$

$$z = \frac{38.3 - 40}{7.8/\sqrt{45}} \approx -1.46$$

$A = 0.4279$, giving *P*-val $= 0.0721 > 0.05 \, (= \alpha)$.

$$\text{Fail to reject } H_0.$$

The sample results do not indicate an exaggerated claim on the part of the manufacturers. ●

Notice that the procedure for this z-score test is very similar to that used for the other z-score test. Now, suppose that you only have time and money to test 20 batteries and no reliable value for σ is known. Then, neither of the tests discussed thus far is appropriate. A new test must be found and this will be discussed in the next chapter.

Two-tailed Tests

When the alternative hypothesis is formulated as $H_1 : \mu < \mu_H$ (as with the battery example), you are only interested in whether or not $\bar{x}$ falls in the left-hand tail. Similarly, if the alternative hypothesis were formulated as $H_1 : \mu > \mu_H$, you would only be interested in whether or not $\bar{x}$ fell in the right-hand tail. Since, with either of the foregoing formulations of H_1, only one-tail presents any interest, they are called **one-tailed tests.** However, suppose the alternative hypothesis were formulated $H_1 : \mu \neq \mu_H$. Then, you should be ready to reject H_0 if $\bar{x}$ falls far enough away from μ_H, *either* in the left-hand tail *or* in the right-hand tail.

With this formulation of H_1, the hypothesis test is called a **two-tailed test.** In a two-tailed test, the application of the P-value is slightly different. To demonstrate this, suppose you are using a significance level α of 0.05. First, one-tailed tests are shown in figures 6–3 and 6–4.

Figure 6–3

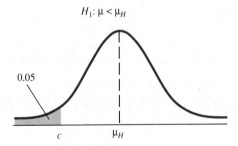

$H_1 : \mu < \mu_H$

0.05

c μ_H

Reject H_0 if $\bar{x}$ falls to the left of c, or P-val < 0.05.

Figure 6–4

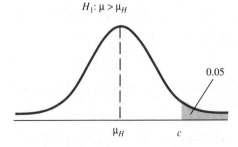

$H_1 : \mu > \mu_H$

0.05

μ_H c

Reject H_0 if $\bar{x}$ falls to the right of c, or P-val < 0.05.

By contrast, figure 6–5 shows a two-tailed test.

Figure 6–5

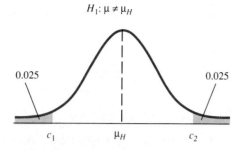

$H_1: \mu \neq \mu_H$

0.025

0.025

c_1 μ_H c_2

Notice that $\alpha = 0.05$ is divided between the two tails. Half of it, or $2\frac{1}{2}\%$, is placed in each tail, and there are two critical values. H_0 is rejected if $\bar{x}$ falls *either* to the left of c_1 *or* to the right of c_2. This amounts to rejecting H_0 if P-val < 0.025.

Summarizing, with a two-tailed test, if everything, including the calculation of P-value, is done in the same way as it was done for a one-tailed test, then the P-value should be compared to $\frac{1}{2}\alpha$ before deciding on the validity of H_0. It is a two-tailed test if $\neq$ appears in H_1. Otherwise it is a one-tailed test.

EXAMPLE 6–6   A statistician working for the manufacturers referred to in example 6–2 weighs the contents of a randomly selected sample of 33 jars of coffee and finds the mean weight to be 16.28 ounces with a standard deviation of 0.5 ounces. Do the sample results indicate that the machine should be adjusted if the statistician uses a 0.01 level of significance?

Solution   Recall that the manufacturers would like to have the machine adjusted if either too much or too little coffee is being dispensed into the jars. A two-tailed test is therefore called for.

$$H_0 : \mu = 16$$

$$H_1 : \mu \neq 16$$

$$\alpha = 0.01$$

Use a z-score test using s, since no σ is known and $n > 30$. Sample results:

$$n = 33$$
$$\bar{x} = 16.28$$
$$s = 0.5$$
$$z = \frac{16.28 - 16}{0.5/\sqrt{33}} \approx 3.22$$

$A = 0.4994$ and P-val $= 0.0006 < 0.005 \; (= \frac{1}{2}\alpha)$.

There is a significant difference at the 1% level between the amount of coffee that ought to be dispensed (16 oz) and the sample results. The machine should be adjusted.   ●

Exercise Set 6–3

1. It is claimed that the mean of a population μ is 70. To test whether this is true or not, a random sample is drawn from the population and reveals that $n = 80$, $\bar{x} = 75$. If σ is assumed to be 27, and a two-tailed z-score test using σ is applied, should the claim be rejected at the 0.05 level?

2. It is believed that the mean of a population μ is 12. A random sample from the population reveals that $n = 100$, $\bar{x} = 14$, $s = 10$. If a two-tailed z-score test using s is applied, should the hypothesized mean be rejected at the 10% level?

3. A supermarket boasts that checkout times for customers are never more than 15 minutes. A random sample of 36 customers reveals a mean checkout time of 17 minutes with a standard deviation of 3 minutes. What can you conclude about the supermarket's boast at the 0.05 level?

4. A high school hires a new football coach. A previous study showed the mean attendance at football games was 550. A random sample of 32 games played after the hiring of the new coach showed a mean attendance of 600 with a standard deviation of 112. Is there reason to believe that attendance has improved since the hiring of this coach at the 0.05 level?

5. A governmental office claims that the average American consumes 100 lb of sugar per year with a standard deviation of 15 lb. To test the validity of this claim, a random sample of 20 Americans were monitored for sugar intake over a year. It found that the mean sugar intake for this sample was 92 lb. Conduct a two-tailed test at the 0.01 level to decide the issue.

6. A sociologist believes that it costs an average of $2,000 to raise a child from birth to age 1. A random sample of 49 families who had a child were selected to see if this figure was too high or too low. The mean expenses for these families revealed a mean of $2,100 with a standard deviation of $98. What can you conclude about the sociologist's belief?

7. Could the standard deviation of the *sample* be used in exercise 5 instead of the population standard deviation given? Why or why not?

8. Is it alright to use the sample standard deviation for the test in exercise 6? Why or why not?

9. A test of hand-eye coordination is designed to have a mean of 100 and a standard deviation of 15. A psychologist wishes to determine whether a group of children she is studying is above or below average. The mean score for a random sample of 12 of these children results in a mean of 112. Are these children significantly different at the 0.01 level?

10. The chamber of commerce of a ski resort attempting to drum up business advertises that the mean rate at hotels there is $30 per person per night. A random sample of 50 skiers reported a mean of $32.50 with a standard deviation of $7. What do these results reveal about the chamber's advertising at the 0.02 level?

11. An organization of marriage counselors believes that the mean length of a marriage in California is 7 years. Forty records chosen at random from county courts all over the state revealed the following lengths of marriages:

6.5	7.4	10.3	9.1	7.0	5.2	3.6	1.8	5.8	9.2	20.1	13.6
2.4	3.7	8.2	7.4	9.6	2.1	0.5	13.6	6.4	7.2	9.3	4.8
5.4	7.2	1.5	7.2	7.5	6.4	8.6	9.1	10.2	12.3	4.6	8.1
7.0	6.5	7.0	8.3.								

Test the organization's claim for over- or underestimate at the 0.04 level.

 12. It is claimed that marijuana users begin smoking at 14 years old or less, on the average. A sociologist interviews 120 smokers revealing these reported ages of first smoking marijuana.

Age	f
12	10
13	25
14	32
15	21
16	12
17	9
18	7
19	4

Test the claim at the 0.05 significance level.

Section 6–4   The z-score Test for Proportions

A Test for Proportions

Often a hypothesis is formed about a proportion in a population, rather than a mean. If someone claims that a certain coin is fair, this means that the *proportion* of heads (or tails) would be $\frac{1}{2}$ if the coin were tossed a large number of times. If a machine in an assembly-line production is supposed to turn out no more than 5% defectives, you can test the machine by examining a random sample of the items to find out what *proportion* are defective. A null hypothesis about proportions is set up as follows:

$$H_0 : P = P_H.$$

As with means, this hypothesis is that a population proportion P is equal to a **hypothesized proportion** P_H. The alternative hypothesis H_1 will be set up similarly: $H_1 : P < P_H$, $H_1 : P > P_H$, or $H_1 : P \neq P_H$, depending on the problem at hand. As with the means, a z-score test can be constructed. The frequency distribution of p, the sample proportion, is centered at P, the population proportion, and has standard deviation $SD_p = \sqrt{PQ/n}$, as well as having a normal distribution. All this was discussed in chapter 5. What is being hypothesized now, under the null hypothesis, is that the population proportion is P_H. Also, define

$Q_H = 1 - P_H$. Then, if it is assumed, as always, that H_0 is true, the standard deviation of this frequency distribution of p is $\sqrt{P_H Q_H / n}$. Thus, if the situation is as shown in figure 6–6, then the area of the tail can be calculated by

Figure 6–6

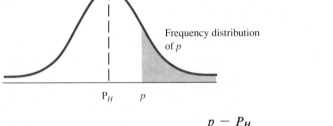

Frequency distribution of p

P_H p

$$z = \frac{p - P_H}{\sqrt{\dfrac{P_H Q_H}{n}}} \qquad (34)$$

and then finding the P-value by using the normal table. Unlike the situation with the means, the standard deviation of the sampling distribution of the proportion under the null hypothesis SD_p is always "known," since P_H and Q_H are given specific values by assumption under the null hypothesis. However, there is a minor problem that crops up here, which did not with the means. It is best illustrated with an example.

EXAMPLE 6–7   You suspect that a certain coin when tossed favors heads. You toss it 50 times and find 31 heads. At the 0.05 significance level, does it favor heads or might it be a fair coin?

Solution   For a fair coin, you *expect* the proportion of heads to be $\frac{1}{2}$ or 0.5. If you suspect the coin favors heads, you wish to test the versus that it doesn't. Since there is no suspicion of favoring tails, a one-tailed test is most appropriate. If you simply wished to test whether a coin were fair or not, a two-tailed test would be more appropriate. Thus,

Here P represents the proportion of heads in the population (*all* tosses of the coin).

$$H_0 : P = 0.5$$

$$H_1 : P > 0.5$$

$$\alpha = 0.05.$$

The sample results show that $n = 50$ and since the number of successes (heads) is 31, the proportion of successes is given by $p = \frac{31}{50} = 0.62$. The question now arises as to whether you should reject the fairness of the coin if p is 0.62 or more. The opposing side to this question is whether you should accept the fairness of the coin if $p = \frac{30}{50} = 0.60$ or less. Since the number of heads is a discrete variable, there is no value of np possible between 30 and 31, and thus no value of p possible between 0.60 and 0.62. The situation is shown in figure 6–7.

Figure 6-7

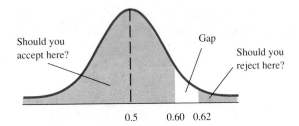

Since the normal distribution is a continuous distribution, this leaves a gap as indicated in figure 6-7. On the horizontal axis, 0.60 and 0.62 are the endpoints of the gap. To remove the gap, take the halfway point between these two, 0.61, and put values to the left of this with the possible acceptance region, and values to the right with the possible rejection region. Having done this, you can ask the question in a new form: Should I reject the fairness of the coin if p is 0.61 or more or accept it if it is less than 0.61? This formulation takes care of the gap. Now, instead of using $p = 0.62$ in formula (34), use $p = 0.61$. This is called applying a **continuity correction** to p. Thus,

$$z = \frac{0.61 - 0.5}{\sqrt{\frac{(0.5)(0.5)}{50}}} \approx 1.56$$

$A = 0.4406$ and P-val $= 0.0594 > 0.05 \; (= \alpha)$.

Therefore, fail to reject H_0 and conclude that the sample results do not indicate that this coin favors heads at the 5% significance level. ●

If the continuity correction were not applied here, then

$$z = \frac{0.62 - 0.5}{\sqrt{\frac{(0.5)(0.5)}{50}}}$$

≈ 1.70, $A = 0.4554$ and P-val $= 0.0446 <$ 0.05 and you *would* have been able to reject H_0.

Whether or not the continuity correction is applied can be important to the outcome. The correct application of formula (34) is *always* to correct p for continuity before calculating it.

Had you originally decided on a two-tailed test (something you might do if you wanted to test whether the coin favored heads *or* tails), the P-value would have been compared to $\frac{1}{2}\alpha = 0.025$ rather than to $\alpha = 0.05$.

EXAMPLE 6-8 It is claimed that at least 10% of the undergraduates at a large university are married. To test this, a random sample of 300 undergrads was selected, and it was found that 22 were married. Do these results dispute the claim at the 10% significance level?

Solution

$$H_0 : P = 0.10$$

$$H_1 : P < 0.10$$

$$\alpha = 0.10$$

We use $<$ in H_1 because the claim is "at least 10%," i.e., 10% or more. The opposite of this is "less than 10%." Sample results:

$$n = 300$$

$$np = 22$$

$$p = \frac{22}{300} \approx 0.0733$$

Now, to correct p for continuity, refer to figure 6–8.

Figure 6–8

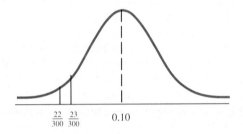

$$\frac{22}{300} \quad \frac{23}{300} \qquad 0.10$$

Since $\frac{22}{300} \approx 0.0733$ is below the center, you need to ask what the chances are that p is $\frac{22}{300}$ or less, as opposed to it being $\frac{23}{300} \approx 0.0767$ or more.

Corrected for continuity, take $p = \dfrac{0.0733 + 0.0767}{2} = 0.075$. Then,

$$z = \frac{0.075 - 0.10}{\sqrt{\dfrac{(0.10)(0.90)}{300}}} \approx -1.44$$

$A = 0.4251$ and P-val $= 0.0749 < 0.10 \ (= \alpha)$.

So, reject H_0 and conclude that less than 10% of the undergrads at this university are married.   ●

The size of the P-value indicates a probability of about $7\frac{1}{2}\%$ that you are committing a type I error.

Notice that in example 6–7, p was corrected for continuity by *reducing* it from 0.62 to 0.61, while in example 6–8, p was corrected for continuity by *increasing* it from 0.0733 to 0.075. An easy way of remembering whether to increase or reduce the value of p when correcting for continuity is to notice that you *always correct p toward the center.*

In example 6–7, p was *above* P_H and was corrected *down*.

In example 6–8, p was *below* P_H and was corrected *up*.

If, by some chance, $p = P_H$, then it doesn't matter, since, without performing any arithmetic, you can say that H_0 should not be rejected.

The whole process of hypothesis testing, and the thinking behind it, is liable to be very new to you. You should, therefore, make sure you are able to do the exercises well before moving on.

Exercise Set 6–4

1. It is hypothesized that the proportion of successes in a population is 0.35. A random sample of 50 produces 22 successes. If a two-tailed z-score test for proportions is applied, what is the resulting P-value?

2. It is claimed that the proportion of successes in a population is at least 75%. A random sample of 100 produces 62 successes. If a one-tailed z-score test for proportions is applied, what is the resulting P-value?

3. A coin is tossed 150 times producing 65 heads. Test the fairness of the coin at the 0.05 level.

4. A die is rolled 300 times producing 33 sixes. Test the hypothesis that the die is biased against sixes at the 0.01 level.

5. An acne specialist has found that 60% of acne sufferers are helped by existing treatments. He decides to test a new treatment on a random sample of 40 acne patients. If 28 of these respond to this treatment, can he conclude that it is significantly better at the 0.05 level?

6. A toy company wishes to put a new toy on the market in time for the Christmas season. It does not want to put it on the market if the manufacturing process produces more than 10% that do not work. In a test run producing 80 of the toys, 10 are found to be defective. This is more than 10%; but is it significantly more at the 0.05 level?

7. A potential political candidate does not wish to run for office if she enjoys less than 45% support at the time of the announcement of her candidacy. If a poll of a random sample of 200 voters reveals that 80 will support her, what should she do if she wants to have less than a 10% chance of committing a type I error?

8. Henry Ford, when he first started producing automobiles, stated that customers could have any color they wanted as long as it was black. Nowadays black cars are much less common. If a random sample of 120 cars included 10 black ones, and the hypothesis that no more than 5% are black is tested, what is the probability of committing a type I error if this hypothesis is rejected?

9. In 1981, a women's group claimed that 60% of a state that had not ratified the Equal Rights Amendment supported its passage. If a poll of 1,575 adults in the state showed 920 who supported ERA, could the group have been correct in its claim at the 0.05 level?

10. It is found that 13% of students enrolled in intermediate algebra at a large university fail the first time they take it. A psychology professor claims that this could be reduced with a program of counseling, and so on. A random sample of 850 students enrolled in intermediate algebra take this program and 99 fail anyway. Is this a significant improvement at the 0.02 level?

Section 6–5   Summary of Terms

alternative hypothesis 226	P-value 229
continuity correction 243	significance level 231
hypothesis 225	two-tailed test 238
hypothesis test 225	type I error 230
hypothesized mean 226	type II error 230
hypothesized proportion 241	z-score test for means, using σ 234
null hypothesis 226	z-score test for means, using s 237
one-tailed test 238	z-score test for proportions 241

Section 6–6   Summary of Symbols

Symbol	Meaning
α	Significance level
H_0	Null hypothesis
H_1	Alternative hypothesis
μ_H	Hypothesized mean
P_H	Hypothesized proportion
Q_H	$1 - P_H$

Section 6–7   Summary of Formulas

Formulas

Use

(32) $z = \dfrac{\overline{x} - \mu_H}{\sigma/\sqrt{n}}$ Formula for use with the z-score test using σ

(33) $z = \dfrac{\overline{x} - \mu_H}{s/\sqrt{n}}$ Formula for use with the z-score test using s

(34) $z = \dfrac{p - P_H}{\sqrt{\dfrac{P_H Q_H}{n}}}$, where p is corrected for continuity.   Formula for use with the z-score test for proportions

Chapter 6 Review Exercises

1. Set up statistical hypotheses suggested by what Sally's daughter is saying in the fifth caption of the first Sally Forth cartoon at the beginning of the chapter.

2. Set up statistical hypotheses suggested by what Sally is saying in the first caption of the second Sally Forth cartoon at the beginning of the chapter.

3. The standard deviation of scores on a test of reaction time is 0.23 seconds. If a random sample of 250 scores on this test shows a mean of 0.92 seconds, test the hypothesis that the mean is 1 second at the 0.05 level.

4. A chess computer is set to respond to a player's move with a mean time of 30 seconds. A sample of 50 moves reveals a mean response time of 32 seconds on the part of the computer. Is it being unusually slow at the 0.01 level if it is assumed the computer's response time has a standard deviation of 5 seconds?

5. The administration of Caesar Hospital claims that the mean annual income of its patients is no more than $14,000. A random sample of 180 of these patients shows a mean income of $14,625 with a standard deviation of $5,200. Is this strong enough evidence against the claim at the 0.05 level?

6. The administration at Vale College claims that its students travel a mean of 5 miles to get to school. A random sample of 110 students reveals that their travel distances to school have a mean of 5.6 miles with a standard deviation of 1.6 miles. Use a two-tailed test to test the administration's claim at the 0.02 level.

7. A manufacturer of fishing line claims it is "8 lb test line," which means it will withstand a force of 8 lb. A random sample of 42 of these lines shows a mean of 7.6 lb with a standard deviation of 0.95 lb. Test the manufacturer's claim at the 0.05 level.

8. The manufacturer in exercise 7 claims that the sample standard deviation quoted there is an underestimate. The population standard deviation is more like 1.3 lb. Would this change your answer to the question?

9. The proportion of students passing an engineering examination is 80%. Students complain that they would do better if they were allowed to use calculators. A random sample of 90 students took this exam with calculators and 75 passed. Is this a significant improvement at the 10% significance level?

10. A casino owner would like to test one of his roulette wheels. He suspects that it favors red. If 3,000 turns produce 1,452 reds, what would you think of his suspicion at the 0.05 level? (*Note:* The wheel has both zero and double-zero, which are considered green.)

11. If it is known that a town has 52% males, what is the probability that in a random sample of 648 people, less than 300 are male?

12. If it is known that the mean income in a town is $12,500, what is the probability that a random sample of 450 residents will have a mean income of more than $13,000 if the standard deviation is assumed to be $6,000?

Hands-on Class/Student Project

Some of the following exercises are a continuation of the exercises for the class/student projects at the end of chapters 4 and 5. So retrieve the data that you collected then and get together in the same groups as you did before.

1. In the class/student projects for chapters 4 and 5, you dealt with a population that you have been calling population H. The mean and standard deviation of this population are known. Take the mean of this population, call it μ_H, and set up hypotheses as if μ_H were merely a claim made about population H. Use a two-tailed test.

2. For each of your 20 samples of size 15, perform a z-score test using σ to test the null hypothesis set up in exercise 1 at a 5% significance level.

3. Of the 20 tests that you performed in exercise 2, how many led you to the conclusion to reject the null hypothesis? (*Note:* Since you are using the actual known population mean, if a sample's results lead you to reject H_0, you are committing a type I error by so doing. Since you were asked to use a 5% significance level, there should be a 5% probability that you reject a true hypothesis. Thus, if the theory works well in practice, 5% [or 1 out of your 20 samples] should lead you to reject the [known to be true] null hypothesis. The other 19 should lead you to fail to reject H_0.)

4. Combine your results of exercise 3 with the corresponding results from all the other groups in the class. Do you get better agreement with the 5% figure when more samples are considered?

5. If you participated in this project after chapter 1, you were asked, when you collected your sample data, to find out the GPA of the students in your sample (variable 5). Before you look back at this data, decide in your group what you think the mean GPA at your college is. Set up hypotheses to test your belief, using a one-tailed test and a 1% significance level.

6. Calculate the mean and standard deviation of your sample of GPA's, and apply the z-score test using s to test the null hypothesis that you set up in exercise 5. Compare your results to other groups' in the class.

7. If you participated in this project after chapter 1, you were asked, when you collected your sample data, to observe the color of eyes of your sample of students (variable 9). Before you look back at this sample, decide in your group what you think the proportion of students who have brown eyes at your college is. Set up hypotheses to test your belief. This time, you decide whether to use a one- or two-tailed test and what significance level to use.

8. Now, look back at your sample of eye colors and figure out the proportion of brown eyes in your sample. Use this proportion and the z-score test for proportions to test the null hypothesis that you set up in exercise 7. Compare your results to those of others in the class.

9. Define a population and make two claims about it: one about a mean of a variable in the population and one about a proportion. Restate these claims in hypothesis notation and decide on appropriate significance levels.

10. In your group, figure out a sampling scheme to test the claims that you formulated in exercise 9, collect the sample data according to your sampling scheme, and use the results to test the null hypotheses that you set up there.

CHAPTER 7

Small-Sample Techniques

". . . makes mighty things from small
beginnings grow."

DRYDEN

Table 4. Gender Differences in Mean Satisfactions with Aspects of Marriage by Family Life-cycle Stage

	Men	Women	t Values
Stage 1, Preparental			
Love	3.24	3.34	1.01
Interest	3.15	3.01	1.44
Help at home	3.34	2.73	5.39**
Treatment by in-laws	3.11	3.07	.37
Time at home	3.35	3.13	2.22**
Spouse's friends	2.88	2.97	.88
Time with children			
Friendship	3.20	3.31	1.45
Sexual gratification	3.35	3.57	2.56**
Stage 2, Preschool Children (up to 6 years)			
Love	3.15	3.08	.76
Interest	2.92	2.74	.89
Help at home	3.34	2.64	7.44**
Treatment by in-laws	3.07	2.79	2.67**
Time at home	3.32	3.04	3.18**
Spouse's friends	2.99	2.95	.47
Time with children	3.53	3.14	2.13**
Friendship	3.30	3.14	2.13*
Sexual gratification	3.48	3.58	1.28
Stage 3, Young Children (6–12 years)			
Love	3.08	3.07	.21
Interest	2.85	2.84	.11
Help at home	3.20	2.60	6.06**
Treatment by in-laws	2.96	2.89	.74
Time at home	3.25	3.10	1.65
Spouse's friends	3.03	2.89	1.59
Time with children	3.34	2.92	4.42**
Friendship	3.13	3.08	.63
Sexual gratification	3.35	3.58	2.94**
Stage 4, Teenaged Children (13–19 years)			
Love	3.06	3.06	0
Interest	2.93	2.85	.88
Help at home	3.31	2.65	6.80**
Treatment by in-laws	3.11	2.81	3.15**
Time at home	3.39	3.15	2.75**
Spouse's friends	3.13	2.95	2.06**
Time with children	3.32	2.86	5.05**
Friendship	3.25	3.04	2.87**
Sexual gratification	3.38	3.49	1.31
Stage 5, Grown Children, All at Home			
Love	3.35	2.83	2.81**
Interest	3.27	2.56	3.02**
Help at home	3.61	2.56	5.38**
Treatment by in-laws	3.08	2.78	1.70
Time at home	3.61	2.93	3.34**
Spouse's friends	3.00	2.93	.33
Time with children	3.50	2.74	4.08**
Friendship	3.42	2.93	2.72**
Sexual gratification	3.38	3.41	.14
Stage 6, Grown Children, Some Launched			
Love	3.07	3.10	.25
Interest	3.11	2.79	2.56**
Help at home	3.34	2.64	5.10**
Treatment by in-laws	3.04	3.06	.17
Time at home	3.36	3.16	1.65
Spouse's friends	3.11	3.26	.44
Time with children	3.21	2.94	2.30*
Friendship	3.49	3.15	3.46**
Sexual gratification	3.59	3.45	1.29
Stage 7, Postparental			
Love	3.30	3.07	1.93
Interest	3.17	3.01	1.36
Help at home	3.44	2.93	4.51**
Treatment by in-laws	3.22	3.13	.78
Time at home	3.42	3.23	1.03
Spouse's friends	3.26	3.15	1.72
Time with children	3.32	3.20	1.21
Friendship	3.35	3.11	2.30*
Sexual gratification	3.39	3.41	.19

*Gender differences significant at the level of .05.   **Gender differences significant at the level of .01.

From Darla Rhyne, "Bases of Marital Satisfaction Among Men and Women" in *Journal of Marriage and the Family,* 43(4):948–949, Nov. 1981. Copyright © 1981 by the National Council on Family Relations, 1910 West County Road B, Suite 147, St. Paul, Minnesota.

....

Chapter Overview

What If the Sample Is Small?

In the previous chapter, two *z*-score tests were introduced to test hypotheses about means. One assumes knowledge of the population standard deviation σ. The other, in not assuming this knowledge, uses *s*, the sample standard deviation, to estimate σ. Unfortunately, this estimate is not always reliable when the size of the sample is smaller than 30. In some research situations, it is impossible or impractical to have as large a sample size as 30. For instance, an anthropologist studying the size of skulls of humans who lived millennia ago only has access to those skulls that are found, and if only 10 have been discovered, 10 has to be the size of the sample. How, then, are CIE's for μ formed or a hypothesis about μ tested?

In forming a CIE for μ, a certain number of standard errors are added to and subtracted from $\bar{x}$, giving an interval of values, which it can be stated, with a certain amount of confidence, includes the true value of μ. The sample standard deviation *s* is used in the calculation of the standard error. If it is felt that the value of *s* obtained is less reliable because the sample is small, logic demands that a *wider* interval be presented as the CIE for μ, so that it can be stated, with an equivalent amount of confidence, that μ is included in the interval. At the moment, for 95% confidence, you add and subtract 2 standard errors. It is likely, then, that when using a small sample and *s*, more than 2 standard errors should be added and subtracted to enjoy the same 95% confidence. But the question is, how many?

By the same token, when testing a hypothesis about μ, shouldn't $\bar{x}$ be further from μ before rejecting the hypothesis to have an equivalently low probability of committing a type I error, when using a small sample and *s?*

This chapter deals with hypothesis testing and CIE procedures for small samples.

....

Section 7–1   Student's *t* Distribution

This was the problem confronted by statisticians at the turn of the century. One of them, W. S. Gosset, found a partial solution. Gosset was an employee of the Guinness Breweries, one of the first companies to hire a statistician. Anxious to keep this a secret from their competition, Guinness asked Gosset to use a pseudonym when publishing his results in scholarly journals and the like. Gosset complied and chose the pseudonym *Student*. In tackling the problem, Gosset made the assumption that all statisticians of the day made, that the population could always be considered to follow a normal distribution. With this assumption, Gosset discovered that the quantity

$$t = \frac{\bar{x} - \mu_H}{s/\sqrt{n-1}} \tag{35}$$

follows a **Student's t distribution.**

What a Student's t distribution is, precisely, will be discussed shortly. Suffice it to say that its name comes from Gosset's chosen pseudonym.

Notice formula (35)'s similarity to formula (33) for z. The only difference is the $n - 1$, in place of n, in the denominator. If you had access to a table for Student's t distribution, as you do for the normal distribution, a decision about the null hypothesis could be reached using formula (35) and its attendant P-value, just as was done with the z-score test. If this were possible, then the problem would be solved for populations whose distribution is close to normal.

Properties of Student's t Distribution

This distribution, like the normal distribution, is a probability distribution. In fact, superficially, it looks very much like a normal distribution. See figure 7–1.

Figure 7–1

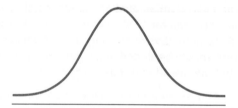

It has the distinctive bell shape, but it is a little flatter at the peak and thicker in the tails than a normal distribution. It has several properties in common with the normal distribution: it is symmetrical about its center, it is asymptotic to the horizontal axis in both tails, and the total area under the curve is 1. However, there is one important point to grasp: there is not merely *one* Student's t distribution, but rather a whole *family* of t distributions. The particular one that is needed to deal with a problem at hand depends on a concept called **degrees of freedom.**

To help you understand the concept of degrees of freedom, consider a story and an example. The story goes that two guys, Alan and Bob, were sitting in a bar arguing about their respective abilities with extrasensory perception, ESP. They decided to settle the matter by staging a friendly competition. They asked the bartender, Carl, to toss a coin and they would attempt to predict the outcome of the toss. Carl tossed the coin and Alan predicted, "the coin will show heads face up." Sure enough, it did. Carl tossed again, Alan predicted, "tails face up," and was right again. Carl tossed for a third time, Alan predicted, "tails face up," and was right yet again. Alan turned to Bob, "that's three right. Now it's your turn." Carl tossed the coin again and Bob predicted, "the coin will show heads face up, tails facing the floor." Sure enough, he was right and claimed, "that's two right." Carl tossed again and Bob predicted, "tails face up, heads facing the floor." Since he turned out to be correct again, he turned to Alan and said, "that's four right, I have beaten you already!" Neither Alan nor, very likely, you would be very impressed with Bob's claim. Why? Because Bob is claiming one more degree of freedom than he actually enjoys. If the coin shows heads face up, it is *forced* to show tails facing the floor, and, likewise, if it shows tails face up, it is *forced* to show heads facing the floor. Bob has not made four *independent* predictions but, rather, two independent predictions and two predictions *dependent* on the first two.

So much for the story. Now for the example. In the calculation of *s,* the sample standard deviation, the original method you studied was first to calculate *d,* the deviations from $\bar{x}$. Thus,

$$
\begin{array}{cc}
x & d = x - \bar{x} \\
2 & -4 \\
3 & -3 \\
5 & -1 \\
8 & 2 \\
\underline{12} & \\
\Sigma x = 30 & \\
n = 5 & \\
\bar{x} = 6 &
\end{array}
$$

The calculation has been interrupted. All but one of the *d*'s have been calculated. What is the sum of the *d*'s so far? It is −6. But it is known what Σd *always* has to be. Do you recall that for deviations about the mean, it is *always* true that $\Sigma d = 0$? For this to continue to be true for this example, the missing *d* is *forced* to be 6. It would matter little which *d* was omitted in the calculation. If all but one of the *d*'s are calculated, the last *d* is not free, it has to be chosen in such a way that $\Sigma d = 0$. Thus, in this sample, there are 4 degrees of freedom, not 5. Likewise, in a sample of size *n,* there are *n* − 1 degrees of freedom, not *n,* when calculating *s.*

That is partly why some statisticians use *n* − 1, rather than *n,* when calculating *s* as an estimate of σ. When the sample is large, the difference in values for *s* thus produced is insignificant. If *n* = 100, whether one divides by 99 or by 100 is a trivial consideration when compared to the different values one can obtain for s due to sampling fluctuations. However, this chapter deals with *small* samples. Thus, it might make a considerable difference to the value obtained for t, were it not for the fact that there is a built-in correction. You can use the two formulas

$$s = \sqrt{\frac{\Sigma f d^2}{n}} \qquad \text{and} \qquad t = \frac{\bar{x} - \mu_H}{s/\sqrt{n-1}}.$$

Those statisticians who use $n - 1$ for the calculation of s use the two formulas

$$s = \sqrt{\frac{\Sigma f d^2}{n-1}} \qquad \text{and} \qquad t = \frac{\bar{x} - \mu_H}{s/\sqrt{n}}.$$

Thus, their formula for t turns out to be exactly the same as formula (33) for z. Can you see that whichever way you go, the *same* value for t will be obtained? In formula (8), division by $\sqrt{n}$ is necessary to calculate s, and then division by $\sqrt{n-1}$ is used in the calculation of t. In the other formulation, first division by $\sqrt{n-1}$ is performed, and then division by $\sqrt{n}$. Either method, of necessity, must produce the same answer.

Using formulas (8) and (35), the complete result by Gossett can be stated as follows:

> If the population is normally distributed, then the quantity
>
> $$t = \frac{\bar{x} - \mu_H}{s/\sqrt{n-1}}$$
>
> follows a Student's t distribution with $n - 1$ degrees of freedom.

The use of "one-sample" is made to distinguish it from the *two-sample* *t-test* discussed in the next chapter.

Armed with this result, a new test can be devised, which will be called a **one-sample t-test.** It will be carried out in exactly the same fashion as the z-score test using s, except formula (35) will be used to calculate t in place of formula (33). This exception means that you can no longer use the normal table to find the P-value to assist you in making a decision about the hypothesis. Assume, for a moment, that you wish to perform a one-tailed test with relevant values of $\bar{x}$ larger than μ_H. Pictorially, the situation might look as is shown in figure 7–2.

Figure 7–2

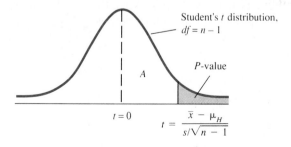

Here *df* is an abbreviation for degrees of freedom. It is necessary to find the area of the shaded tail, which will equal the relevant *P*-value. When performing a *z*-score test, *A* was found using a normal table. To find *A* under a *t* distribution, a separate table is needed for *each t* distribution (that is, for $df = 1$, $df = 2$, . . . , $df = 29$). This means 29 different tables are required! (When $df = 30$, $n = 31$, the sample can be considered large, and the normal table, table 1, is used.) To save space, a table has been created, table 2 in appendix B, giving values of *t* up to 3.9, but only to one decimal place (rather than to two places, as was done for *z*), and all values of degrees of freedom up to 29. As with table 1, the values in the body of table 2 give values for *A*, the area between the central line and the value of *t*. To obtain the *P*-value, subtract from 0.5. To obtain *P*-values of *t* that are specified to two decimal places, the method of **linear interpolation** can be used. For $df = 19$, $t = 1.4$, $A = 0.4112$ can be read from the table. Suppose you wanted *A* for $t = 1.47$. Here is the information for $df = 19$.

Warning: Linear interpolation only works well in the tails of the distribution. Closer to the center it is not so accurate.

t	*A*
1.40	0.4112
1.47	?
1.50	0.4250

The difference between the values of *A* for $t = 1.40$ and $t = 1.50$ is $0.4250 - 0.4112 = 0.0138$. Now 1.47 is $\frac{7}{10}$ of the distance between 1.40 and 1.50 and

$$\frac{7}{10}(0.0138) = 0.00966 \approx 0.0097.$$

Adding this to the value of *A* for $t = 1.40$, find

$$0.4112 + 0.0097 = 0.4209.$$

Thus, for $t = 1.47$, $A \approx 0.4209$ and *P*-val $\approx 0.5 - 0.4209 = 0.0791$. The *exact* *P*-value for $t = 1.47$ is 0.0790, and the approximation is very good.

EXAMPLE 7-1 It is assumed that "average" systolic blood pressure is about 120. A random sample of 20 junior high school teachers, each having taught for at least 10 years, shows their mean systolic blood pressure is 123 with standard deviation 7.9. Test, at a 5% significance level, to see if junior high school teachers have above average systolic blood pressure.

Solution

$$H_0 : \mu = 120$$

$$H_1 : \mu > 120$$

$$\alpha = 0.05$$

Accepting H_0 is to say junior high school teachers are no different than average as far as systolic blood pressure goes.

Rejecting H_0 and, by implication, accepting H_1 is to say that junior high school teachers have systolic blood pressure significantly above average.

Use the one-sample t-test.

Sample results:

$$n = 20$$

$$\bar{x} = 123$$

$$s = 7.9$$

Assuming systolic blood pressures follow a normal distribution,

$$t = \frac{123 - 120}{7.9/\sqrt{19}} \approx 1.66, \, df = 19$$

$$A = 0.4370 + \tfrac{6}{10}(0.0103) = 0.4432$$

$$P\text{-val} = 0.5 - 0.4432 = 0.0568 > 0.05 \ (= \alpha).$$

Fail to reject H_0.

On the basis of this sample's results, it cannot be concluded at the 5% level of significance that junior high school teachers have above normal systolic blood pressure. ●

Had z been used (incorrect on the basis of such a small sample), A = 0.4554 (for z = 1.70) and P-val = 0.0446 and it would have been possible to reject H_0 at the 5% significance level. In using the t-test, $\bar{x}$ has to be further away from μ_H to be able to reject H_0 with an equal probability of committing a type I error.

EXAMPLE 7–2   When a manufacturer of light bulbs first started production, it was found that a mean lifetime of 1,000 hours could be claimed. Two years later, a random sample of 15 of the bulbs was tested with the following results:

962   1,008   914   952   992   886   1,092   942   928   986   1,012
961   978   811   912.

Can the manufacturer still claim that the bulbs last 1,000 hours on the average, or has the mean lifetime gone up or down? Use a 5% significance level.

Solution

$$H_0 : \mu = 1,000$$

$$H_1 : \mu \neq 1,000$$

$$\alpha = 0.05$$

A two-tailed test is used here, since the manufacturer wishes to know if the mean has changed *either way*. Check for yourself that the sample results give

$$n = 15$$

$$\bar{x} = 955.7$$

$$s = 62.0.$$

Assume lifetimes of light bulbs are normally distributed and use the *t*-test.

$$t = \frac{955.7 - 1,000}{62.0/\sqrt{14}} \approx -2.67, \, df = 14$$

$$A = 0.4895 + \tfrac{7}{10}(0.0019) = 0.4908$$

$$P\text{-val} = 0.5 - 0.4908 = 0.0092 < 0.025(= \tfrac{1}{2}\alpha).$$

Reject H_0. On the basis of this sample, you can be at least 95% confident that the mean lifetime of these bulbs has gone down.   ●

Notice, as was done with the *z*-score test, that the *P*-value is compared to $\tfrac{1}{2}\alpha$ for a two-tailed test. Had α been chosen as 0.01, $\tfrac{1}{2}\alpha = 0.005$, and the *P*-value, being larger than this, would have indicated nonrejection of H_0. For $\alpha = 0.02$, however, H_0 would be rejected.

Exercise Set 7-1

You may assume that the populations in these exercises follow a normal distribution.

1. It is hypothesized that the mean of a population is 18. A random sample from this population reveals that $n = 18, \bar{x} = 16.2, s = 5.1$. Use a two-tailed *t*-test to test the hypothesis at the 0.05 level.

2. It is claimed that the mean of a population is at least 25. A random sample drawn from the population reveals that $n = 23, \bar{x} = 21, s = 8.2$. Use a *t*-test to test the claim at the 0.02 level.

3. An anthropologist believes that the diameter of the skull of a primitive man is more than 7 inches. After many years search, 6 skulls of this type are found. The mean diameter of these is found to be 6.8 inches with a standard deviation of 1.2 inches. What can you conclude about the anthropologist's belief at the 0.05 level?

4. The label on a high nutrition cereal claims that it contains no more than 23 grams of carbohydrate per serving. A random sample of 25 servings reveals a mean carbohydrate content of 24.3 grams with a standard deviation of 1.8 grams. Do these results substantiate or refute the label at the 0.01 significance level?

5. Perpetrators of a crime, which carries a maximum sentence of 10 years, are believed by a district attorney to be sentenced to no more than 7 years on the average. A random sample of 20 of this crime's offenders reveals a mean sentence of 7.6 years with a standard deviation of 2.4 years. Might the D.A. be correct in his suspicions at the 0.04 level?

6. A spokesperson for a real estate company claims that the mean selling price for 3-bedroom houses in a certain community is no more than $110,000. A random sample of 24 3-bedroom houses in the community reveals a mean price of $112,300 with a standard deviation of $9,200. Is this consistent with the real estate company's claim at the 5% level?

7. An examination in an elementary algebra class is supposed to have a mean of 75. A class of 23 students score the following grades on this exam:

71   78   83   84   70   92   68   100   95   72   85   87   75   79
62   83   80   90   69   83   78   74   91.

At a 1% significance level, does this appear to be an above-average class?

8. It is claimed by a union of mine workers that the life span of its members is no more than 55 years. A random sample of miners who have died recently produced these ages at death.

56   48   34   62   49   52   65   42   72   61

How would you evaluate the union's claim?

9. An educational psychologist working with mentally retarded children gives a sample of 15 of them a test. She will consider her treatment a success if the mean on this test is significantly greater than 30 at the 2% significance level. The results were as follows:

32   28   25   35   31   30   26   34   30   22   35   36   24   32   33.

Has the treatment been a success?

10. In exercise 9, what is the largest mean score (instead of 30) that the psychologist could use to claim successful treatment of those 15 children if everything else is the same?

Section 7–2 Modification of CIE's

In chapter 5, formula (25), giving a CIE for μ, was adapted, when σ was unknown, to:

$$\bar{x} \pm zSE_{\bar{x}} \qquad \text{or} \qquad \bar{x} \pm z\frac{s}{\sqrt{n}}.$$

This CIE for μ is valid only if the sample size n is larger than 30. For samples with n smaller than 30, the estimate for σ is not a sufficiently reliable substitute for σ. However, if the underlying population follows a normal distribution, the Student's t-distribution can be used. The calculation of z is replaced by a similar calculation of t.

To see where the formula comes from, compare formulas (25), (33), and (35).

Assuming that the population is normally distributed, a CIE for μ is given by

$$\bar{x} \pm t\frac{s}{\sqrt{n-1}}, \tag{36}$$

where the value of t depends on the amount of confidence required and the number of degrees of freedom.

When working with CIE's for μ with known σ or $n \geq 30$, the value of z was found by using the normal table (table 1) in reverse. Since table 2 gives values of t to only one decimal place, table 3 has been constructed, giving values of t for several selected levels of confidence. If a level of confidence is required for which t values are not presented in table 3, table 2 can still be used in reverse, using the linear interpolation method.

EXAMPLE 7-3   For $df = 19$, which value of t is used for 95% confidence?

Solution (a)   From table 3, $df = 19$ gives $t = 2.09$ for a 0.95 level of confidence.   ●

Solution (b)   For 95% confidence, $A = 0.4750$. Using table 2, look under $df = 19$. For $A = 0.4700$, $t = 2.0$; for $A = 0.4753$, $t = 2.1$. Thus,

t	A		**Differences**	
2.00	0.4700			
?	0.4750	0.0050		0.0053.
2.10	0.4753			

Thus, the required $t = 2.00 + \dfrac{0.0050}{0.0053}(0.10)$

$$= 2.00 + 0.0943$$

$$\approx 2.09.$$ ●

This second solution demonstrates that table 3 is much easier to apply.

EXAMPLE 7-4   Using the data of example 7-2, find a 90% CIE for the mean lifetime of this manufacturer's light bulbs.

Solution   Recalling the sample results: $n = 15$, $\bar{x} = 955.7$, $s = 62.0$.
     Since $n = 15$, $df = 14$ and, using table 3, for 90% confidence, $t = 1.76$. Thus,

$$\text{a 90\% CIE for } \mu \text{ is } 955.7 \pm 1.76\frac{62.0}{\sqrt{14}}$$

$$\approx 955.7 \pm 29.2$$

$$= 926.5 \text{ to } 984.9.$$

You can state, with 90% confidence, that the mean lifetime of these bulbs is somewhere between 926.5 and 984.9 hours.   ●

Notice that, had it been (incorrectly) decided to use the normal table, for 90% confidence, $z = 1.65$. By using $t = 1.76$ instead, a wider interval is obtained, a necessary precaution given the small sample size and the consequent unreliable estimate for σ.

In summary, Student's t distribution, together with P-values and critical t-scores derived from it, may be used instead of the z-score tests and CIE's discussed earlier. *Do not* forget, however, that these techniques are only appropriate if the population can be assumed to be normally distributed. If it is known that the population is not normally distributed, other methods are available. These will be discussed in chapter 9.

The table at the beginning of the chapter examines the differences in satisfaction with nine aspects of marriage between men and women during seven stages of marriage. The last column presents values from Student's t distribution. As the footnotes indicate, one asterisk denotes significance at the 0.05 level, while two asterisks indicate significance at the 0.01 level.

Notice that for the stage 2 categories "time with children" and "friendship," the last column shows identical t-values of 2.13 but that these are tagged with *different* significance levels. Why do you suppose that is? The answer to this question is that P-values associated with t-values depend not only on the size of t but also on the number of degrees of freedom. You can conclude, therefore, that when comparing time with children, the researchers must have used different-sized samples than when they were comparing friendship. The different-sized samples produce a different number of degrees of freedom and, hence, a different conclusion as to how significant the results are.

The t-values presented in this table must, of necessity, have been obtained using two sample techniques. You will learn about these in the next chapter.

Exercise Set 7–2

In exercises 1 and 2, find the critical t-value by (a) using table 3 and (b) using table 2 and interpolation.

1. Find t for $n = 18$ and 95% confidence.

2. Find t for $n = 27$ and 99% confidence.

3. If, for a random sample from a normal population, $n = 28$, $\bar{x} = 19.1$, $s = 3.7$, find a 95% CIE for μ, the mean of the population.

4. If, for a random sample from a normal population, $n = 14$, $\bar{x} = 7,840$, $s = 2,910$, find a 99% CIE for μ, the mean of the population.

 5. A certain hormonal deficiency produces dwarfism in adults. If their heights follow a normal distribution and a random sample of 10 of these adults produced the following heights:

 58   56   59   53   48   51   50   55   54   52,

find a 90% CIE for the true mean height of all adults with this deficiency.

 6. A department store is researching its credit card customers. A random sample of 25 of these customers shows a mean outstanding balance of $42.50 with a standard deviation of $6.18. Construct a 95% CIE for the mean outstanding balance of all customers.

Section 7–3 Summary of Terms

degrees of freedom 252
linear interpolation 255
one-sample *t*-test 254

Student's *t* distribution 252
t distribution 251

Section 7–4 Summary of Symbols

Symbol	Meaning
t	Standard score for a variable following the *t* distribution
df	Degrees of freedom

Section 7–5 Summary of Formulas

Formula	Use
(35) $t = \dfrac{\bar{x} - \mu_H}{s/\sqrt{n-1}}$	Formula for use with the one-sample *t*-test
(36) $\bar{x} \pm t \dfrac{s}{\sqrt{n-1}}$	CIE for μ when sample size is less than 30

Chapter 7 Review Exercises

 1. A chemist wishes to test the hypothesis that the boiling point of a certain substance is 572° C. She boils the substance 8 times and obtains the values

570°   572°   573°   571°   575°   571°   573°   574°.

Assuming these measurements follow a normal distribution, what conclusion can be drawn about the boiling point of this substance at the 5% significance level?

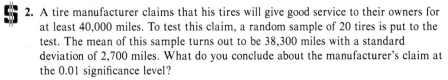 **2.** A tire manufacturer claims that his tires will give good service to their owners for at least 40,000 miles. To test this claim, a random sample of 20 tires is put to the test. The mean of this sample turns out to be 38,300 miles with a standard deviation of 2,700 miles. What do you conclude about the manufacturer's claim at the 0.01 significance level?

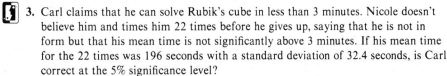 **3.** Carl claims that he can solve Rubik's cube in less than 3 minutes. Nicole doesn't believe him and times him 22 times before he gives up, saying that he is not in form but that his mean time is not significantly above 3 minutes. If his mean time for the 22 times was 196 seconds with a standard deviation of 32.4 seconds, is Carl correct at the 5% significance level?

4. Part of the net weight of canned fruit salad is the fruit itself and part is the juice. A spokesperson for the company producing the fruit claims that the 18 oz cans of fruit contain at least 16 oz of drained fruit. A random sample of 15 cans had the following weights of drained fruit:

    15.8   16.3   15.2   16.1   15.7   14.9   15.6   15.7   16.0   15.8   15.5
    16.5   15.4   15.4   15.3.

What conclusion can be drawn about the company's claim at the 2% significance level?

 5. In a study of time spent on the telephone by its employees, a large office monitors a random sample of 27 telephone calls and finds the mean length of these calls to be 6.3 minutes with a standard deviation of 1.4 minutes. Construct a 95% CIE for the mean length of all calls made by these employees.

 6. A student finds that she has to spend $224 for 8 textbooks this semester and that the standard deviation for this sample is $2.39. Assuming that this is a random sample of all the textbooks for sale in the bookstore, find a 90% CIE for the mean price of all textbooks.

 7. A random sample of 18 cars of a certain make were each filled with exactly 12 gallons of gasoline and then driven until they ran out of gas. Following are the number of miles driven:

    238   245   222   232   249   252   218   235   196   256   240   239
    243   245   231   238   245   208.

Find a 98% CIE for the number of miles per gallon that can be expected from this make of car.

 8. To estimate the mean length of the longest whiskers on Siamese cats, a random sample of 25 cats produced these lengths in centimeters.

    7.9 8.3   9.2 6.4 10.1   8.5 7.9 10.6 9.5 7.5 8.2 10.2 11.1 6.9 7.4
    9.3 9.5 10.2 9.4   8.0 10.0 6.4   8.5 8.6 9.2

Construct a 95% CIE for the mean length of the longest of a Siamese cat's whiskers.

Hands-on Class/Student Project

The following exercises involve a large number of repetitive calculations, which would preferably be performed with the aid of a computer. Also, quite a few of these exercises are almost identical to exercises at the end of chapters 5 and 6, but they are now performed using the t distribution, rather than the normal distribution. If you worked in groups before, work with these same groups again.

1. In exercise 5 of the class/student project in chapter 4, you were asked to select 20 samples, each of size 15, from population H and to find $\bar{x}$, the sample mean of each. Now, in addition, calculate the standard deviation of each of these samples.

2. For each of the 20 samples, use its mean and standard deviation, which you calculated in exercise 1, to form a 95% confidence interval estimate (CIE) for the mean of population H. (*Note:* Since you are now using sample standard deviations and the size of each sample is only 15, you must use an appropriate t-value instead of a z-value.)

3. Of your 20 CIE's, how many do include the mean of population H (μ)? (*Note:* For 95% confidence, there ought to be a 95% probability that any one of your CIE's does contain μ. So, if the theory worked perfectly in practice, you should find that 95% [or 19 out of 20] of your CIE's do include μ.)

4. Combine your results of exercise 3 with the corresponding results from all the other groups in the class. Do you get better agreement with the 95% figure when more samples are considered?

5. The mean and standard deviation of population H are known. Take the mean of this population, call it μ_H, and set up hypotheses as if μ_H were merely a claim made about population H. Use a two-tailed test.

6. For each of your 20 samples of size 15, perform a t-test to test the null hypothesis set up in exercise 5 at a 5% significance level.

7. Of the 20 tests that you performed in exercise 6, how many led you to the conclusion to reject the null hypothesis? (*Note:* Since you are using the actual known population mean, if a sample's results lead you to reject H_0, you are committing a type I error by so doing. Since you were asked to use a 5% significance level, there should be a 5% probability that you reject a true hypothesis. Thus, if the theory works well in practice, 5% [or 1 out of your 20 samples] should lead you to reject the [known to be true] null hypothesis. The other 19 should lead you to fail to reject H_0.)

8. Combine your results of exercise 7 with the corresponding results from all the other groups in the class. Do you get better agreement with the 5% figure when more samples are considered?

9. If you participated in this project after completing chapter 1, in your sample you collected data in answer to the question, "How much do you expect your annual income to be in your first job after you have completed your formal education?" (variable 12). Before looking back at this data, discuss in your group what you think that the mean of this population is. Set up hypotheses to test your belief.

10. From your data, choose 25 responses to the question at random and perform a t-test to test the hypotheses that you set up in exercise 9. Compare your results to those in other groups.

Two-Sample Techniques

....

"Comparisons are odorous."

....

Much Ado About Nothing　　　　　　　　　**SHAKESPEARE**

Prison-Issued Release Clothing

Of course, the releasees do not have to rely on prison-issued clothing exclusively, and may even reject it entirely. Prisoners may have civilian clothing sent them when they are nearly ready for release, and it is then issued to them just before they depart from the prison. Sometimes inmates are met at the institution by relatives or friends who bring the inmate a change of clothing to wear home. When confinement is not for a long period and the inmate does not change greatly in weight, his preprison wardrobe may still be usable if someone has taken care of it for him. At Terre Haute the inmates were charged points for clothing sent in for them, thus reducing the amount of prison-issued clothing they received, so that funds for release clothing could be extended to provide more liberally for those without such outside resources.

Since satisfaction with the release clothing is an aspect of the inmate's immediate economic resources on release, as well as a possible index of his attitude, we asked all releasees whom we interviewed how they utilized the clothing issued in prison. With the returned violators and successful releasees, the median time between release from prison and our interview was about two years. Their reports on their utilization of their release clothing are summarized in table 14.14.

It will be seen from table 14.14 that over half the returned violators discarded their release apparel as soon as possible. About a third of the successful releasees also discarded this clothing. The difference in these percentages suggests that the success of the latter group may have been, in part, a consequence of their more cautious husbanding of such economic resources. Nevertheless, the distinctive feature of these findings is that a large proportion of both groups failed to retain clothing which the institution made great effort to procure for them.

In our interviews it became apparent that there was considerable diversity of opinion on the prison-issued clothing in both the successful and the violating groups. Some releasees were quite satisfied with this clothing and used it until it wore out. At the time of our interviews several of the successful releasees were wearing the clothing that they received at the prison. Several who were interviewed at home brought the prison-issued clothing forth from their closets. In other cases releasees justified discarding the clothing on grounds that the clothes reminded them of prison. Men released from Leavenworth repeatedly told of a used-clothes dealer who met the bus bringing them to Kansas City, spotted the released prisoners readily, and talked them into trading in their prison clothing for other used clothing which he sold cheaply. There was some suspicion that he merely sewed commercial labels on prison-issued clothing which he purchased from other releasees.

Table 14.14   Use of Prison-Issued Release Clothing by Returned Violators and Successful Releasees

Use of Clothing	Returned Violators	Successful Releasees
Discarded as soon as possible	53%	34%
Retained and used	45%	58%
Retained but did not use	2%	8%
No. of cases covered	308	250

. . . .

Chapter Overview
Making Comparisons

Despite what Shakespeare says, comparisons are often made in life and statistics. Commercials on TV often claim that one headache remedy is better than another, or a particular toothpaste results in better checkups than others. Not only commercials, but researchers, too, find it necessary to make comparisons sometimes. In this chapter, you will learn methods for constructing CIE's for the differences between parameters of two populations and procedures for testing hypotheses comparing two populations. To accomplish these objectives, it will be necessary to draw two samples—one from each population. The material in this chapter will give you some tools and methods for dealing with hypotheses that make comparisons between two populations. They come under the general heading of two-sample tests. Hence the title of the chapter. The next chapter will present you with some alternative methods for both one- and two-sample tests.

. . . .

Section 8-1   Estimating Differences

Sometimes researchers wish to **estimate the difference** between parameters of *two* populations by random sampling. If μ_1 is the mean of one population and μ_2 the mean of a second, you might be more interested in estimating $\mu_1 - \mu_2$ than estimating μ_1 and μ_2 separately—similarly for proportions, medians, or standard deviations. For instance, you might be interested in how much greater the vocabulary of fifth-graders from a wealthy suburb is when compared with the vocabulary of fifth-graders from the inner city. In such a case, you could draw random samples from each of the two populations, calculate the mean number of words recognized by each of the two groups, and use the difference between the two sample means as an estimate for the difference between the corresponding population means. As with all estimation, it is desirable to attach a standard error to this difference so that you can judge the accuracy of the estimate of the difference as well as set up CIE's for the true difference.

Difference Between Means

If you wish to estimate $\mu_1 - \mu_2$, the difference between means of separate populations, the logically best point estimate is $\bar{x}_1 - \bar{x}_2$. This should be true since $\bar{x}_1$ is the best point estimate for μ_1 and $\bar{x}_2$ for μ_2. From what you have learned so far, you could probably guess this without being told. What is not so obvious, or easy to guess, is what the standard error of this estimate is. It will be designated by $SE_{\bar{x}_1 - \bar{x}_2}$. In fact, to discover the formula for $SE_{\bar{x}_1 - \bar{x}_2}$, a theorem from advanced statistics is needed. This theorem is similar to the central limit theorem and, as with the CLT, will be stated without proof. This theorem concerns the frequency distribution of $\bar{x}_1 - \bar{x}_2$, which is another frequency distribution of a statistic (see section 4–2). The way it would be formed, in practice, is to draw all possible samples of size n_1 from the first population (with mean μ_1 and standard deviation σ_1), all possible samples of size n_2 from the second population (with mean μ_2 and standard deviation σ_2), and then calculate all possible differences between sample means $\bar{x}_1 - \bar{x}_2$. A very large number (depending on the sizes of the two populations) of these differences would be obtained, and thus a frequency distribution can be formed. This frequency distribution of $\bar{x}_1 - \bar{x}_2$, according to the aforementioned theorem, possesses three properties not unlike the three parts of the CLT.

1. The frequency distribution of $\bar{x}_1 - \bar{x}_2$ has mean $= \mu_1 - \mu_2$.

2. If the two samples are separately and independently drawn, then the variance (abbreviated: Var) of the difference in their statistics is equal to the sum of their separate variances. In symbols,

$$\text{Var}_{\bar{x}_1 - \bar{x}_2} = \text{Var}_{\bar{x}_1} + \text{Var}_{\bar{x}_2} .$$

3. The frequency distribution of $\bar{x}_1 - \bar{x}_2$ has a normal distribution if the separate frequency distributions of $\bar{x}_1$ and $\bar{x}_2$ are normal. The CLT guarantees that these separate distributions *are* normal, providing the sample sizes are not too small.

Results 1 and 3 probably come as no surprise to you. Result 2 is the one that, most likely, demands explanation. Variance, if you recall, is the square of the standard deviation. Why, you may wonder, is the variance of a difference not the *difference* between the separate variances? If this were indeed so, some mighty strange results could develop. It is quite possible that the two separate frequency distributions have identical, or almost identical, variances. If the difference between them were then calculated, a value near to zero would be obtained. Variance, like standard deviation, is a measure of spread. To say that $\text{Var}_{\bar{x}_1 - \bar{x}_2} = 0$ is to say that the frequency distribution of $\bar{x}_1 - \bar{x}_2$ has no spread at all. All the differences would have to be the same—a highly improbable occurrence! Since *all* possible differences $\bar{x}_1 - \bar{x}_2$ are being calculated, some of these differences may be quite large, while others are close to zero. Here is a very much simplified example illustrating the point: Suppose the first distribution consists of the three numbers 2, 3, 6, while the second consists of the four numbers 1, 5, 6, 8. The set of *all* differences consists of the twelve numbers $1, -3, -4, -6, 2, -2, -3, -5, 5, 1, 0, -2$. Using the ideas of chapter 2, check for yourself that

$$\text{Var}_{\bar{x}_1} = 2.89, \text{Var}_{\bar{x}_2} = 6.50, \text{Var}_{\bar{x}_1 - \bar{x}_2} = 9.39.$$

In this example, it *is* true that, $\text{Var}_{\bar{x}_1 - \bar{x}_2} = \text{Var}_{\bar{x}_1} + \text{Var}_{\bar{x}_2}$. The theorem, quoted above, asserts that this result is *always* true for separately and independently drawn samples. Suppose you are convinced. How does this result help in arriving at a formula for $SE_{\bar{x}_1 - \bar{x}_2}$? Read on.

Since the variance is the square of the standard deviation, result 2, above, can be rewritten as follows:

$$(SD_{\bar{x}_1 - \bar{x}_2})^2 = (SD_{\bar{x}_1})^2 + (SD_{\bar{x}_2})^2$$

or, by taking the square root of each side,

$$SD_{\bar{x}_1 - \bar{x}_2} = \sqrt{(SD_{\bar{x}_1})^2 + (SD_{\bar{x}_2})^2}.$$

Now, from part 2 of the CLT,

$$SD_{\bar{x}_1} = \frac{\sigma_1}{\sqrt{n_1}} \quad \text{and} \quad SD_{\bar{x}_2} = \frac{\sigma_2}{\sqrt{n_2}}.$$

Incorporating these into the above equation,

$$SD_{\bar{x}_1 - \bar{x}_2} = \sqrt{\frac{\sigma_1^2}{n_1} + \frac{\sigma_2^2}{n_2}}.$$

Finally, since σ_1 and σ_2 will, in general, be unknown, it is necessary to estimate them by s_1 and s_2, giving,

$$SE_{\bar{x}_1 - \bar{x}_2} = \sqrt{\frac{s_1^2}{n_1} + \frac{s_2^2}{n_2}}. \tag{37}$$

Using this formula for the standard error of the difference between two means, it can be stated:

A CIE for $\mu_1 - \mu_2$ is $(\bar{x}_1 - \bar{x}_2) \pm zSE_{\bar{x}_1 - \bar{x}_2}$.

> **Warning:** Although it is true that for positive quantities a and b,
>
> $\sqrt{a^2} = a$ and $\sqrt{b^2} = b$,
>
> it is not true that
>
> $\sqrt{a^2 + b^2} = a + b$.
> $\sqrt{3^2 + 4^2} = 5$, *not* 7.
>
> Don't attempt to cancel the squares and square root.

It must be stressed, here, that the theorem that allowed the derivation of formula (37) is *only* true if the two samples are **separately and independently drawn samples.** The previous example of the two samples of fifth-graders would fit this criterion. But when is formula (37) inapplicable? Here is an example: A fifth-grade teacher may want to judge the effectiveness of his/her teaching by giving the class a vocabulary test at the beginning of the school year and again at the end. Assuming that this class forms a random sample of all fifth-graders at the school, formula (37) still cannot be applied to attach a standard error to the difference in vocabulary knowledge at the beginning and end of the school year. Why? Because the *same* students are involved, albeit at different times, and the vocabulary scores at these two times constitute what are known as **dependent samples.** Statistical theory does have a method of dealing with dependent samples; but it is, of necessity, different. This will be discussed at the end of this section. First, examine an example of the use of formula (37).

EXAMPLE 8–1   Suppose random samples of fifth-graders from a wealthy suburb and an economically deprived neighborhood were shown a vocabulary list of 100 words and asked their meanings. Identifying all fifth-graders from the wealthy suburb as population 1 and all fifth-graders from the economically deprived neighborhood as population 2, the following results are obtained:

Random sample from population 1	Random sample from population 2
$n_1 = 75$	$n_2 = 54$
$\bar{x}_1 = 83.2$	$\bar{x}_2 = 76.4$
$s_1 = 12.7$	$s_2 = 14.8$

These results mean that 75 children from population 1 identified correctly an average of 83.2 of the 100 words, with a standard deviation of 12.7 words; and 54 children sampled from population 2 identified an average of 76.4 of the 100 words. Notice that the sample sizes are not equal (75 and 54), nor do they need to be. Formula (37) can be applied even with unequal sample sizes. It is essential, however, that the samples are drawn *randomly* and *independently*.

Find a 90% CIE for the difference in mean numbers of words known by the population of fifth-graders in the two communities.

Solution   For 90% confidence, $z = 1.65$.

$$SE_{\bar{x}_1 - \bar{x}_2} = \sqrt{\frac{(12.7)^2}{75} + \frac{(14.8)^2}{54}} \approx 2.49$$

A 90% CIE for $\mu_1 - \mu_2$ is $(83.2 - 76.4) \pm (1.65)(2.49)$

$$\approx 6.8 \pm 4.1$$

$$\approx 2.7 \text{ to } 10.9.$$

The interpretation of this result is that fifth-graders in the wealthy community know, on the average, between 2.7 and 10.9 more of the 100 words than their more disadvantaged counterparts; and you can have 90% confidence that this is correct.   ●

Difference Between Proportions

In chapter 5, you saw how to use an indicator variable to "convert" proportions into means. By very similar reasoning, a formula can be developed for $SE_{p_1 - p_2}$, the standard error of a difference between proportions. It would be an estimate for $SD_{p_1 - p_2}$, the standard deviation of the frequency distribution of $p_1 - p_2$. Because of the link between proportions and means, for separately and independently drawn samples,

$$\text{Var}_{p_1 - p_2} = \text{Var}_{p_1} + \text{Var}_{p_2},$$

which leads to

$$SD_{p_1 - p_2} = \sqrt{(SD_{p_1})^2 + (SD_{p_2})^2}$$

or, by invoking formula (28),

$$SD_{p_1 - p_2} = \sqrt{\frac{P_1 Q_1}{n_1} + \frac{P_2 Q_2}{n_2}}.$$

The population proportions are necessarily unknown. Thus, they are estimated by the corresponding sample proportions, giving,

$$SE_{p_1 - p_2} = \sqrt{\frac{p_1 q_1}{n_1} + \frac{p_2 q_2}{n_2}}. \tag{38}$$

Having the formula for the standard error, a confidence interval can be constructed in the usual way:

$$\text{A CIE for } P_1 - P_2 \text{ is } (p_1 - p_2) \pm z SE_{p_1 - p_2}.$$

EXAMPLE 8-2 A random sample of 100 students from your college reveals that 30% attend only at night while a sample of 75 students from a campus across town shows that 18 of them attend only at night. Find a 99% CIE for the true difference in proportions of students who attend only at night.

Solution Check the following figures:

$$n_1 = 100 \qquad n_2 = 75$$

$$p_1 = 0.30 \qquad p_2 = \tfrac{18}{75} = 0.24$$

$$q_1 = 0.70 \qquad q_2 = 0.76$$

$$SE_{p_1 - p_2} = \sqrt{\frac{(0.30)(0.70)}{100} + \frac{(0.24)(0.76)}{75}} \approx 0.06732.$$

A 99% CIE for $p_1 - p_2$ is $(0.30 - 0.24) \pm 2.58(0.06732)$

$$= 0.06 \pm 0.174$$

$$= -0.114 \text{ to } 0.234.$$

The difference in proportions of night-only students between the two campuses is somewhere between 11.4% in favor of the other campus and 23.4% in favor of your campus. ●

Note: A negative sign in the CIE means that there would be a higher proportion in the second population.

Dependent Samples

Returning now to the situation when the two samples cannot be considered as separately and independently drawn, if two samples are dependent, the individual members of the samples can be paired in some natural way. Having done this, *individual* differences can be calculated, and the mean of these differences used as an estimate of the mean difference in the population. The two samples are thus reduced to one sample (of differences). Since you are now dealing with the mean of *one* sample, the CLT can be applied directly.

If the individual differences are labeled d, the mean of the sample of differences $\bar{d}$, and the standard deviation of differences s_d, then the true mean of differences can be estimated by $\bar{d}$, and to it attached a standard error of this estimate given by

$$SE_d = \frac{s_d}{\sqrt{n}}.$$

Thus, a CIE for the true mean of differences in the population is given by

$$\bar{d}, \pm zSE_d.$$

Consider the sample of fifth-graders tested at the beginning and end of a school year. What you would do is record the difference between Nicole's score at the beginning of the year and her score at the end, and similarly for Fred, Michelle, Charles, and so on, until differences for the whole class had been recorded. The mean $(\bar{d})$ and standard deviation (s_d) of these differences could then be calculated, and a CIE for the true mean difference could be calculated as well. Once the sample of differences has been established, the techniques that are applied parallel exactly the techniques in chapter 5.

If n, the size of each of the samples, is smaller than 30, then s_d is apt to be an unreliable estimate for the true standard deviation of the differences. If it can be assumed that the population of individual differences follows a normal distribution, the t distribution can be used to determine a CIE for the true mean difference:

$$\bar{d} \pm t\frac{s_d}{\sqrt{n-1}}, \tag{39}$$

where, again, t depends on the level of confidence required and the number of degrees of freedom, $n - 1$.

EXAMPLE 8–3   A random sample of eleven fifth-graders is selected at the beginning of the school year. They are given a test of vocabulary then and again at the end of the school year to estimate the difference. Here are the results.

Child	1	2	3	4	5	6	7	8	9	10	11
Starting score	63	72	54	82	72	85	83	67	75	80	75
Ending score	80	79	69	85	72	91	81	78	78	84	77
Difference d	17	7	15	3	0	6	−2	11	3	4	2

Find a 95% CIE for the true average difference in vocabulary scores between the end and the beginning of the school year of all the fifth-graders at this school.

Solution   Check for yourself that $\bar{d} = 6.0$ and $s_d = 5.8$. For 95% confidence and $df = 10$, table 3 gives $t = 2.23$. Thus, a 95% CIE for the true mean difference is

$$6.0 \pm 2.23\frac{5.8}{\sqrt{10}}$$

$$= 6.0 \pm 4.1$$

$$= 1.9 \text{ to } 10.1.$$

You can be 95% confident that the true mean difference in vocabulary scores of all fifth-graders at this school between the end and the beginning of the school year is an increase of between 1.9 and 10.1 points.   ●

Formula (39) and the above method is only applicable when you are dealing with small, dependent samples.

Exercise Set 8–1

1. (Calculator problem) Verify the figures given in the text example on page 268 to justify $\text{Var}_{\bar{x}_1 - \bar{x}_2} = \text{Var}_{\bar{x}_1} + \text{Var}_{\bar{x}_2}$.

2. (Calculator problem) Suppose one population of proportions consists of 0.02, 0.18, 0.34, while a second consists of 0.01, 0.09, 0.24, 0.46. Show that the formula $\text{Var}_{p_1 - p_2} = \text{Var}_{p_1} + \text{Var}_{p_2}$ holds for these figures by calculating the three quantities involved in the equation.

3. Two separately and independently drawn random samples produce the following results: $n_1 = 50$, $\bar{x}_1 = 10.3$, $s_1 = 5$ and $n_2 = 72$, $\bar{x}_2 = 8.2$, $s_2 = 6$. Construct a 95% CIE for the difference in population means $\mu_1 - \mu_2$.

4. Two separately and independently drawn random samples produce the following results: $n_1 = 100$, $\bar{x}_1 = 12.7$, $s_1 = 5$ and $n_2 = 100$, $\bar{x}_2 = 19.3$, $s_2 = 12$. Construct a 68% CIE for the difference in population means $\mu_1 - \mu_2$.

5. Two community colleges are to be compared. A random sample of 80 students from the first college is found to have mean age 25 with a standard deviation of 3.7 years, while a random sample of 60 students from the second college has a mean age of 27 with a standard deviation of 4.6 years. Construct a 90% CIE for the difference in mean ages between the two colleges.

6. Two brands of low-tar cigarettes are to be compared for tar content. A random sample of 90 cigarettes of the first brand was shown to have a mean tar content of 9.6 mg with a standard deviation of 1.6 mg, while a random sample of 90 of the second brand had a mean of 8.7 mg with a standard deviation of 1.9 mg. Find a 98% CIE for the difference in mean tar content between the two brands.

7. Two separately and independently drawn random samples produce the following results: $n_1 = 100$, $p_1 = 0.8$ and $n_2 = 100$, $p_2 = 0.9$. Construct a 68% CIE for the difference in population proportions $P_1 - P_2$.

8. Two separately and independently drawn random samples produce the following results: $n_1 = 350$, $p_1 = 0.65$ and $n_2 = 250$, $p_2 = 0.58$. Construct a 95% CIE for the difference in population proportions $P_1 - P_2$.

9. After a history test had been graded, it was found that 126 of the 180 male students and 90 of the 120 female students had answered a particular question correctly. Find a 98% CIE for the difference in percentages of male and female performances in general on this test question, assuming that these samples represent all male and female students.

10. Of 300 United States citizens tested, 138 were found to have blood type O, and of 250 Europeans, 85 were found to have blood type O. Construct a 95% CIE for the difference in proportions of O blood types between the two populations.

11. In a study of dependent differences, a random sample of 28 differences showed a mean of $\bar{d} = 4.8$ with a standard deviation of 1.7. Find a 99% CIE for the true difference in the population, if the difference can be considered to follow a normal distribution.

12. A typing teacher wishes to examine the effectiveness of a new method of instruction. After using this method for several weeks, he tests the speed of the students and compares it to their speeds before instruction started. Here are the results.

Student	1	2	3	4	5	6	7	8	9	10
Before	43	51	70	36	48	55	68	48	47	51
After	48	60	79	62	72	80	63	65	69	68

The numbers represent speeds in words per minute. Construct a 90% CIE for the true difference in typing speed (after minus before).

🎧 **13.** Users of a method of reducing weight are offered a new, improved method. Their weight losses in pounds over 3-month periods with the old and new methods were as follows:

User	1	2	3	4	5	6	7	8	9
Old	8	5	4	12	7	9	14	16	10
New	12	6	3	15	8	7	19	16	11.

Find a 95% CIE for the true difference (new minus old) in losses of weight by the two methods.

🎧 **14.** Use a 5% significance level to test that the mean difference in losses of weight, μ_d, is zero, using the data in exercise 13. In other words, is the new method no different from the old, or is the new treatment a significant improvement over the old?

Section 8–2 Two-Sample Tests

Experimental and Control Groups

A new product might be launched if it is found to be more effective than an existing one. To test this, the proposed new product could be used on one sample, while the old one was used on another. If a significant difference in the effects of the two products is shown, the researchers can act accordingly. The sample that is subjected to an old procedure (or an existing one) is referred to as the **control group**, while the sample subjected to the new procedure is referred to as the **experimental group**.

When researchers were investigating the Salk vaccine for the prevention of polio, a large sample of children were given the vaccine (the experimental group), while another large sample were injected with a placebo (the control group). A placebo is anything that simulates medication but is a neutral substance doing neither harm nor good to the recipient.

To test a proposed new gasoline additive to improve an automobile's gas mileage, a sample of cars might be run on gasoline alone (the control group). Another sample of cars might be run on gasoline with the additive (the experimental group). Then it can be seen if there is a significant improvement in gas mileage.

On the other hand, comparisons between populations are often made. A friend of yours, at another college, may claim that the mean age of students at your college is less than the mean age of students at his/hers. To test this, a random sample from each college might be drawn and the difference between the sample means investigated. This difference would then indicate whether or not there is a difference in population means. Or you might claim that the proportion of exclusively evening students is higher at your college than at your friend's. Again, the difference in sample proportions could be analyzed to test the validity of this claim.

Setting Up Hypotheses

Hypotheses (or claims) that make comparisons are almost always stated in a very vague fashion. "This headache remedy is better than another" has no mention of *how much* better. "The mean age of students at one college is less than the mean age of another" makes no claim as to *how much* less. As with hypotheses about one population, you need the null hypothesis to be definite (not vague) so that you know where the center of the relevant distribution is.

To get around this problem, vague hypotheses are often turned around so that the null hypothesis expresses the fact that there is *no* difference between the two populations. The *rejection* of the null hypothesis would then indicate the *acceptance* of the original, while the *acceptance* of the null hypothesis would indicate the *rejection* of the original. *Note:* The original hypothesis becomes the alternative H_1.

EXAMPLE 8–4   Set up the hypotheses to test "Headache remedy A is better than headache remedy B." The method of testing will be as follows: one sample of headache sufferers will be given remedy A, while another sample will be given remedy B. Then the proportions of the headache sufferers who claim they got relief will be computed for each remedy.

Solution   Suppose P_A is the proportion of all headache sufferers who are helped by remedy A, while P_B is the proportion of all headache sufferers who are helped by remedy B. Then the hypotheses would be set up as follows:

$$H_0 : P_A - P_B = 0$$

$$H_1 : P_A - P_B > 0.$$

What H_0 is saying, in effect, is that there is no difference between the remedies, while H_1 says that remedy A is better. Eventual rejection of H_0, implying acceptance of H_1, would mean remedy A appears to be better, based on the sample results. Eventual acceptance of H_0 would mean the remedies are about equal. ●

Setting up the hypotheses in this fashion allows you to postulate that if H_0 is true, the frequency distribution of $p_A - p_B$ is centered at 0, a definite number. Had H_0 and H_1, been interchanged, you would not know where to center the distribution under the null hypothesis.

In this problem, rejection of H_0 would imply that remedy A helps a higher proportion of headache sufferers. It would *not* imply that remedy A is better for everyone than remedy B. Even though a higher proportion may be helped by remedy A, a certain individual's body chemistry may be such that remedy B helps more. You should always make clear, to yourself and readers of your research, what the results of your hypothesis testing do, and do *not*, imply.

EXAMPLE 8–5   Set up the hypotheses to test "The mean age of students at college A is less than the mean age of students at college B."

Solution   Suppose μ_1 is the mean age of students at college A and μ_2 the mean age at college B. Then the hypotheses would be stated

$$H_0 : \mu_1 - \mu_2 = 0$$

$$H_1 : \mu_1 - \mu_2 < 0.$$

As with example 8–4, the claim made in H_0 is that the means are equal, while the claim in H_1 is that μ_1 is less than μ_2. Rejection of H_0 would imply acceptance of the hypothesis as stated in the example, and failure to reject H_0 would imply rejection of the original hypothesis.   ●

To test these types of hypotheses, two samples are required: either one from each population or a control group and an experimental group from the same population, as described above. Tests of these types of comparative hypotheses are called **two-sample tests.** In previous chapters, the tests you have studied have been **one-sample tests.**

As with one-sample tests, it is always the case with two-sample tests that the null hypothesis has an equals sign ($=$) in it, while the alternative hypothesis will use one of the symbols $<$, $>$, or $\neq$. If either of the symbols $<$ or $>$ is present in H_1, this will be an indication of a one-tailed test, while the presence of the symbol $\neq$ in H_1 indicates a two-tailed test.

In the vast majority of cases requiring two-sample tests, the null hypothesis will include $= 0$. Occasionally, numbers different from 0 can appear.

Do not confuse the concepts of one- and two-*tailed* tests on the one hand, and one- and two-*sample* tests on the other. These are separate and distinct concepts, and neither one has anything to do with the other.

EXAMPLE 8–6   Set up the hypotheses to test "The mean age of students at college A is at least 2 years less then the mean age of students at college B."

Solution   Using μ_1 and μ_2 as was done in example 8–5:

$$H_0 : \mu_1 - \mu_2 = -2$$

$$H_1 : \mu_1 - \mu_2 < -2.$$

Since college A is supposed to have the younger students, the difference (A minus B) should be negative, which explains the presence of -2 in the hypotheses. Rejecting H_0, and thereby accepting H_1, would imply that the original hypothesis, as stated in the example, is true. On the other had, failing to reject H_0 would imply that the difference in mean ages of the students at the two colleges is 2 years or less. In this example, under the assumption that H_0 is true, you would center the frequency distribution of $\bar{x}_1 - \bar{x}_2$ at -2, rather than the more usual 0. ●

Exercise Set 8–2

In these exercises, state the null and alternative hypotheses to set up the problem.

1. It is claimed that the mean of one population is larger than the mean of a second.

2. It is claimed that the proportion of successes in one population is different from the proportion in a second.

3. The mean score on a mathematical aptitude test is lower at one elementary school than at another.

4. The proportion of rats who successfully negotiate their way through a maze is lower for one strain of rats than another.

5. Two states are to be compared regarding amount spent on utilities. One question to be answered is whether the average amount spent by families on electricity is the same or different in each of the states.

6. Two countries with different cultures are to be compared. It is believed that the average age at which men first get married in the first country is higher than the corresponding age in the second country.

7. A Los Angeles Dodger fan claims that the percentage of Los Angelenos who consider themselves baseball fans is higher than the percentage of New Yorkers who do. A New York Yankee fan disputes this.

8. An ABC spokesman claims to have had a bigger percentage of the national TV audience watching their latest miniseries than CBS had watching theirs. A CBS spokesperson denies this.

9. An automobile battery manufacturer introduces a new and improved battery onto the market, claiming that, despite its higher cost, it will last at least 6 months longer, on the average, than the old model. The two types of batteries, old and new, are to be compared, for length of wear, by a consumer's group.

10. A sociologist believes that children today spend at least 10% more of their waking hours watching TV than they did 20 years ago.

Section 8–3 The Two-Sample z-score Test for Means

First of all, suppose the null hypothesis is stated as

$$H_0 : \mu_1 - \mu_2 = d_H.$$

The symbol d_H is used to stand for hypothesized difference. As indicated previously, d_H will often be 0, but not always. The discussion in section 8–1 derived the following result: If all possible samples of size n_1 are drawn from population 1, all possible samples of size n_2 are drawn from population 2, and all possible differences $\bar{x}_1 - \bar{x}_2$ calculated, then the frequency distribution of $\bar{x}_1 - \bar{x}_2$ will tend to be a normal distribution centered at $\mu_1 - \mu_2$ with a standard deviation

$$SD_{\bar{x}_1 - \bar{x}_2} = \sqrt{\frac{\sigma_1^2}{n_1} + \frac{\sigma_2^2}{n_2}}.$$

In actual practice, one random sample of size n_1 is drawn from population 1, and a (separately and independently drawn) random sample of size n_2 from population 2. Under the usual assumption that H_0 is true, the situation is illustrated in figure 8–1, with the actual difference $\bar{x}_1 - \bar{x}_2$ occurring somewhere in this

Figure 8–1

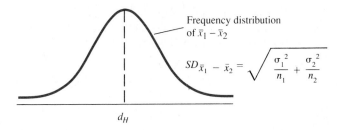

normal distribution. In an analogous way to the one-sample tests, if this difference $\bar{x}_1 - \bar{x}_2$ is far enough away from the center, you would like to reject H_0. This can be decided by converting to a z-score and using the normal table to establish a P-value. As usual, the relevant z-score can be found by taking the actual difference, subtracting the mean, and dividing by the standard deviation, thus:

$$z = \frac{(\bar{x}_1 - \bar{x}_2) - d_H}{\sqrt{\frac{\sigma_1^2}{n_1} + \frac{\sigma_2^2}{n_2}}}.$$

This is one possible version of the two-sample z-score test. However, it is a rarely used one, since, more often than not, σ_1 and σ_2, the two population standard deviations, are unknown. When the population standard deviations are unknown, they are estimated by sample values s_1 and s_2, and the standard error of the difference between the two means is given by the formula

$$SE_{\bar{x}_1 - \bar{x}_2} = \sqrt{\frac{s_1^2}{n_1} + \frac{s_2^2}{n_2}}$$

as an estimate for $SD_{\bar{x}_1 - \bar{x}_2}$. If n_1 and n_2 are both over 30, s_1 and s_2 can be considered good estimates of σ_1 and σ_2, and $SE_{\bar{x}_1 - \bar{x}_2}$ will be a close approximation of $SD_{\bar{x}_1 - \bar{x}_2}$. Therefore, if $n_1 > 30$ and $n_2 > 30$, z can be calculated as

$$z = \frac{(\bar{x}_1 - \bar{x}_2) - d_H}{\sqrt{\frac{s_1^2}{n_1} + \frac{s_2^2}{n_2}}}. \tag{40}$$

Then a P-value can be found using the normal table, and a decision can be reached.

EXAMPLE 8–7   Suppose you wish to test the hypothesis that students at college A are, on the average, younger than students at college B. A random sample of 80 students is drawn from college A and it is found that their mean age is 22.1 with a standard deviation of 2.6, while a random sample of 64 students from college B is found to have a mean of 23.5 with a standard deviation of 2.3. The samples do indicate that college A's students are younger than college B's; but is the difference significant at the 1% level?

Solution

$$H_0 : \mu_1 - \mu_2 = 0$$

$$H_1 : \mu_1 - \mu_2 < 0$$

$$\alpha = 0.01$$

Sample results:

$n_1 = 80$	$n_2 = 64$
$\bar{x}_1 = 22.1$	$\bar{x}_2 = 23.5$
$s_1 = 2.6$	$s_2 = 2.3$

$$z = \frac{(22.1 - 23.5) - 0}{\sqrt{\frac{(2.6)^2}{80} + \frac{(2.3)^2}{64}}} \quad \text{(See figure 8–2.)}$$

$$z \approx -3.42$$

$$A = 0.4997$$

$$P\text{-val} = 0.0003 < 0.01 \ (= \alpha)$$

Reject H_0.

It can be stated, with a probability of less than 1% of committing a type I error, that the students at college A *are* younger, on the average, than the students at college B.   ●

Figure 8-2

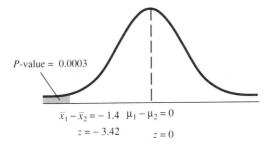

$$\bar{x}_1 - \bar{x}_2 = -1.4 \quad \mu_1 - \mu_2 = 0$$
$$z = -3.42 \qquad z = 0$$

P-value = 0.0003

Exercise Set 8-3

1. It is hypothesized that the mean of one population is larger than the mean of a second. A random sample from the first population reveals that $n_1 = 50$, $\bar{x}_1 = 34.2$, $s_1 = 8.1$, while a random sample from the second gives $n_2 = 75$, $\bar{x}_2 = 31.8$, $s_2 = 9.3$. What conclusion can you draw at the 0.05 significance level?

2. It is hypothesized that the mean of one population is different from the mean of a second. A random sample from the first population reveals that $n_1 = 44$, $\bar{x}_1 = 806$, $s_1 = 54.2$, while a random sample from the second gives $n_2 = 40$, $\bar{x}_2 = 840$, $s_2 = 46.3$. What conclusion can you draw at the 0.01 significance level?

3. The average monthly amount spent on electricity by a random sample of 60 families from one state was found to be $42.07 with a standard deviation of $8.93, while a random sample of 50 families from a second state spent, on the average, $39.24 with a standard deviation of $9.16. Do these results indicate that families in the first state spend significantly more on electricity, on the average, than families in the second? Use a 0.05 significance level.

4. A random sample of 80 married men from one country reported a mean age of 24 when they first married with a standard deviation of 3.2, while a random sample of 100 married men from a second country revealed the mean and standard deviation to be 26 and 2.8, respectively. Do these results indicate a difference in the populations of married men in the two countries at the 1% significance level?

5. In a long-range study of toothpaste, a random sample of 40 children use a paste without fluoride for a period of 10 years, resulting in a mean of 5.65 cavities with a standard deviation of 1.3. Meanwhile, a random sample of 40 children use a paste with fluoride during the same period, resulting in a mean of 4.75 cavities with a standard deviation of 1.1. Do these results indicate a significantly smaller number of cavities in the fluoride users at the 2% significance level?

6. A psychologist devises a test of problem-solving skills and presents it to random samples of 45 males and 65 females. There is no significant difference in the mean scores of the two groups at the 5% level. It has long been suspected that an existing test is biased against women. To show this, this test is presented to the same two samples with these results.

	$\bar{x}$	s
Males	68.9	12.2
Females	62.1	14.3

Do these results bear out the psychologist's suspicion at the same significance level?

 7. A company has two branches in Chicago and Detroit. The sales of the two branches were recorded for a random sample of 60 days with these results.

	Total sales (Σx)	Σx^2
Chicago	$18,240	6,300,000
Detroit	$19,310	7,200,000

Do these results indicate a significant difference in sales between the two branches at the 1% level?

 8. The Federal Trade Commission is testing two brands of "ultra light" cigarettes for nicotine content, in milligrams, with the following results:

	Mean	Variance	Sample size
Brand S	0.43	0.0029	35
Brand M	0.46	0.0031	35.

Do these results indicate that there is a significant difference in nicotine content between the two brands at the 0.01 level?

 9. In a test of the psychological effect of music on the buying habits of supermarket shoppers, a market research company conducts the following test: Random samples of adults who usually do the shopping for their families are observed doing their normal shopping. Of these, two samples are chosen on the basis that there is no significant difference in the mean amount spent on food at the 20% significance level. These two samples are then requested to do their shopping at the same store, one sample with music, one without. Here are the results.

	Sample size	Mean amount spent	s
With music	45	$84.30	$9.40
Without music	55	$79.70	$8.20

Is there a significant difference in the amounts spent by the two groups at the 5% significance level?

10. If you did exercise 9 correctly, you should have found that there is a significant difference. What is the smallest value for the mean of the sample who did their shopping without music, which would result in no significant difference between the groups, if everything else is the same?

Section 8-4   The Two-Sample z-score Test for Proportions

Tests involving the difference between two proportions are similar in many respects. The null hypothesis would be

$$H_0 : P_1 - P_2 = 0.$$

Here, it will be assumed that $= 0$ always appears in H_0. The modifications necessary when a number other than 0 appears in H_0 are minor.

If all possible samples of size n_1 are drawn from population 1, all possible samples of size n_2 from population 2, and all possible differences $p_1 - p_2$ calculated, then the frequency distribution of $p_1 - p_2$ will be a normal distribution centered at $P_1 - P_2$ ($= 0$ under the null hypothesis) with a standard deviation

$$SD_{p_1 - p_2} = \sqrt{\frac{P_1 Q_1}{n_1} + \frac{P_2 Q_2}{n_2}} .$$

A possible formula for p^* is

$$p^* = \frac{n_1 p_1 + n_2 p_2}{n_1 + n_2} .$$

Of course, P_1, Q_1, P_2, Q_2 are unknown; otherwise, no test would be needed. They might be estimated by p_1, q_1, p_2, q_2, the values obtained from the actual *two* samples drawn in practice. However, there is a better estimate. Under H_0, the null hypothesis, P_1 and P_2 are assumed to be equal, so $P_1 - P_2 = 0$. If H_0 is true, then $P_1 = P_2$. Call their common value P^*; and since $Q_1 = Q_2$, call their common value Q^*. Now, estimate P^*, Q^*, by p^*, q^*. These latter two values can be calculated by finding the proportions in the two samples *combined*. Thus, for example, if p_1 is the proportion of smokers in a sample of 50 from population 1, and p_2 is the proportion of smokers in a sample of 40 from population 2, then p^* would be the proportion of smokers in the combined sample of 90 coming from both populations. As usual, $q^* = 1 - p^*$. $SD_{p_1 - p_2}$ can then be estimated by

$$SE_{p_1 - p_2} = \sqrt{\frac{p^* q^*}{n_1} + \frac{p^* q^*}{n_2}}$$

$$= \sqrt{p^* q^* \left(\frac{1}{n_1} + \frac{1}{n_2}\right)}.$$

As with the means, a z-score can now be formed,

$$z = \frac{p_1 - p_2}{\sqrt{p^* q^* \left(\dfrac{1}{n_1} + \dfrac{1}{n_2}\right)}}, \tag{41}$$

and the test completed in the usual way.

It is preferable to use the *two* values, p^*, q^*, rather than the *four* values, p_1, q_1, p_2, q_2, since the estimates p^*, q^* are based on a larger sample.

EXAMPLE 8–8   It is claimed that the proportions of exclusively evening students at two colleges, A and B, differ. To test this, the random samples of 80 from college A and 64 from college B are reexamined and it is found that 25 of the 80 and 14 of the 64 are exclusively evening students. Test to see if that represents a significant difference at the 5% level.

Solution

$$H_0 : P_1 - P_2 = 0$$

$$H_1 : P_1 - P_2 \neq 0$$

$$\alpha = 0.05$$

A two-tailed test is chosen here, since it is of interest if the proportions differ in *either* direction.
Sample results:

$$n_1 = 80 \qquad\qquad n_2 = 64$$

$$p_1 = \frac{25}{80} \approx 0.31 \qquad p_2 = \frac{14}{64} \approx 0.22$$

$$q_1 = 0.69 \qquad\qquad q_2 = 0.78$$

Using the formula

$$p^* = \frac{80(0.31) + 64(0.22)}{80 + 64}$$

To calculate p^*, note that there are $25 + 14 = 39$ exclusively evening students out of a combined sample of $80 + 64 = 144$ students.

Figure 8–3

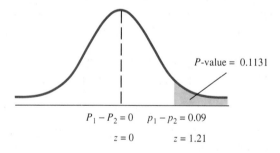

Thus,

$$p^* = \frac{39}{144} \approx 0.27$$

and

$$q^* \approx 1 - 0.27 = 0.73.$$

Thus,

$$z = \frac{0.31 - 0.22}{\sqrt{(0.27)(0.73)\left(\dfrac{1}{80} + \dfrac{1}{64}\right)}}$$

$$z \approx 1.21 \text{ (See figure 8-3.)}$$

$$A = 0.3869$$

$$P\text{-val} = 0.1131 > 0.025 \ (= \tfrac{1}{2}\alpha).$$

Fail to reject H_0.

Although the proportion of exclusively evening students in the sample from college A was 31%, while the proportion in the sample from college B was 22%, this did not represent a significant difference at the 5% level. It must be concluded that the actual proportions of exclusively evening students at the two colleges could be identical. ●

Exercise Set 8-4

1. It is hypothesized that the proportion of successes in one population is smaller than the proportion in a second. A random sample from the first population reveals 35 successes in a sample of 100, while a random sample from the second has 64 successes out of 120. If a two-sample *z*-score test for proportions is conducted, what P-value results?

2. It is believed that there is a difference between proportions of successes in two populations. If random samples of 200 from each population produce, respectively, 30 and 34 successes, what *P*-value is produced by an application of the *z*-score test for proportions?

3. To determine whether men have nightmares as often as women, 28 out of 88 randomly chosen men and 30 out of 100 randomly chosen women reported having nightmares at least once a month. Do these figures indicate a significant difference in the frequency of nightmares between the two sexes at the 0.05 significance level?

4. To see if experimentation with marijuana has changed in 10 years, a sociologist asked a random sample of adults in their mid-twenties whether they tried it in junior high school. Of this sample, 21 of the 55 said that they had. Of a random sample of 75 junior high school students, 34 said that they had. Test the hypothesis that more junior high school students these days experiment with marijuana than junior high school students of 10 years ago. Use a 5% significance level.

5. Brown and Green are running for political office. An opinion poll shows Brown with 48% support in Brownstown based on a random sample of 152 voters, while another shows him only enjoying 42% support in Greenstown based on a random sample of 100 voters. Is there a significant difference in the amount of support Brown enjoys in the two towns at the 5% level?

6. The two samples in exercise 5 showed Green with 37.5% support in Brownstown and 49% in Greenstown. Does Green have significantly more support in Greenstown than in Brownstown at the 5% level?

 7. A doctor has discovered what he believes to be a better treatment for acne. He treats a random sample of 32 of his patients with this new method and finds 18 respond well, while a random sample of 36 patients treated with existing methods only showed a 50% success rate. Do these results show that the new treatment is significantly better than the old at the 10% significance level?

 8. Two firms submit bids to supply a contractor with pipes for the construction of a new high-rise office building. Firm A's bid is lower than firm B's, but the contractor wishes to know about the quality difference. A random sample of 120 of firm A's pipes show 108 to be acceptable, while a random sample of 140 of firm B's pipes reveals that 95% are acceptable. The contractor will go with firm A unless it can be shown that firm B has a significantly higher percentage of acceptable pipes at the 10% significance level. Who should the contractor give the contract to?

9. In a two-sample z-score test for proportions, if the two sample sizes are equal, p^* can be calculated from p_1 and p_2 by substituting n for n_1 and n_2 and simplifying the formula for z to

$$z = \frac{p_1 - p_2}{\sqrt{p^*q^* \cdot \frac{2}{n}}} = \frac{\sqrt{n}\,(p_1 - p_2)}{\sqrt{2p^*q^*}}, \text{ where } n \text{ is the size of each sample.}$$

If such a test is conducted as a one-tailed test, find the smallest value of n that would produce a significant result at the 5% significance level if $p_1 = 26\%$ and $p_2 = 20\%$.

10. Redo exercise 9 if $p_1 = 60\%$ and $p_2 = 72\%$.

Section 8–5 The Two-Sample t-Test

Suppose you wish to test a hypothesis concerning the difference between two population means in the form $H_0 : \mu_1 - \mu_2 = d_H$. Suppose that the population standard deviations σ_1 and σ_2 are unknown, and that only small samples are available. For this two-sample test with small samples, the Student's t distribution can be used.

Recall that the one-sample t-test is valid only if the population being tested can be considered to have a normal distribution. You should also be aware that the **two-sample t-test,** a widely used and practical test, should only be used when certain conditions are satisfied:

Assumption 1: Both populations are normally distributed.
Assumption 2: The two populations are equally dispersed. That is, $\sigma_1 = \sigma_2$.

If you suspect that either or both of these assumptions are violated by the populations under study, then the two-sample t-test should *not* be used, since it can give misleading results.

The remaining discussion in this section pertains to hypothesis testing only for populations in which both of the above assumptions are satisfied.

$$\text{Recall that } SD_{\bar{x}_1 - \bar{x}_2} = \sqrt{\frac{\sigma_1^2}{n_1} + \frac{\sigma_2^2}{n_2}}.$$

Under assumption 2, $\sigma_1 = \sigma_2$; call their common value σ^*. Then,

$$SD_{\bar{x}_1 - \bar{x}_2} = \sqrt{\frac{\sigma^{*2}}{n_1} + \frac{\sigma^{*2}}{n_2}}$$

$$= \sqrt{\sigma^{*2}\left(\frac{1}{n_1} + \frac{1}{n_2}\right)}$$

$$= \sigma^*\sqrt{\frac{1}{n_1} + \frac{1}{n_2}}.$$

As is usually true, σ^* is unknown and must be estimated by s^*, where s^* is computed from the sample data.

Even if $\sigma_1 = \sigma_2$, s_1 and s_2 could be quite different, especially with small samples. How can s^* be estimated from s_1 and s_2? You are again asked to accept a result from more advanced statistical theory that states that the best estimate is

$$s^* = \sqrt{\frac{n_1 s_1^2 + n_2 s_2^2}{n_1 + n_2 - 2}},$$

a weighted average of s_1 and s_2. The quantity in the denominator $(n_1 + n_2 - 2)$ comes from the number of degrees of freedom in the first sample, $n_1 - 1$, added to the number of degrees of freedom in the second sample, $n_2 - 1$.

$$(n_1 - 1) + (n_2 - 1) = (n_1 + n_2 - 2)$$

This quantity represents the total number of degrees of freedom for this test—a fact that you need to know when you look up the *P*-value. When s^* is used to estimate σ^*, the revised formula for $SE_{\bar{x}_1 - \bar{x}_2}$ for the two-sample *t*-test is

$$SE_{\bar{x}_1 - \bar{x}_2} = s^*\sqrt{\frac{1}{n_1} + \frac{1}{n_2}},$$

where

$$s^* = \sqrt{\frac{n_1 s_1^2 + n_2 s_2^2}{n_1 + n_2 - 2}}$$

and t, with $(n_1 + n_2 - 2)$ degrees of freedom, is

$$t = \frac{(\bar{x}_1 - \bar{x}_2) - d_H}{s^*\sqrt{\frac{1}{n_1} + \frac{1}{n_2}}}. \tag{42}$$

Table 2 can now be used to compute the relevant *P*-value, and a decision about H_0 can be made.

EXAMPLE 8–9   There is a psychological eating disorder called anorexia nervosa. Sufferers eat very little, believing they are too fat even when they are little more than skin and bones. Suppose a psychiatrist devises a new treatment for this disease, and a sample of 21 women suffering from anorexia nervosa is selected for a test to determine whether or not the new treatment is more effective than the existing one. The women are randomly divided into two groups; 10 are treated with the current prevailing method, while the other 11 comprise the experimental group that is treated with the new method.

A measure of success can be judged by weight gain. Suppose that for the 10, the mean weight gain after four weeks is 12 lb with a standard deviation of 2.4 lb, while for the 11, the mean weight gain, in the same time period, is 15 lb with a standard deviation of 3.2 lb. The new method seems to be more effective, based on average changes in these two small samples; but can you infer that the difference is significant at the 5% level? If so, you can be 95% confident that the difference between sample results is not due merely to random differences caused by sampling (i.e., sampling error), and that the new treatment would really be more effective on average than the current treatment if applied to the population as a whole.

Solution   Weight gain is assumed to follow a normal distribution. There is also no reason to believe that the two population standard deviations of weight gain are radically different simply because treatment differs. Assumptions 1 and 2 appear to be justified in this test, and a two-sample t-test is therefore appropriate. The test seeks to determine whether or not the new method is *more* effective than the old. Since there would be no reason to change treatment methods unless the new method is *more* effective, researchers are not interested in testing the possibility that the new method is *less* effective. Therefore, a one-tailed test is chosen here, and the hypotheses are stated as

$$H_0 : \mu_1 - \mu_2 = 0$$
$$H_1 : \mu_1 - \mu_2 < 0$$
$$\alpha = 0.05,$$

where population 1 is the population mean for the new method and population 2 refers to persons treated by the current mehtod. Sample results are

$$n_1 = 10 \qquad\qquad n_2 = 11$$
$$\bar{x}_1 = 12 \qquad\qquad \bar{x}_2 = 15$$
$$s_1 = 2.4 \qquad\qquad s_2 = 3.2$$

The s's are squared, while the n's are *not*. Be careful!

$$s^* = \sqrt{\frac{10(2.4)^2 + 11(3.2)^2}{10 + 11 - 2}}$$

$$\approx 2.99.$$

Figure 8-4

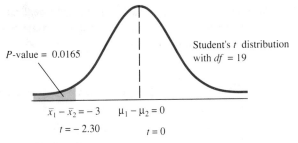

$$t = \frac{(12 - 15) - 0}{2.99\sqrt{\dfrac{1}{10} + \dfrac{1}{11}}}$$

$$\approx -2.30 \text{ (See figure 8-4.)}$$

$$df = 10 + 11 - 2 = 19$$

$$A = 0.4835 \text{ from table 2}$$

$$P\text{-val} = 0.0165 < 0.05(= \alpha)$$

Reject H_0.

There is a significant difference in weight gains at the 5% level, indicating with 95% confidence that the new method is more effective.   ●

A CIE for a Difference in Means with Small Samples

In section 8-1, a CIE for a difference in means was developed. It was

$$\text{a CIE for } \mu_1 - \mu_2 \text{ is } (\bar{x}_1 - \bar{x}_2) \pm z\sqrt{\frac{s_1^2}{n_1} + \frac{s_2^2}{n_2}}.$$

This formula is only really valid if the sample sizes n_1 and n_2 are each larger than 30. For small sample sizes, the formula can be adapted using t if assumptions 1 and 2 (cited above) are maintained:

$$\text{a CIE for } \mu_1 - \mu_2 \text{ is } (\bar{x}_1 - \bar{x}_2) \pm ts^*\sqrt{\frac{1}{n_1} + \frac{1}{n_2}}, \qquad \textbf{(43)}$$

where there are $(n_1 + n_2 - 2)df$, and where s^* is given in formula (42) and t can be found from table 3 (or table 2 and linear interpolation if an unusual confidence level is required).

EXAMPLE 8–10   Find a 95% CIE for the true difference in weight gains of all anorexia nervosa sufferers, given the data in example 8–9.

Solution   In example 8–9, it was calculated that $s^* = 2.99$ and $df = 19$. For 95% confidence, $t = 2.09$.

Thus, a 95% CIE for $\mu_1 - \mu_2$ is $(12 - 15) \pm (2.09)(2.99) \sqrt{\dfrac{1}{10} + \dfrac{1}{11}}$

$$= (-3) \pm 2.73$$
$$= -5.73 \text{ to } -0.27.$$

Thus, it can be stated with 95% confidence that the new method of treatment will provide anorexia nervosa patients between 0.27 lb and 5.73 lb more gain in weight than the old method in a period of four weeks. ●

Exercise Set 8–5

1. Random samples from two normal populations produce the following results:

	n	$\bar{x}$	s
Sample 1	15	10.2	3.1
Sample 2	12	7.6	2.6

If the two-sample t-test is applied, what is the resulting P-value?

2. Random samples from two normal populations produce the following results:

	n	$\bar{x}$	s
Sample 1	8	141.3	18.1
Sample 2	13	136.5	19.6

If the two-sample t-test is applied, what is the resulting P-value?

3. A corporation uses two computers on a trial basis for a month. The time during which a computer is not operating properly due to a malfunction is called downtime. The corporation executives would like to choose the computer that would give them the lowest total downtime. Fourteen days are chosen at random from the month and the mean downtime calculated for each computer. Here are the results, in hours.

	n	$\bar{x}$	s
Computer 1	14	1.1	0.62
Computer 2	14	0.57	0.56

Do these results indicate that there is a significant difference between the mean downtimes for the two computers at the 5% level?

4. There is a psychological test called the Incomplete Sentences Blank (ISB), which measures maladjustment. Higher scores on it indicate more maladjustment. Random samples of heavy marijuana users and nonusers were given the ISB test with these results.

	n	$\bar{x}$	s
Heavy users	14	146.1	17.4
Nonusers	12	138.2	15.2

Do these results indicate that heavy marijuana smokers are more maladjusted (according to the ISB test) than nonusers at the 5% significance level?

5. There has been much controversy over the teaching of "new math." A random sample of 25 students was randomly divided into two groups. Of the 25, 12 were taught by "new" math methods, while the other 13 were taught by more "traditional" methods. After one year, these 25 students were all given the same test with these results.

	$\bar{x}$	s
New	87.8	17.4
Traditional	82.1	12.3

Do these results show any significant difference between the results of students taught by the two methods at the 10% level?

6. How big a difference should there be between the two mean scores in exercise 5 to show significance at the 5% level, if the standard deviations and sample sizes are the same?

7. Construct a 95% CIE for the difference between population means, given the data in exercise 1.

8. Construct a 99% CIE for the difference in scores on the ISB test between heavy and nonusers of marijuana, given the data in exercise 4.

9. A sample of 20 male college students were found to have a mean height of 70.3 inches with a standard deviation of 3.2 inches, while a random sample of 18 female students had a mean height of 64.7 inches with a standard deviation of 2.6 inches. Find a 90% CIE for the difference in mean heights of all male and female students on campus.

10. The two samples in exercise 9 were found to have the following results for their weights, in pounds:

	$\bar{x}$	s
Males	175.2	27.5
Females	142.1	26.0

Construct a 90% CIE for the difference as you did in exercise 9. How valid do you think your result is?

11. Sociologists have developed a social distance scale to measure how people relate to each other socially. The higher the score, the less social distance that person has to the group under test. Here are the results for random samples from two ethnic groups on how they relate to a third.

Group A: 1.8   2.4   3.2   1.7   1.5   2.8   3.0
Group B: 1.4   2.7   2.9   1.2   3.3   1.6   1.8   2.9   2.7   1.0   1.1

Find a 95% CIE for the difference in attitude by these two groups toward the third.

12. It is believed that Catholics have larger families than Protestants. To test this, a sociologist investigates samples of Catholic and Protestant families with these results.

Catholic family sizes: 3   6   2   4   7   6   5   4   3   7   5   5   8   6
Protestant family sizes: 2   3   3   4   2   5   2   3   5   4   6

Construct a 95% CIE for the difference in mean family sizes between the two religions.

13. Using the data of exercise 12, test the hypothesis that the mean size of Catholic families is at least 1 more than Protestant families at the 5% significance level.

....

Section 8–6   Summary of Terms

control group 275
dependent samples 269
estimating differences 267
experimental group 275
one-sample tests 277

separately and independently drawn
samples 269
two-sample tests 277
two-sample t-test 286
two-sample z-score test 279

Section 8–7   Summary of Symbols

Symbol	Meaning
d_H	Hypothesized difference
P^*	When two populations are being compared, this is the proportion that the two population proportions are assumed to be equal to.
p^*	This is an estimate of P^*. It is found by combining the results of samples drawn from the two populations being compared.
Q^*	$1 - P^*$
q^*	$1 - p^*$
σ^*	When using the two-sample t-test, this is the common value of the standard deviations of the two populations being compared.
s^*	This is an estimate of σ^*. It is found by computing a kind of weighted average of the two standard deviations of the two samples drawn from the populations being compared.
$SD_{p_1 - p_2}$	The standard deviation of a frequency distribution of a difference in sample proportions
$SD_{\bar{x}_1 - \bar{x}_2}$	The standard deviation of a frequency distribution of a difference in sample means
$SE_{p_1 - p_2}$	The standard error of a difference in sample proportions
$SE_{\bar{x}_1 - \bar{x}_2}$	The standard error of a difference in sample means
Var	Abbreviation for variance

Section 8–8   Summary of Formulas

Formula	Use
(37) $SE_{\bar{x}_1 - \bar{x}_2} = \sqrt{\dfrac{s_1^2}{n_1} + \dfrac{s_2^2}{n_2}}$	To find the standard error when estimating the difference between the means of two populations
(38) $SE_{p_1 - p_2} = \sqrt{\dfrac{p_1 q_1}{n_1} + \dfrac{p_2 q_2}{n_2}}$	To find the standard error when estimating the difference between proportions in two populations

Note: Formulas (37) and (38) should only be used if the samples used are separately and independently drawn from their respective populations.

(39) $\bar{d} \pm t \dfrac{s_d}{\sqrt{n-1}}$

CIE for a dependent difference when the sample size is less than 30

(40) $z = \dfrac{(\bar{x}_1 - \bar{x}_2) - d_H}{\sqrt{\dfrac{s_1^2}{n_1} + \dfrac{s_2^2}{n_2}}}$

Formula used in the two-sample z-score test for means

(41) $z = \dfrac{p_1 - p_2}{\sqrt{p^* q^* \left(\dfrac{1}{n_1} + \dfrac{1}{n_2}\right)}}$,

Formula used in the two-sample z-score test for proportions

where $p^* = \dfrac{n_1 p_1 + n_2 p_2}{n_1 + n_2}$

(42) $t = \dfrac{(\bar{x}_1 - \bar{x}_2) - d_H}{s^* \sqrt{\dfrac{1}{n_1} + \dfrac{1}{n_2}}}$,

where

$s^* = \sqrt{\dfrac{n_1 s_1^2 + n_2 s_2^2}{n_1 + n_2 - 2}}$

and $df = n_1 + n_2 - 2$

Formulas used in the two-sample t-test

(43) $(\bar{x}_1 - \bar{x}_2) \pm t s^* \sqrt{\dfrac{1}{n_1} + \dfrac{1}{n_2}}$

Formula giving a CIE for $\mu_1 - \mu_2$ when n_1, n_2 are small, the populations are normally distributed, and $\sigma_1 = \sigma_2$, they being unknown

Chapter 8 Review Exercises

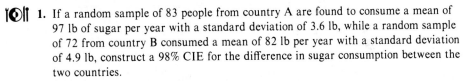 **1.** If a random sample of 83 people from country A are found to consume a mean of 97 lb of sugar per year with a standard deviation of 3.6 lb, while a random sample of 72 from country B consumed a mean of 82 lb per year with a standard deviation of 4.9 lb, construct a 98% CIE for the difference in sugar consumption between the two countries.

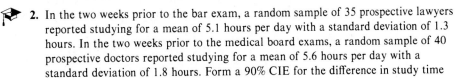 **2.** In the two weeks prior to the bar exam, a random sample of 35 prospective lawyers reported studying for a mean of 5.1 hours per day with a standard deviation of 1.3 hours. In the two weeks prior to the medical board exams, a random sample of 40 prospective doctors reported studying for a mean of 5.6 hours per day with a standard deviation of 1.8 hours. Form a 90% CIE for the difference in study time prior to important exams between the two groups.

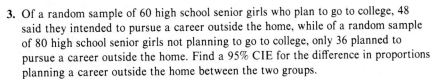 **3.** Of a random sample of 60 high school senior girls who plan to go to college, 48 said they intended to pursue a career outside the home, while of a random sample of 80 high school senior girls not planning to go to college, only 36 planned to pursue a career outside the home. Find a 95% CIE for the difference in proportions planning a career outside the home between the two groups.

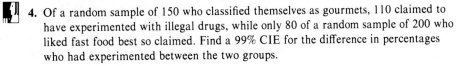 **4.** Of a random sample of 150 who classified themselves as gourmets, 110 claimed to have experimented with illegal drugs, while only 80 of a random sample of 200 who liked fast food best so claimed. Find a 99% CIE for the difference in percentages who had experimented between the two groups.

5. In a test of new fish food, a random sample of 10 baby goldfish were measured, fed this food for a month, and then measured again. Here are the results in centimeters.

Fish	1	2	3	4	5	6	7	8	9	10
At the start	2.4	2.7	3.0	2.9	4.2	3.6	3.4	3.1	5.0	2.5
After 1 month	3.6	3.9	4.2	3.6	5.3	4.8	4.7	4.8	5.2	3.5

Find a 90% CIE for the mean difference in lengths after feeding the fish this food for a month.

 6. A random sample of 12 married couples participate in a study. Here are their ages.

Husband	37	20	25	31	26	18	44	35	36	22	27	52
Wife	27	19	23	28	32	18	43	30	38	23	24	48

Construct a 95% CIE for the mean difference in ages (husband's minus wife's) in the population from which the sample was drawn.

 7. A community college instructor experiments with two different methods of teaching, by trying them on random samples of students, with these results.

	n	$\bar{x}$	s
First method	14	71.8	13.2
Second method	16	79.2	8.7

Is there a significant difference between the results obtained from the two teaching methods at the 5% level?

 8. If the sample sizes in exercise 7 are each increased to 32 and the same means and standard deviations result, is the answer to the question the same or different as the answer using the small samples?

 9. Of a random sample of 132 blue-collar workers, 60 liked a new TV show, while of a random sample of 108 white-collar workers, 58 liked it. Is there a significant difference between the appeal of the show among blue- and white-collar workers at the 10% significance level?

 10. Random samples of 400 males and 350 females were tested for color-blindness. Of these, 6 males and 1 female were found to be color-blind. Does this data allow us to conclude that the proportion of color-blind people is lower for females than for males at the 5% significance level?

 11. Do college graduates subscribe to more magazines than noncollege graduates? A survey was made with these results.

	n	$\bar{x}$	s
College graduates	20	3.0	0.87
Noncollege grads	23	2.5	0.73

How would you answer the question at the 5% significance level?

12. If you did exercise 11 correctly, you should have used a t-test with 41 degrees of freedom. At that large a number of degrees of freedom, the P-value can safely be computed using the normal table. You might wonder why the z-score test could not have been used from the beginning even though it is incorrect to apply it with sample sizes less than 30. If the z-score test is incorrectly applied to the same data, is the conclusion the same? If so, would the conclusion necessarily always be the same?

13. To investigate reaction time, the DMV conducts this test. Two random samples of drivers were tested and it was found that there was no significant difference between the groups. Then one sample was given three alcoholic drinks each, with these results.

Drivers with no alcohol: 2.1   3.5   1.8   1.2   2.4   2.2   3.0
Drivers with alcohol: 4.2   3.2   2.4   4.1   5.4   2.6

Are drivers who had the drinks significantly slower at the 5% level?

 14. Reaction times are typically right-skewed, making the application of the *t*-test in exercise 13 invalid. What would have been a better experimental design to use to test the hypothesis of exercise 13?

15. To test the hypothesis $H_0 : \mu_1 = \mu_2$, using samples of 10 and 12, a two-sample *t*-test can be used when certain assumptions are satisfied. If a one-tailed test at a 5% significance level is used, the *t*-score thus obtained could be compared to a critical *t*-value. If this critical value is used to form a CIE for $\mu_1 - \mu_2$, how much confidence do you have?

16. Redo exercise 15 if the sample sizes are 7 and 8 and 1% level is used.

 17. Using the data given in the article at the beginning of the chapter, test the hypothesis that the proportion of returned violators who discard their prison-issued release clothing as soon as possible is the same as the proportion for successful releasees. Use a 0.01 level.

 18. Using the data given in the article at the beginning of the chapter, test the hypothesis that the proportion of returned violators who retain and use their prison-issued release clothing is the same as the proportion for successful releasees. Use a 0.05 level.

Hands-on Class/Student Project

Follow similar directions as found for the corresponding project exercises in previous chapters for these exercises:

1. If you haven't already done so, divide the sample you collected in the project exercises of chapter 1 into two samples, one of male students and the other of female students. Find the mean and standard deviation of each of these samples' GPA (variable 5).

2. Form a 95% CIE for the difference in mean GPA between the male and female students at your college and compare your result to the results of other groups in your class.

3. If the samples you used in exercise 2 are smaller than 30, you should have used the formulas found in section 8–5 rather than those found in section 8–1. However, to use those formulas, it is required that the corresponding populations are normally distributed. Whether or not you used small samples, discuss in your group or with your instructor if you think GPA's at your college are normally distributed.

4. Set up hypotheses to test the claim that there is a higher proportion of blonds (natural or not) among females than there is among males for students on your campus.

5. Using your data from chapter 1 (variable 8), test the hypotheses you set up in exercise 4 using a 5% significance level. (*Note:* Your samples might be quite small for testing a hypothesis about proportions, so you might wish to return to your campus and get larger samples. Since you can use observation rather than inquiry, this might not take as long as you think.)

6. Compare your conclusion to exercise 5 with those of other groups in the class.

7. Set up hypotheses to test the claim that males on your campus have owned, on the average, more different cars than females.

8. Using your data from chapter 1 (variable 11), test the hypotheses you set up in exercise 6 using a 5% significance level.

9. Compare your conclusion to exercise 8 with those of other groups in the class.

10. Make up your own claim about a comparison between two populations, set up hypotheses to test this claim, select an appropriate test and significance level to test the claim, gather data and conduct the test.

Nonparametric Tests

"By different methods different men excel;"

CHARLES CHURCHILL

Education
How to Promote Sophomore Surge

The trick worked like magic. Two researchers informed a group of Duke freshmen who were worried about their below-average grades that academic difficulties in the freshman year are usually temporary. And lo, for reasons the researchers still find mysterious, two semesters later those students' grades had improved, while the grades of equally worried students in a control group had stayed the same.

The 40 psychology students in the study had averages of about 2.7, somewhere between a B and C. All indicated by ratings that they were more worried than their peers about their grades, felt themselves intellectually inferior to almost half their classmates, and thought that they had not done as well as they could have in their first semester. Half of them learned, from reading a report on the academic histories of upper-classmen and seeing videotaped interviews with several students, that grades usually improve in the sophomore year.

A modest effect became apparent immediately in scores on two versions of a six-item reading comprehension test. The students took the test immediately after learning about the sophomore improvement, and again a week later. Those who had heard the good news got an average of 4.18 answers correct (70 percent), while their uninformed peers got an average of only 3.5 answers right (58 percent).

The major effect emerged over a year later, when the researchers compared the grades that the worried students had achieved in their first freshman semesters with their grades in their second sophomore semesters. The uninformed students averaged a slight decrease (.05 points), suggesting that they may have been too worried and too far behind to share in the second-year improvement. But the informed students averaged an increase of .34 points on the four-point scale used for grading. Moreover, the researchers discovered that five uninformed students had left school—but only one informed student had left.

The idea of telling worried students that their academic problems would be temporary came from studies in "attribution therapy" that suggest that people have a better chance of solving a problem if they think their difficulty comes from a tough situation rather than from their character. —*Harris Dienstfrey*

The researchers are psychologists Timothy D. Wilson, at the University of Virginia, Charlottesville, Va. 22901, and Patricia W. Linville, at Carnegie-Mellon University. Their study appeared in *The Journal of Personality and Social Psychology,* Vol. 42, No 2.

Chapter Overview
What If Things Aren't Normal?

In the preceding chapters, the discussion has centered on tests of hypotheses, both of the one-sample and the two-sample variety. When the population standard deviations were known, the z-score test was found to be completely satisfactory. However, in most practical instances, population standard deviations are unknown, and this fact forces you to use sample standard deviations as estimates. If the sample sizes are large enough (30 or over), this again presents no problem, since z-score tests with s's replacing σ's are again satisfactory. When sample sizes are small *and* population standard deviations are unknown, z-score tests become unreliable, and you can fall back on t-tests.

However, as you have seen, t-tests should only be used when the population is normally distributed. For some types of tests, assumptions concerning the population variance(s) must also hold. What can be done when you have only small samples available, the population standard deviation(s) are unknown, and you *cannot* assume the population(s) are normal? None of the tests discussed so far is appropriate.

Use of a z-score test requires some knowledge about a *parameter*, namely the population standard deviation, with σ either known or estimated by s. Use of a t-test requires that the population or populations are normally distributed; and the normal distribution, as you know, depends on the two *parameters* μ and σ. It is for this reason that these tests are sometimes referred to as parametric tests. There is another group of tests for which the assumptions made do *not* involve knowledge of, estimation of, or information concerning the sampling distributions of population parameters. They are, therefore, referred to as nonparametric tests. The first test you will study, for instance, is called the sign test. The assumption necessary for the use of this test is simply that the population is symmetric. This is a nonparametric assumption, since no *specific* distribution, such as the normal distribution for the t-test, is mentioned. It is also not so strict an assumption as is required for the t-test, since it allows for a much wider variety of distributions for the population. The population could be distributed as a normal, triangular, uniform, bell-shaped, and non-normal, or *any* other distribution that is not skewed . Since no specific distribution is attributed to the population, no parameters can be involved. As you study the tests in this chapter, examine carefully the assumptions necessary for each one's application. You should satisfy yourself, in each case, that these assumptions are nonparametric in nature, leading to the test being labeled nonparametric.

Section 9–1   The Sign Test

The sign test is a one-sample test that assumes that the population being tested has a symmetric distribution.

Suppose that the null hypothesis is

$$H_0 : \mu = \mu_H.$$

A random sample is drawn from the population to test this hypothesis. If (a) the population is symmetric, (b) H_0 is true, and (c) the sample is random, then the proportion of values in the sample that are larger than μ_H should be 0.5 or 50%, and similarly for the proportion of values that are smaller. To see that this must be true, let P = proportion of values in the *population* that are larger than μ_H. If H_0 is true and the population is symmetric, the situation might be as shown in figure 9–1.

Figure 9–1

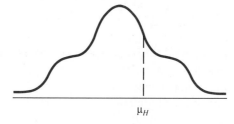

Notice that exactly 50% of the population must be larger than μ_H. Thus $P = 0.5$.

If, on the other hand, either the population is symmetric and H_0 is false, or if the population is asymmetric and H_0 is true, you might have one of the situations shown in figures 9–2 and 9–3.

Figure 9–2

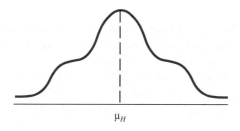

Figure 9–3

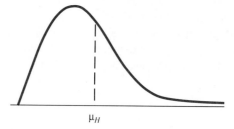

In either of the cases shown here, less than 50% of the population would be larger than μ_H. Thus $P < 0.5$. Situations could also be constructed where $P > 0.5$, when either or both of assumptions (a), (b) are not satisfied. If (a) and (b) are true, is it not logical to suppose that a *random* sample from the population would contain in it roughly half of the values larger than μ_H and half smaller? If the proportion of values in the sample that are larger than μ_H is substantially different from 0.5, then it will be necessary to conclude that either the population is not symmetric or that H_0 is false and should be rejected. If you know *for sure* that the population can be assumed to be symmetric, and p, the sample proportion larger than μ_H, is significantly different from 0.5 (at some predetermined significance level), then this would lead to the inescapable conclusion to reject H_0.

To conduct the sign test, then, the previous null hypothesis is transformed into

$$H_0 : P = 0.5.$$

When the sample is drawn, a plus sign is attached to every value above μ_H and a minus sign to every value below μ_H. (This is where the test gets its name.) The proportion of plus signs in the sample p is then computed. A z-score test for proportions is then conducted with P_H replaced by 0.5. Thus, a test about a mean is transformed into a test about a proportion, which you have already studied. *Note:* To carry out this test, do not forget the continuity correction (see section 6–4).

Note: If the hypothesis is made about the median $\bar{\mu}_H$ instead, the assumption of symmetry is not necessary, since the definition of the median guarantees that 50% of the population should be above it.

EXAMPLE 9–1 A magazine claims that teenagers watch TV for no more than 20 hours per week. A random sample of teenagers was selected and a record kept of how many hours of TV were watched over a period of a few weeks. The average number of hours of TV watched per week by each teenager was then recorded. Here are the results.

21.2	18.6	15.3	25.5	19.1	24.1	29.7	10.6	12.3
27.4	26.5	22.3	18.1	9.9	22.8	23.5	25.2	23.7
25.3	27.6	31.2	24.5	16.7	29.1	22.2		

Test to see if the magazine's claim is believable at the 5% significance level.

Solution

$$H_0 : \mu = 20$$
$$H_1 : \mu > 20$$
$$\alpha = 0.05$$

If it is decided to use the sign test, the hypotheses would be rewritten

$$H_0 : P = 0.5$$
$$H_1 : P > 0.5$$
$$\alpha = 0.05,$$

where P is the proportion in the population of teenagers who watch more than 20 hours of TV per week. The reason $>$ appears in H_1 is that if, as is stated in the original H_1, the actual value of μ is greater than 20, then the situation would look as shown in figure 9–4 and there would be a proportion of *more* than 50% of teenagers who watch more than 20 hours of TV.

Figure 9–4

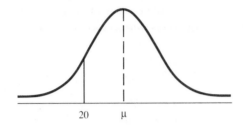

20   μ

Now, to conduct a sign test:

If one of the sample values is actually equal to μ_H, no sign can be attached. Usually, then, this value is ignored and the sample size is reduced by one. The situation can usually be avoided by precise measurement.

Sample value, x	Sign	Sample value, x	Sign
21.2	+	9.9	−
18.6	−	22.8	+
15.3	−	23.5	+
25.5	+	25.2	+
19.1	−	23.7	+
24.1	+	25.3	+
29.7	+	27.6	+
10.6	−	31.2	+
12.3	−	24.5	+
27.4	+	16.7	−
26.5	+	29.1	+
22.3	+	22.2	+
18.1	−		

Here, a plus sign was noted if a sample value was more than $\mu_H (= 20)$, and a minus sign if it was less.

There are 17 plus signs and $n = 25$. Thus,

$$p = \frac{17}{25} = 0.68.$$

This is more than 0.5, but is it significantly more?

The z-score test for a proportion is given by

$$z = \frac{p - P_H}{\sqrt{\dfrac{P_H Q_H}{n}}},$$

where p is corrected for continuity.

Since 0.68 is larger than $P_H (= 0.5)$, the continuity correction should be made in the *down*ward direction. The next possibility down is 16 out of 25, which would give $p = 0.64$. Halfway between this and 0.68 is 0.66. Thus,

$$z = \frac{0.66 - 0.5}{\sqrt{\dfrac{(0.5)(0.5)}{25}}}$$

$$z = 1.60$$

$$A = 0.4452$$

$$P\text{-val} = 0.0548 > 0.05\ (= \alpha)$$

Fail to reject H_0.

On the basis of this sample's results, the magazine could be correct in its claim that teenagers watch no more than 20 hours a week of TV.   ●

Exercise Set 9–1

1. Here is some data that is the result of examining a random sample.

   | 28 | 31 | 34 | 38 | 27 | 33 | 33 | 35 | 32 | 26 | 21 | 28 | 29 | 36 |
   | 38 | 37 | 35 | 29 | 34 | 39 | 27 | 31 | 35 | 31 | 28 |

   Use this to test the hypothesis that the mean of the population from which the sample was drawn is 30. Use a 5% significance level and the sign test.

For exercises 2 and 3, you will need to be aware of the marginal note next to the solution of example 9–1.

2. Test the claim that the mean of a population is at least 12, using the sign test and a 2% significance level if the sample results are

   | 8 | 10 | 12 | 13 | 7 | 5 | 6 | 4 | 8 | 11 | 10 | 13 | 14 | 11 | 10 |
   | 9 | 12 | 9 | 9 | 10 | 15 | 14 | 3 | 7 | 8 | 19 | 18 | 9 | 9 | 7. |

 3. To test the claim that the average high temperature in a Southern California city during October is at least 71° F, a meteorologist observes the temperature on 15 randomly chosen days during October getting

   | 79 | 78 | 74 | 68 | 72 | 70 | 77 | 76 | 72 | 67 | 71 | 69 | 70 | 66 | 78. |

   Do you think the claim is correct at the 5% significance level?

4. The meat department of a supermarket sells hamburger meat in packages marked "32 oz." To test whether consumers are being treated fairly, a random sample of these packages are selected and the meat weighed, with these results.

   | 31.8 | 32.1 | 31.5 | 32.2 | 31.3 | 32.4 | 31.8 | 31.9 | 31.6 | 32.2 | 31.3 |
   | 31.4 | 30.9 | 31.7 | 32.6 | 31.9 | 31.8 | 32.1 |

   Test to see if the meat department is selling customers the amount of meat it says it is. Use $\alpha = 0.05$.

5. A factory produces nails in boxes of 200 each. The manufacturers are anxious not to have more than an average of 5% defective nails per box. To test this a random sample of 12 boxes is examined, and the number of defectives in each box was found to be

15   3   0   2   19   16   11   8   20   3   5   12.

Use the sign test to see if the boxes being turned out conform to the manufacturer's guideline at the 10% level.

 6. A secretarial school advertises that, in a two-month course, it can train students to type at least 50 words per minute. To test this claim, a random sample of 30 students who have completed the course are timed, with these results.

52   62   48   57   40   45   38   52   47   49   52   47   41   46
48   48   36   45   52   41   48   49   34   68   45   61   42   48
40   52

Is the school being as effective as it claims it is? Use $\alpha = 0.05$.

 7. A random sample of adults is assumed to dream 110 minutes per night measured by a number of measures including rapid eye movements. This same sample was asked to smoke a marijuana cigarette before they went to sleep to test whether this stimulus increased or decreased dream time. The same measures were used to produce the following data:

115   120   100   125   90   112   123   100   114   130   128   108.

What conclusion can you draw at the 5% significance level?

8. If a testing problem has the hypotheses $H_0 : \mu = \mu_H$, $H_1 : \mu > \mu_H$ and a random sample of size 10 and the sign test is used, what is the smallest number of the data values in the sample that should be larger than μ_H that would allow you to reject H_0 at the 5% significance level?

Section 9–2   The One-Sample Wilcoxon Test

The sign test's big advantage is the ease with which it can be carried out. However, it is not a very satisfactory test. Suppose in example 9–1 that the 17 plus signs all corresponded to values in the 30s, while the 8 minus signs corresponded to values at 18 or 19. Wouldn't you, then, be more inclined to reject the magazine's claim? Yet the arithmetic involved would be *exactly* as it is above, and, on the basis of the sign test, the same conclusion of *failing* to reject H_0 would be reached. Statisticians and others, therefore, searched for an alternative to the sign test. It was in 1945 that a chemist, Frank Wilcoxon, discovered such an alternative, and it was, naturally, named after him.

The principle behind the one-sample Wilcoxon test is to give a heavier emphasis to those values in the sample that are further from μ_H than those that are closer. The process by which this is accomplished is to rank the absolute value of the differences between the sample values and μ_H. If H_0 is true and the population is symmetric, the sum of the ranks corresponding to values in the sample above μ_H should be approximately equal to the sum of the ranks corresponding to values in the sample below μ_H. If these two sums are substantially different, it may be an indication to reject H_0. If they are close to each other, this would be an indication to fail to reject H_0.

EXAMPLE 9-2   Rework example 9-1 using the Wilcoxon test.

Solution

$$H_0 : \mu = 20$$

$$H_1 : \mu > 20$$

$$\alpha = 0.05$$

Table 9-1

x	$|x - \mu_H|$	Rank	Sign of Rank
21.2	1.2	2	+
18.6	1.4	3	−
15.3	4.7	13	−
25.5	5.5	16	+
19.1	0.9	1	−
24.1	4.1	11	+
29.7	9.7	23	+
10.6	9.4	22	−
12.3	7.7	20	−
27.4	7.4	18	+
26.5	6.5	17	+
22.3	2.3	6	+
18.1	1.9	4	−
9.9	10.1	24	−
22.8	2.8	7	+
23.5	3.5	9	+
25.2	5.2	14	+
23.7	3.7	10	+
25.3	5.3	15	+
27.6	7.6	19	+
31.2	11.2	25	+
24.5	4.5	12	+
16.7	3.3	8	−
29.1	9.1	21	+
22.2	2.2	5	+ .

The creation of table 9-1 is as follows: All the sample values are listed under x, then μ_H ($= 20$) is subtracted from each of these values and the *sign is dropped* to determine the "absolute value" of each difference. These differences are then ranked with 1 going to the smallest, 2 to the next smallest, and so on, until 25 is assigned to the largest (25, here, because the size of the sample $n = 25$). The last column is created in the same way as for the sign test: + if x is above μ_H, and − if it is below.

Next, the sum of the ranks corresponding to the pluses is computed and labeled V^+. Check for yourself that, in this example, $V^+ = 230$. Similarly, for the minuses, label the sum of ranks as V^-; here $V^- = 95$. These sums, V^+ and V^-, are different, but are they significantly different? To decide this for samples of $n \leq 10$, you can refer to a set of tables, one for each value of n, giving P-values associated with the smaller of the two sums V^+ and V^-. When n is larger than 10, a very accurate P-value can be obtained by using a normal approximation; for these larger samples, V^+ and V^- each follow a normal distribution with

$$\left. \begin{array}{c} \text{Mean of } V = \dfrac{1}{4}n(n + 1) \\[2mm] SD_V = \sqrt{\dfrac{1}{24}n(n + 1)(2n + 1)} \end{array} \right\} . \qquad \textbf{(44)}$$

Figure 9–5

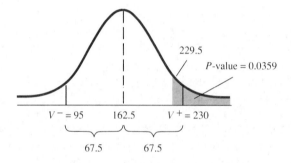

These formulas are derived with the help of college algebra, and you are again asked to accept them on trust. With their help, you can convert either V^+ or V^- to a z-score and use the normal table to obtain a P-value. For example 9–2,

$$\text{Mean of } V = \frac{1}{4} \cdot 25 \cdot 26 = 162.5$$

$$SD_V = \sqrt{\frac{1}{24} \cdot 25 \cdot 26 \cdot 51} \approx 37.2.$$

Notice that $V^+ = 230$ is 67.5 above the mean, while $V^- = 95$ is 67.5 below the mean (see figure 9–5). It makes no difference, therefore, which you convert to a z-score. The same value, but opposite in sign, will be obtained. Since the sign of z is irrelevant to the P-value, it does not matter which V you use.

However, since the values of V will be discrete, you do have to use the continuity correction. For V^+, which is above the mean, correct down. Thus,

$$z = \frac{229.5 - 162.5}{37.2}$$

$$z \approx 1.80$$

$$A = 0.4641$$

P-val $= 0.0359 < 0.05 \ (= \alpha)$. (See figure 9–5.)

The Wilcoxon test indicates that, with less than 5% probability of committing a type I error, the magazines's claim is incorrect. •

Here is a summary of the **one-sample Wilcoxon test.**

---

1. The population must be assumed to be symmetric.
2. The absolute values of the differences between the sample values and μ_H are computed and ranked.
3. The sum of the ranks corresponding to positive differences is computed and called V^+; or the sum of the ranks corresponding to negative differences is computed and called V^-.
4. If a Wilcoxon table is available, a P-value can be obtained from the smaller of V^+, V^-, and n. If such a table is not available, but $n > 10$, either V^+ or V^- can be converted to a z-score by using the formula

$$z = \frac{V - \text{mean of } V}{SD_V},$$

where mean of V and SD_V are found from the formulas at (44) and V is corrected for continuity. The test can then be completed using the normal table.

The ranking process is performed from the smallest to the largest. If there are any ties, they are treated as was discussed in chapter 2 on page 88. However, if there are many ties, then the formula for SD_V at (44) becomes unreliable. Therefore, when you intend to use the Wilcoxon test, and you are studying continuous data, the number of ties should be minimized by measuring as accurately as possible. By measuring weights to the nearest ounce, for example, rather than merely the nearest pound, it is less likely that any two values (or differences from μ_H) will be the same, so ties in rank are less likely as well.

Notice that using the same data, the Wilcoxon test and the sign test resulted in different decisions regarding the null hypothesis. For this sample, with no ties in the rankings, the Wilcoxon test is more dependable than the sign test.

You now have *five one-sample tests* in your arsenal for testing a hypothesis about a mean. They are: the z-score test using σ, the z-score test using s, the t-test, the sign test, and the Wilcoxon test. Each is appropriate under certain conditions. However, there is still one set of circumstances where none of them is appropriate. This is when σ is unknown, n is smaller than 30 with a larger sample unavailable, and the population is believed to be skewed. A possible solution to this problem is to reword the hypotheses to be about $\tilde{\mu}$, the median of the population, rather than the mean, and use the sign test or Wilcoxon test. It is always true, even for skewed distributions, that P, the proportion of values above the *median*, is 0.5, by the *definition* of median.

Exercise Set 9–2

1. If $n = 10$, $V^+ = 8$, a Wilcoxon table of P-values gives that the P-value $= 0.0244$. What is the value of the P-value obtained by using the normal approximation?

2. The exact P-value for $n = 12$, $V^- = 9$ is 0.008. What is the approximate P-value obtained by using the normal approximation?

 3. A farmer is currently raising a mean of 100 units of a crop per acre. He wants to test the effect of using a growth hormone and applies this to a random sample of 15 acres, obtaining these yields.

    108   120   95   102   90   105   110   92   122   109   95   112
    114   107   97

Do these results indicate that the hormone is effective in increasing growth at the 5% significance level?

 4. At the beginning of a college algebra class, the instructor gives the students a test of their knowledge of intermediate algebra. The usual mean on this test is 75. Here are the results for this class.

    69   78   79   84   89   100   92   75   64   83   91   60   68   82
    89   78   76   80   81   66   74   77   81   79   70   76

Does this data show that this is an above average class at the 5% significance level?

 5. It is believed that, in a certain city, people who buy new cars keep them for a mean length of $3\frac{1}{2}$ years. An association of automobile dealers would like to see if recent set-backs in the economy have caused new-car buyers to keep their cars longer. By inspecting a random sample of records on new-car buyers, the following data on the lengths of time new cars are kept is obtained.

    3.6   3.0   2.8   3.9   4.2   3.4   4.8   4.7   4.0   3.8   3.3   3.4   4.6
    4.1   4.2   4.5   4.8   5.2   3.7   3.2

Test the hypothesis at the 0.02 significance level.

6. A random sample of women admitted to a hospital over the past year to have their first babies were reported to have these ages.

19   24   23   20   21   19   25   18   26   29   30   28   24   17

Is this data consistent with the claim that the mean age of women in the community served by the hospital is no more than 21 when they have their first babies? Use $\alpha = 0.05$.

7. When using the Wilcoxon one-sample test with a sample of size 12, what is the smallest possible value of V^+ before applying the continuity correction and assuming no ties in the ranks that would produce significance at the 5% level? Assume a one-tailed test and give the exact P-value that this value of V gives.

8. Answer the question in exercise 7 if the sample size is 18 and a 1% significance level is used.

Section 9–3 The Two-Sample Wilcoxon Test

Frank Wilcoxon also proposed a nonparametric alternative to the two-sample t-test that, like the one-sample Wilcoxon test, uses ranks. Two years after Wilcoxon's initial idea, H. B. Mann and D. R. Whitney helped develop this test in its final form, which is sometimes referred to as the **Mann–Whitney test.**

Like the two-sample t-test, the two-sample Wilcoxon test is used to test hypotheses of the form $H_0 : \mu_1 - \mu_2 = 0$. Here, as there, two assumptions must be satisfied to assure reliability of the test:

1. The two populations under study should have the same shape, regardless of what that shape is.
2. The two populations under study should have the same amount of dispersion, that is, $\sigma_1 = \sigma_2$.

The first assumption is much less strict than the first assumption for the two-sample t-test. To be able to use the two-sample Wilcoxon test, both populations may have a normal distribution, or they may both have a uniform distribution, or they may both be right-skewed, and so on. Since the second assumption is exactly the same as the second assumption for the two-sample t-test, the two-sample Wilcoxon test is always applicable when the two-sample t-test is, but there are other situations when it is applicable and the two-sample t-test is not.

The second assumption may seem to you to be a parametric assumption, since it involves the parameters σ_1 and σ_2. However, just as the first assumption simply states that the shapes should be the same, without specifying the shapes, the second assumption states that σ_1 and σ_2 should have the same size, without specifying the size.

The null hypothesis for this test of the difference between two means states that $\mu_1 = \mu_2$. Therefore, if H_0 is true and assumptions 1 and 2 apply, then the two populations have the same mean, the same shape, and the same standard deviation. If these three facts are true, then the two populations may as well be one. This is the key fact that should be borne in mind when applying the two-sample Wilcoxon test: Under the null hypothesis, the two random samples that are drawn from two populations can be considered as though they were drawn from the same population.

Suppose you have also pooled the two samples and ranked them as if they formed a single combined sample. Each of the samples should be expected to have some high and some low ranks. The sums of the ranks for each of the two samples should then turn out to be roughly equal if $n_1 = n_2$, and otherwise proportionate to their respective sample sizes. The more different these sums turn out to be, the less likely it is that the null hypothesis is true; the closer they are, the more inclined you would be to accept the null hypothesis.

The two-sample Wilcoxon test can also be used when two random samples are independently drawn from the same population and then subjected to different treatments. The two samples become the control and experimental groups. Consider an example illustrating how the two-sample Wilcoxon test works.

Figure 9–6

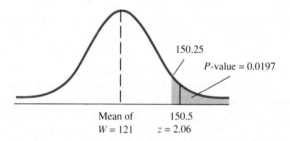

EXAMPLE 9–3   Redo example 8–9 in chapter 8 concerning anorexia nervosa patients. Usually, weight gains are skewed rather than normally distributed. There is no reason, however, to believe that weight gains would be skewed differently according to treatment. Thus, the Wilcoxon two-sample test might be appropriate here.

Solution

Table 9–2 shows the individual weight gains together with their ranks.

Table 9–2

Control Group		Experimental Group	
Gain	*Rank*	*Gain*	*Rank*
15.7	16	20.2	21
8.5	1	9.8	3
10.0	4	10.5	6
11.9	7	16.5	17
13.2	10	17.9	19
14.5	13	14.2	12
10.2	5	13.3	11
12.3	8.5	15.4	15
9.0	2	18.3	20
14.7	14	12.3	8.5
		16.6	18

The sum of the ranks in the first sample is 80.5 and in the second, 150.5. You would expect the first sample to have a smaller sum since the sample is smaller; but would you expect it to be that much smaller? This can be tested using a normal approximation as was done for the one-sample Wilcoxon test. If W is the sum of the ranks of the *second* sample, then the mean and standard deviation of W are given by

$$\left. \begin{aligned} \text{Mean of } W &= n_2\left(\frac{n_1 + n_2 + 1}{2}\right) \\ SD_W &= \sqrt{n_1 n_2 \cdot \left(\frac{n_1 + n_2 + 1}{12}\right)} \end{aligned} \right\} . \qquad (45)$$

Then W, again corrected for continuity, can be converted to a z-score in the usual way, and a P-value can be found using the normal table.

For this example, $W = 150.5$ and

$$\text{Mean of } W = 11\left(\frac{10 + 11 + 1}{2}\right) = 121$$

$$SD_W = \sqrt{10 \cdot 11 \left(\frac{10 + 11 + 1}{12}\right)} \approx 14.2.$$

As you can see, W, for this example, is larger than the mean of W averaged across both samples. This is the correct direction for possible rejection of H_0, since you would like to reject H_0 if the weight gains due to the new method of treatment are significantly more than with the existing treatment. If W had been smaller than the mean of W, you could have stopped immediately. You are testing to see if the new treatment is an improvement on the old, not vice versa. It is always a good idea, at this point, to see if W is on the correct side of its mean for possible rejection. Of course, with a two-tailed test, it doesn't matter. Also, the fact that W is larger than its mean, indicates that the continuity correction should be made in the downward direction. Had there been no ties, the next possible value for W below 150.5 is 150. Hence, the continuity corrected value for W is 150.25 and (see figure 9–6)

$$z = \frac{150.25 - 121}{14.2}$$

$$z \approx 2.06$$

$$A = 0.4803$$

$$P\text{-val} = 0.0197 < 0.05 \ (= \alpha)$$

Reject H_0.

The same conclusion has been reached as with the two-sample t-test, although, here, the P-value is slightly larger.   ●

As with the one-sample Wilcoxon test, if there are too many ties, the two-sample Wilcoxon test becomes unreliable. So accurate measurement is highly desirable when using continuous data with this test. When there are few or no ties, this test is to be preferred over the two-sample t-test, unless you are absolutely sure that the populations are normally distributed. It is conceivable that different decisions could result depending on which test is chosen. An unscrupulous researcher may run both tests and present only the one that "proves" the result desired. An ethical researcher should choose the test in *advance* of collecting the sample data, and the choice should be based on which assumptions it is felt can be legitimately made, and *not* based on the outcome.

Here is a summary of the **two-sample Wilcoxon test.**

---

1. The assumptions for its use are that both populations have the same shape and the same amount of dispersion.
2. The two random samples are ranked as if they were one sample.
3. The sum of the ranks of the second sample is computed and assigned to W. At this point, make sure W is on the correct side of its mean.
4. W is converted to a standard score using

$$z = \frac{W - \text{mean of } W}{SD_W},$$

where mean of W and SD_W are found from the formulas at (45) and W is corrected for continuity. The test can then be completed using the normal table.

The ranking process is performed from the smallest to the largest. Ties are treated as before. However, if there are many ties, the test becomes unreliable.

Exercise Set 9–3

1. If a two-sample Wilcoxon test is applied to samples of size $n_1 = 16$, $n_2 = 18$, find the mean of W and SD_W.

2. If a two-sample Wilcoxon test is applied to samples of size $n_1 = 13$, $n_2 = 19$, find the mean of W and SD_W.

3. A local convenience store would like to test the effect of a display on selling a certain item. From the next 6 months, the manager chooses 10 weeks at random to put on the display and 14 weeks when the display is not put on. Here are the weekly sales results.

Without display: 72 90 84 65 83 70 75 65 55 62 63
71 80 91

With display: 80 70 95 108 100 72 85 75 69 101

Does the display appear to have the desired effect of increasing sales if a 5% level of significance is used?

4. Fishermen can buy various strengths of fishing line. For instance, "6 lb test line" means it can stand a force of at least 6 lb before breaking. To see if two different brands of 6 lb test line are significantly different, two samples of the line, one from each brand, were tested with these results.

Brand A: 6.3   5.7   6.8   5.4   6.2   6.7   6.1   6.0   6.6   5.8   5.9

Brand B: 6.1   6.4   5.7   6.9   7.2   7.0   6.3   6.5   6.8

Is there a significant difference between the brands at the 1% significance level?

5. To see if supermarket shoppers spend more when they are hungry, a random sample of 30 shoppers were randomly divided into two groups of 15 each. One group was given a big meal, while the other group was left hungry. Both groups were then given identical lists and told they could buy anything else they thought would be a good buy. Here is the data that consists of the bills of the 30 shoppers.

Hungry shoppers: 52.95   59.70   49.84   65.18   66.24   72.18   54.16   55.00
48.81   52.16   54.50   63.80   60.05   58.20   50.18

Full shoppers: 53.17   48.12   45.14   55.62   54.21   40.30   42.40   52.86
53.14   62.12   47.81   45.26   53.25   51.60   47.18

What can you conclude at a 5% significance level?

6. During the controversy over busing in a city, it was desired to determine whether black children in integrated schools did better than those in segregated schools. A preliminary analysis consisted of taking random samples of 20 black children from a segregated school and 24 from an integrated school and giving them a standardized test under the same conditions. Here are the results.

From segregated school: 525   564   510   492   508   490   482   532   546   471
483   491   501   504   481   472   503   522   509   515

From integrated school: 518   560   574   530   565   550   541   535   521   508
492   538   481   493   521   581   572   502   500   540
544   539   520   519

What can you conclude at a 1% significance level?

7. If it is intended to apply a two-sample Wilcoxon test to random samples with sizes $n_1 = 4$, $n_2 = 3$, what is the smallest P-value that could be obtained?

8. If it is intended to apply a two-sample Wilcoxon test to random samples with sizes $n_1 = 6$, $n_2 = 8$, what is the smallest P-value that could be obtained?

9. In the article at the beginning of the chapter, of 40 students with a mean GPA of about 2.7, almost half were informed, in various ways, that grades usually improved in their sophomore year. The others, forming the control group, remained uninformed. One year later, all 40 students' GPA's were reexamined. Here are some hypothetical figures that fit the facts of the article.

GPA's in Sophomore Year

Informed students: 2.18 2.25 2.36 2.42 2.59 2.67 2.78 2.82 2.91 2.95
                     3.04 3.12 3.13 3.14 3.26 3.33 3.45 3.48 3.49 3.52
                     3.54 3.60

Uninformed students: 1.88 1.97 2.29 2.37 2.45 2.56 2.58 2.65 2.73 2.77
                     2.79 2.82 2.83 2.85 2.95 3.02 3.11 3.15

Conduct a two-sample Wilcoxon test on this data to test whether the informed students have a significantly higher mean GPA than the uninformed students at a 0.01 level.

10. Repeat exercise 9, using a two-sample *t*-test to see if a similar conclusion is obtained.

Section 9–4   The Runs Test

Much emphasis has been placed on the necessity for having *random* samples when performing any kind of statistical inference. How can you be sure, especially if you have not been involved in the selection of the sample, that a sample in front of you is random? Sometimes a sample consists of only two different values of the response variable, such as heads and tails when tossing a coin. If the sample size were 50 and you were presented with 25 heads followed by 25 tails, or, at the other extreme, strictly alternating heads and tails, would you not be a little suspicious of its purported randomness? With a run of 25 heads followed by a run of 25 tails, you would have exactly 2 runs. With strictly alternating heads and tails, you would have 25 separate runs of heads and 25 separate runs of tails, giving a total of 50 runs. These two extreme examples should demonstrate to you that too few runs or too many runs should cause you to suspect the purported randomness of the sample. In the example of 50 tosses of a coin, a truly random sample would have neither as few runs as 2 nor as many runs as 50, but rather a number of runs somewhere between these extremes.

A very simple test has been devised, called the **runs test,** that enables you to put the alleged randomness of a sample to the test. If you are examining a sample involving only two characteristics, the number of times the first characteristic occurs, n_1, the number of times the second characteristic occurs, n_2, and the total number or runs, R, can be counted. If n_1 and n_2 are both less than 20, tables are available that tell you which two values R should be between, given the values of n_1 and n_2, so that the sample can be considered to be random at a given significance level. As an illustration, one such table, for a 5% level of significance, is given in the appendix. It is table 4. A separate table would be necessary for each different level of significance, but space limitations prevent the inclusion of these.

Example 9–4 illustrates an application of the runs test for qualitative data with only two possible values. As examples 9–5 and 9–6 illustrate, the runs test can also be used to assess the randomness of samples involving continuous, numerical data. In the latter application, n_1 is the number of observations below the median; unless there are ties at the median, $n_1 = n_2$.

EXAMPLE 9-4   Suppose you are shown this sample of 15 tosses of a coin: T, H, T, T, H, H, H, H, H, H, H, T, T, H, T. Is it random?

Solution   There are 9 heads and 6 tails. Take the smaller to be n_1 and larger n_2. Therefore, $n_1 = 6$, $n_2 = 9$. Check for yourself that the number of runs is 7. Thus, $R = 7$. Now look at the table. For $n_1 = 6$, $n_2 = 9$, find the box

$\boxed{\begin{array}{c} 4 \\ 13 \end{array}}$. This means that if 4 or fewer or 13 or more runs are observed, then

the hypothesis of randomness will be rejected at the 5% significance level. Since 7 runs appear in the sample, conclude that the sample could indeed be random.   ●

The runs test can also be used with quantitative data. If discrete data is being studied, the data can be arbitrarily divided into two categories such as odd and even, and the number of odds, number of evens, and total number of runs of odd and even counted. If the data is continuous, the usual method is to divide the data into the two categories above and below the median and to conduct the runs test with these categories.

EXAMPLE 9-5   Use the runs test to test the randomness of the first sample of example 9–3.

Solution   The sample in order of size is: 8.5, 9.0, 10.0, 10.2, 11.9, 12.3, 13.2, 14.5, 14.7, 15.7. The median is, therefore, given by

$$\tilde{x} = \frac{11.9 + 12.3}{2} = 12.1.$$

The sample in original order is: 15.7, 8.5, 10.0, 11.9, 13.2, 14.5, 10.2, 12.3, 9.0, 14.7. If a and b are used for above and below the median, respectively, then this translates into

$$a, b, b, b, a, a, b, a, b, a.$$

There are 5 above and 5 below. Thus, $n_1 = 5$, $n_2 = 5$, and the relevant box in

the table is $\boxed{\begin{array}{c} 2 \\ 10 \end{array}}$. Since, here, $R = 7$, the hypothesis of randomness cannot be

rejected at the 5% significance level.   ●

It is important to realize that not rejecting the hypothesis of randomness does *not* prove that the sample is random because there may be some other reason, apart from runs, why it is not random. You have more of a "proof" when you *can* reject the hypothesis of randomness as a result of the runs test. Then, at least at a certain significance level, you can express your suspicion of the alleged randomness of the sample.

If the sample is of such a size that either n_1 or n_2, or both, is larger than 20, a normal approximation, such as the ones used for the Wilcoxon tests, is available. To use the approximation, you will need the following formulas:

$$\left. \begin{aligned} \text{Mean of } R &= \frac{2n_1 n_2}{n_1 + n_2} + 1 \\[2ex] SD_R &= \sqrt{\frac{(2n_1 n_2)(2n_1 n_2 - n_1 - n_2)}{(n_1 + n_2)^2(n_1 + n_2 - 1)}} \end{aligned} \right\} . \qquad (46)$$

With the actual R corrected for continuity, you can then test using

$$z = \frac{R - \text{mean of } R}{SD_R},$$

finding the *P*-value from the normal table in the usual way.

EXAMPLE 9–6   Return to the sample of 38 female student weights in chapter 1. The weights were allegedly drawn in this order.

130	108	135	120	97	110	130	112	123	117	170	124	120
133	87	130	160	128	110	135	115	127	102	130	89	135
87	135	115	110	105	130	115	100	125	120	120	120	

Use the runs test to test the randomness of this sample.

Solution   In chapter 2, the median of this sample, using the frequency distribution, was found to be 120.75. Using a and b again, the sample translates into

a, *b*, *a*, *b*, *b*, *b*, *a*, *b*, *a*, *b*, *a*, *a*, *b*, *a*, *b*, *a*, *a*, *a*, *b*, *a*, *b*, *a*, *b*, *a*, *b*, *a*, *b*, *a*, *b*, *b*, *b*, *a*, *b*, *b*, *a*, *b*, *b*, *b*.

There are 17 above and 21 below the median. Thus, set $n_1 = 17$ and $n_2 = 21$. Since n_2 is larger than 20, table 4 cannot be used. Also, here, $R = 28$. Using the formulas at (46),

$$\text{Mean of } R = \frac{2 \cdot 17 \cdot 21}{17 + 21} + 1 \approx 19.8$$

$$SD_R = \sqrt{\frac{(2 \cdot 17 \cdot 21)(2 \cdot 17 \cdot 21 - 17 - 21)}{(17 + 21)^2(17 + 21 - 1)}} \approx 3.0.$$

Since the actual value of R is above the mean of R, the correction for continuity is made downward to 27.5 giving

$$z = \frac{27.5 - 19.8}{3.0}$$

$$z \approx 2.57$$

$$A = 0.4949$$

$$P\text{-val} = 0.0051.$$

The sample of weights was actually a random sample, but the order of drawing was not in the order given. It was altered so that you could see a runs test that led to the rejection of the hypothesis of randomness.

This is a highly significant P-value and should lead you to the rejection of randomness at a 5%, or even a 1%, significance level. ●

In summary, the runs test is a test for the randomness of a sample. Like all tests discussed in this chapter, it is a nonparametric test, since there is no assumption made about the population's shape. It can be used whenever the sample is, or can be, split up into two characteristics, such as heads/tails, odd/even, or above/below the median. Failing to reject the hypothesis of randomness does not necessarily enable you to conclude that a sample *is* random, for there might be other reasons for its lack of randomness. Rejecting the hypothesis of randomness due to an application of the runs test, however, is cause to suspect the sample's randomness, provided you are sure that you have the sample in the order in which it was drawn.

Exercise Set 9–4

Use the runs test and a 5% significance level to test the randomness of the samples given in exercises 1–6, assuming the samples are presented in the order drawn.

1. Tossing a coin: H H H T T T T T H H H H H T T T T
   T T T

2. Rolling a die: 3 6 1 2 4 1 6 5 4 6 5 2 3 1 4 5 2 3
   4 5 2 4 3 6 4 1 6 2
   Use odd/even as the criterion.

3. Continuous data: 5.7 11.4 10.3 3.2 7.6 12.2 7.7 2.1 2.8 11.7
   10.0 9.0

4. Gender: M F M F M F F F M M F F M F F M F M F M
   F F M F F

5. Continuous data: 13.1 20.2 19.2 24.6 10.4 11.9 17.5 21.4 23.5 11.2
   5.3 7.8 8.6 20.7 23.3 17.0 14.8 16.1 14.6 13.6
   25.2 21.5 19.5 3.8 8.3 7.2 13.3 16.5 27.2 17.9
   11.3 10.2 5.0 20.3 21.6 21.2 17.3 19.1 8.5 11.0
   14.2 16.9 13.8 19.4

6. Continuous data: 29.3 31.5 20.5 24.6 31.8 32.6 33.3 29.5 29.8 28.5
27.4 27.0 27.3 25.2 22.3 24.9 26.5 28.1 30.0 32.3
35.5 24.4 28.2 27.4 26.2 24.8 29.1 32.7 33.4 32.4
27.5 27.6 26.3 22.5 26.9 28.0 28.7 31.7 28.3 29.4
30.5 25.5 22.1 27.1 27.6 28.8 27.8 26.7 25.1 24.2

7. When performing a runs test, it is found that $n_1 = 31$, $n_2 = 35$. What are the smallest and largest values of R that would permit rejection of the hypothesis of randomness at the 5% significance level?

8. When performing a runs test, it is found that $n_1 = 18$, $n_2 = 40$. Answer the question posed in exercise 7, but using a 1% significance level.

···

Section 9-5   Summary of Terms

Mann–Whitney test 309	runs test 314
nonparametric tests 299	sign test 299
one-sample Wilcoxon test 307	two-sample Wilcoxon test 312

Section 9-6   Summary of Symbols

Symbol	Meaning
a	Above the median in the application of the runs test
b	Below the median in the application of the runs test
R	Number of runs
SD_R	Standard deviation of the frequency distribution of R
SD_V	Standard deviation of the frequency distribution of V
SD_W	Standard deviation of the frequency distribution of W
V	A sum of ranks in the one-sample Wilcoxon test
V^+	The sum of positive ranks
V^-	The sum of negative ranks
W	The sum of ranks of the second sample in the two-sample Wilcoxon test

Section 9-7   Summary of Formulas

Formula   Use

(44) Mean of $V = \dfrac{1}{4}n(n + 1)$

$SD_V = \sqrt{\dfrac{1}{24}n(n + 1)(2n + 1)}$

Formulas for use in the one-sample Wilcoxon test

(45) Mean of $W = n_2\left(\dfrac{n_1 + n_2 + 1}{2}\right)$

$SD_W = \sqrt{n_1 n_2\left(\dfrac{n_1 + n_2 + 1}{12}\right)}$

Formulas for use in the two-sample Wilcoxon test

(46) Mean of $R = \dfrac{2n_1n_2}{n_1 + n_2} + 1$

$$SD_R = \sqrt{\dfrac{(2n_1n_2)(2n_1n_2 - n_1 - n_2)}{(n_1 + n_2)^2(n_1 + n_2 - 1)}}$$

> Formulas for use in the runs test when either or both n_1, n_2 are larger than 20

Chapter 9 Review Exercises

In an investigation of the age when human infants first learn to walk on their own, samples of infants were chosen for the experiment. The control group consisted of infants who were left to their own devices, while the experimental group consisted of infants who were given special exercises and encouragement. The age (rounded to the nearest tenth of a month) when each child first started to walk without assistance was then recorded. Allegedly random samples were then drawn from each group with these results.

Control group: 12.3   11.3   11.7   13.2   14.5   10.6   11.8
            12.6   12.9   14.0   13.4   12.2   10.4   12.7

Experimental group: 11.4   11.0   10.3   10.8   12.2   12.5   9.0
            10.9   11.6   12.1   13.2   10.6   9.2   13.6   12.8

 1. Use the runs test to test the control group for randomness.

 2. Use the runs test to test the experimental group for randomness.

 3. Use the data of the control group and the sign test to test the claim that children left to their own devices will begin walking on their own only after their first birthdays. Use $\alpha = 0.05$.

 4. Use the data of the experimental group and the sign test to test the claim that children who are given special exercises and encouragement will begin walking before their first birthdays. Use $\alpha = 0.05$.

 5. Repeat exercise 3 using the one-sample Wilcoxon test.

 6. Repeat exercise 4 using the one-sample Wilcoxon test.

7. Use a two-sample Wilcoxon test to see if special exercises and encouragement will lead to a lower age when children start walking alone. Use $\alpha = 0.05$.

8. To see if the same or a different conclusion is obtained, repeat exercise 7 using a two-sample t-test.

Hands-on Class/Student Project

Follow similar directions as found for the corresponding project exercises in previous chapters for these exercises.

1. Retrieve your sample of male GPA's that you considered in project exercises 1 and 2 of chapter 8 and perform a runs test on it to check randomness.

2. Retrieve your sample of female GPA's that you considered in project exercises 1 and 2 of chapter 8 and perform a runs test on it to check randomness.

3. Perform a sign test on your sample of male GPA's to test the hypothesis that the mean male GPA at your college is at least 2.0 using a 5% significance level.

4. Perform a sign test on your sample of female GPA's to test the hypothesis that the mean female GPA at your college is at least 2.0 using a 5% significance level.

5. Repeat exercise 3 using the one-sample Wilcoxon test.

6. Repeat exercise 4 using the one-sample Wilcoxon test.

7. Set up hypotheses to test the claim that males on your campus have owned, on the average, more different cars than females.

8. Using your data from chapter 1 (variable 11), test the hypotheses you set up in exercise 6 using a 5% significance level and the two-sample Wilcoxon test.

9. Compare your conclusion to exercise 8 with those of other groups in the class.

10. Repeat project exercise 10 of chapter 8, but this time use a two-sample Wilcoxon test.

10

The Chi-Square Test

"The art of being wise is the art of knowing what to overlook."

WILLIAM JAMES

Law

The Curious Case of
The Mitchell-Stans Trial

The 1974 trial of the *United States v. John Mitchell and Maurice Stans* was replete with oddities and contradictions, including the verdict: acquittal for the two former Cabinet members on charges of conspiracy to impede the investigation of financier Robert Vesco.

Legal scholar Hans Zeisel and psychologist Shari Seidman Diamond called the verdict "an unlikely event" because it fooled the odds. Research finds that a jury's first ballot is a reliable indication of the final verdict. In the Mitchell-Stans trial, the first ballot was eight for conviction and four for acquittal, making acquittal a 19-to-one long shot (see table).

In the film *Twelve Angry Men,* juror Henry Fonda held out for acquittal against a first ballot that stood 11-to-one for conviction, and finally convinced the other jurors of the defendant's innocence. The Mitchell-Stans trial found its Fonda in the person of Andrew Choa, a vice-president of the First National City Bank. . . .

The 12 jury members were a bank teller, a steel cutter, a telephone installer, a Western Union messenger, a mailroom supervisor, a Post Office supervisor, a city-highway-department employee, a shipping-clerk foreman, a subway conductor, two elderly women (one retired and the other an insurance company clerk), and Choa, a bank vice-president.

Once the jury began to deliberate, Choa became its dominant force. According to Zeisel and Diamond, "he persuaded the jury at several points to have testimony as well as the judge's instructions reread.

We do not know what testimony, but we are told that these requests were drafted by Choa for the forelady's signature." Eventually, the unlikely happened. Eight jurors changed their minds and both defendants were acquitted on all counts.

Zeisel and Diamond believe the defense attorneys used the public-opinion survey effectively. The eight jurors who were persuaded to change their original guilty verdict were perfectly selected for such persuasion. The fact that they all came from the lower and middle social strata, the authors maintain, "made it easier for the one non-profile juror who came from the upper class and was at home in the world of high finance to explain to his fellow jurors that the evidence in the case did not warrant conviction."

If the trial were written up as a mystery story, it might be called *The Case of the 13th Juror.* But there's another good title, *The Case of the Short Alphabet.* Zeisel and Diamond unearthed the puzzling fact that 95 percent of the original 196-person jury panel had names starting with the letters A through J. The chance that this would happen in a list of names chosen at random from voter lists is about one in 100 million, according to the authors. While there is no reason to believe this mistake changed the outcome of the trial, it represented one more oddity in this curious case.   —Jack Horn

Hans Zeisel is emeritus professor of law and sociology, University of Chicago. Shari Seidman Diamond is assistant professor of criminal justice and psychology, University of Illinois at Chicago Circle. Their article appeared in the *American Bar Foundation Research Journal,* Vol. 1976, No. 1.

First Ballot and Verdict

Number of guilty votes on first ballot	0	1–5	6	7–11	12
Likelihood that verdict will be:					
Not guilty	100%	91%	50%	5%	—
Hung jury	—	7%	—	9%	—
Guilty	—	2%	50%	86%	100%
Total	100%	100%	100%	100%	100%
Share of cases	(12%)	(18%)	(4%)	(47%)	(19%)

Table adapted from *The American Jury,* Harry Kalven Jr., and Hans Zeisel, University of Chicago Press.

Chapter Overview
Making Multiple Comparisons

In previous chapters, when dealing with qualitative data, you have learned how to test hypotheses about proportions using *z*-score tests. However, these *z*-score tests are only applicable if the data can be classified into two, mutually exclusive categories, commonly designated by success and failure. Sometimes it is desired to test something about proportions of categories when the number of categories involved is more than two and there is no obvious way of reducing this number to just two.

When attempting to decide the fairness of a coin, a *z*-score test *can* be used since there are only two categories: heads and tails. What about deciding the fairness of a die? As you know, a die has 6 faces. Now, there may be some perfectly good reason why you suspect an abnormality due to one of the faces. If so, a *z*-score test might still be conducted using the two categories "that face" and "not that face." Alternatively, a *z*-score test might be run using pairs of categories such as odd and even; or 1,2,3, and 4,5,6. On the other hand, there may be no indication as to which face, if any, is suspect, and splitting the sample space into any kind of pairs may prove unsatisfactory as far as deciding the issue of the fairness of the die.

To determine whether a die is fair or loaded, without any prior information about which numbers are favored, requires a procedure capable of testing the relative frequencies of all 6 faces at the same time. None of the tests discussed thus far is adequate for this purpose. The subject of this chapter is the chi-square test, a test that is capable of testing proportions of more than two categories simultaneously.

Section 10-1 Deriving the Chi-Square Test

Chi is a Greek letter that looks like this: χ. The chi-square test was first derived by the English statistician Karl Pearson. Before considering the chi-square test, example 10–1 illustrates a problem that might be tackled by using it.

EXAMPLE 10–1   The distribution of the four human blood types, O, A, B, and AB, in the United States is approximately as follows:

Blood type	O	A	B	AB	Total
Proportion	45%	40%	11%	4%	100%

A hospital in a small town in the United States is anxious to have its blood bank well stocked and is interested in knowing whether the distribution of blood types in the community that it serves is the same as, or different from, the rest of the country. To determine this, a random sample of 120 residents of this community is selected, and their blood types determined. Here are the results.

Blood type	O	A	B	AB	Total
Number	41	59	17	3	120

To make a meaningful comparison between this sample's results and the distribution of blood types in the United States as a whole, the sample results need to be translated into proportions. Do this for the above sample.

Solution   Recall from chapter 5 that if p is the *proportion* of a category in a sample of size n, then np is the *number* of times the category occurred in the sample. To convert from number to proportion, it is necessary to divide by n, since $\dfrac{np}{n} = p$. To express this proportion as a percentage, merely multiply by 100. Thus, the *proportion* of persons in the sample with blood type O, for example, is $\frac{41}{120} \times 100\% = 34.17\%$. The above results from the sample could then be expressed thus:

Blood type	O	A	B	AB	Total
Proportion	34.17%	49.17%	14.17%	2.5%	100.01%

The reason that the total is not exactly 100%, as it should be, is due to round-off error occurring in the calculations.   ●

Notice that there is quite a difference between these sample proportions and the United States proportions; but are the differences significant enough to be able to conclude that the distribution of blood types in this community is substantially different from the United States as a whole?

To tackle this problem, Pearson's derivation of the χ^2-test is presented: Suppose the theoretical proportions are labeled $P_1, P_2, \ldots, P_c$, while the corresponding sample proportions are labeled $p_1, p_2, \ldots, p_c$. Here, c is the number of categories. In example 10–1, $c = 4$, because there are 4 blood types. For a χ^2-test, c can be as small as 3 or larger depending on the number of categories involved. Also, in example 10–1, $P_1 = 45\%$ while $p_1 = 34.17\%$, and so on for the rest of the proportions.

If you simply wanted to test whether the theoretical and sample proportions were significantly different for the *first* category, you could run a z-score test, which would mean calculating the following ratio:

$$z_1 = \frac{p_1 - P_1}{\sqrt{\dfrac{P_1 Q_1}{n}}},$$

and for the second category, the ratio would be

$$z_2 = \frac{p_2 - P_2}{\sqrt{\dfrac{P_2 Q_2}{n}}},$$

and so on, up to the *last* category, when the ratio would be

$$z_c = \frac{p_c - P_c}{\sqrt{\dfrac{P_c Q_c}{n}}}.$$

Some of these z's may be positive, others negative. Squaring each makes all the ratios positive and removes the square roots.

$$z_1^2 = \frac{(p_1 - P_1)^2}{\dfrac{P_1 Q_1}{n}}, \ z_2^2 = \frac{(p_2 - P_2)^2}{\dfrac{P_2 Q_2}{n}}, \ \ldots \ , \ z_c^2 = \frac{(p_c - P_c)^2}{\dfrac{P_c Q_c}{n}}$$

These expressions are called complex fractions, since there are two fraction bars. To simplify each expression, multiply the numerator and the denominator of the main fraction by n^2. (Since the numerator is a squared expression, n^2 may be used instead of n.) For z_1^2, for instance,

$$z_1^2 = \frac{n^2 (p_1 - P_1)^2}{n^2 \left(\dfrac{P_1 Q_1}{n} \right)}$$

or $\quad z_1^2 = \dfrac{(np_1 - nP_1)^2}{nP_1 Q_1} \quad$ since $a^2(b - c)^2 = (ab - ac)^2$ and one pair of n's divide out in the denominator.

Exactly the same thing can be done for each of the other z's. Now, if any of the sample proportions differs substantially from its corresponding theoretical proportion, a large value of its associated z will result, and this may lead to rejection of the hypothesis that states they are essentially the same. Alternatively, all of the sample proportions may differ slightly from the theoretical proportions, but the cumulative effect of these differences may warrant rejection of the null hypothesis. In either of these situations, a good measure of the differences in *all* pairs of proportions can be obtained by adding the z^2's.

$$z_1^2 + z_2^2 + \cdots + z_c^2 = \frac{(np_1 - nP_1)^2}{nP_1 Q_1} + \frac{(np_2 - nP_2)^2}{nP_2 Q_2} + \cdots + \frac{(np_c - nP_c)^2}{nP_c Q_c}$$

Pearson examined this expression and noticed that there are a lot of terms of the form np or nP that, as was indicated earlier, represent a *number* of successes (rather than a proportion). The expression np_1 represents the number of occurrences of the first category *observed* in the sample, while the expression nP_1 represents the number of occurrences of the first category *expected* in the sample if the theoretical proportion P_1 were correct. The same holds true for the other categories. What is spoiling things is the occurrence of a Q in the denominator of each of the fractions. At this point, Pearson had a stroke of genius. As the quotation at the beginning of the chapter says, he was wise enough to know what to overlook. By simply erasing the Q's, he discovered that the expression remaining followed a χ^2 distribution. Thus,

$$\chi^2 = \frac{(np_1 - nP_1)^2}{nP_1} + \frac{(np_2 - nP_2)^2}{nP_2} + \cdots + \frac{(np_c - nP_c)^2}{nP_c}$$

follows a χ^2 **distribution,** a known right-skewed probability distribution. The graph of a χ^2 distribution is shown in figure 10–1.

Figure 10–1

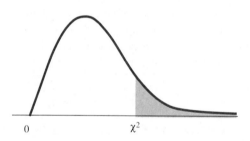

If all the sample proportions were equal to all of the corresponding theoretical proportions, χ^2 would equal zero. If there are any differences, χ^2 becomes positive (because of the squaring, χ^2 can never be negative). The larger the differences, the larger χ^2 becomes and, at some point, will cause the rejection of the null hypothesis. With suitable tables, the value of χ^2 can be translated into a P-value, which is the area of the tail shaded in figure 10–1. The smaller this P-value is, the more inclined will you be to reject the null hypothesis. As with previously discussed hypothesis tests, you can compare this P-value to a pre-chosen significance level to decide the issue.

Here O is for observed frequencies and is a letter, *not* a zero.

Look again at the expression for χ^2. There are no annoying Q's left. To make things clearer, the formula for χ^2 will be rewritten in a simpler way: Let $O_1, O_2, \ldots, O_c$ represent the **observed number** occurring in the sample for each category. Also, let $E_1, E_2, \ldots, E_c$ represent the **expected number** that would have occurred in the sample for each category had the theoretical proportions been exactly correct. Then, because np or nP represent the *numbers* O and E, the expression for χ^2 can be rewritten as

$$\chi^2 = \frac{(O_1 - E_1)^2}{E_1} + \frac{(O_2 - E_2)^2}{E_2} + \cdots + \frac{(O_c - E_c)^2}{E_c}.$$

This, in turn, can be rewritten as

$$\chi^2 = \Sigma\frac{(O - E)^2}{E}, \tag{47}$$

where the summation is taken over all categories. To obtain a value for χ^2 using formula (47), perform the following calculation for each category: Take the observed frequency (O), subtract the expected frequency (E), square the difference, and divide by E. Then, sum over all categories.

EXAMPLE 10-2   Calculate χ^2 for the data of example 10-1.

Solution

Table 10-1

Blood Type	O	A	B	AB	Total
1. Theoretical proportion	45%	40%	11%	4%	100%
2. Observed number (O)	41	59	17	3	120
3. Expected number (E)	54.0	48.0	13.2	4.8	120.0
4. $O - E$	−13.0	11.0	3.8	−1.8	0.0
5. $(O - E)^2$	169.0	121.0	14.44	3.24	N/A
6. $\dfrac{(O - E)^2}{E}$	3.130	2.521	1.094	0.675	$7.420 = \chi^2$ ●

The lines of table 10-1 have been numbered so that each can be identified readily and explained.

1. The theoretical proportions must be known in advance of drawing the sample. They are the values associated with the null hypothesis. These values are either known or given to you, or can be calculated from theoretical considerations that are inherent in the null hypothesis.
2. The observed numbers (O) in the sample are the frequencies, not proportions, of occurrences in each category. The total of the numbers (O) in this line should be the sample size.
3. Each expected frequency (E) is calculated by multiplying the sample size (the total of line 2) by the corresponding theoretical proportion in line 1. The total of the numbers (E) in this line should also equal the sample size.
4. This is the difference between lines 2 and 3. The total of the numbers in this line should always be zero.
5. This is the squares of the numbers in line 4. The total of the numbers in this line is not meaningful for this analysis.
6. This is line 5 divided by line 3. The total of the numbers in this line gives you the value of χ^2.

After χ^2 has been calculated, its corresponding P-value needs to be determined. The important result that Pearson established is that this value of χ^2 falls in a χ^2 distribution. Chi-square, like Student's t, consists of a family of distributions that are associated with the number of degrees of freedom. To determine the P-value, you need to know the degrees of freedom. In table 10–1, it is always true that the total of line 3 is n, the sample size, and the total of line 4 is zero. Knowing the numbers corresponding to all the categories except one in either line 3 or line 4 would completely determine the number in the remaining category in each case. Therefore, the number of degrees of freedom present is precisely one less than the number of categories, that is, $df = c - 1$.

Table 5 in the appendix, like table 2 for t, gives P-values for values of χ^2, to one decimal place, corresponding to degrees of freedom values between 2 and 10, inclusive. In the blood-type example, $c = 4$, and so, $df = 4 - 1 = 3$. Consulting table 5, find the following for $df = 3$:

$$\left. \begin{array}{ll} \chi^2 = 7.4 & P\text{-val} = 0.0602 \\ \\ \chi^2 = 7.5 & P\text{-val} = 0.0576 \end{array} \right] \quad \text{difference} = 0.0026.$$

Thus, for $\chi^2 = 7.42$, which is $\frac{2}{10}$ of the way from 7.4 to 7.5,

Subtract in the calculation of the P-value since as χ^2 gets larger, the P-value gets smaller.

$$P\text{-val} = 0.0602 - \frac{2}{10}(0.0026)$$

$$= 0.0597.$$

At a 5% significance level, you could not reject the null hypothesis and would conclude that the distribution of blood types in this community might be the same as the rest of the United States.

Before summarizing this χ^2-test, you should be aware that this is only one form of the χ^2-test. You will be examining another χ^2-test in section 10–2. To distinguish χ^2-tests, this one is referred to as a χ^2-**test for a simple hypothesis.** (It is also referred to, by some statisticians, as a test of a *multinomial* distribution. If you compare this name to that of the binomial distribution, which concerned just 2 categories, you can see where it came from.) Another very common name for this test is the χ^2-test for **goodness-of-fit,** since it tests how well the observed numbers *fit* the theoretical proportions.

Summary of the χ^2-test for a Simple Hypothesis (Goodness-of-fit Test)

1. The null hypothesis is set up by stating that the population under study follows a certain distribution of proportions, which are either given or due to some theoretical considerations.
2. After the sample data has been collected, a table such as table 10–1 is constructed to calculate $\chi^2 = \Sigma \dfrac{(O - E)^2}{E}$, where O stands for observed number, E expected number, and the summation is taken over all categories.
3. Table 5 is consulted to determine the P-value and the test is completed in the usual way. The number of degrees of freedom is determined by $df = c - 1$, where c is the number of categories.

Note: There is no need to specify the alternative hypothesis for this test, since this is invariably a one-tailed test. The value of χ^2 will always be either zero or positive. Rejecting H_0 implies that the population under study does not fit the given distribution of proportions.

Examine table 10–1 in the solution to example 10–2. Notice that the biggest contribution to χ^2 comes from the blood type O category.

EXAMPLE 10–3 Using the sample from example 10–2, conduct a hypothesis test to determine whether or not the blood type O in this community is statistically significantly lower (at a 5% level) than in the United States population as a whole. Comment on the validity of the result.

Solution

$$H_0 : P = 0.45$$

$$H_1 : P < 0.45$$

$$\alpha = 0.05$$

Here P is the true proportion of blood type O in the whole of this community. According to the null hypothesis, P for the community equals the proportion for the United States population as a whole from the previous examples. Sample results:

$$n = 120$$

$$np = 41$$

$$p = \frac{41}{120} \approx 0.342$$

Since this is below the hypothesized proportion of 0.45, the continuity correction should be made upwards. The next possibility above 41 is 42, and $\frac{42}{120}$ $= 0.350$. Thus, the continuity corrected value of p is $\dfrac{0.342 + 0.350}{2} = 0.346$ and

$$z = \frac{0.346 - 0.45}{\sqrt{\dfrac{(0.45)(0.55)}{120}}}$$

$$z = -2.29$$

$$A = 0.4890$$

$$P\text{-val} = 0.0110 < 0.05\ (= \alpha)$$

Reject H_0.

You might then conclude that blood type O is underrepresented in this community, relative to the country as a whole. However, this is a completely *invalid* result. Sherlock Holmes was once supposed to have said, "It is a capital mistake to theorize before one has data." In this situation, it is a capital mistake to theorize *after* seeing the data.

Hypotheses should always be set up before the sample results are examined. In this example, there was no logical reason for choosing to isolate blood type O rather than any of the other three blood types. The reason blood type O was chosen was that it contributed the largest amount to the value of χ^2. But, remember the χ^2-test result itself did not lead to the conclusion that the distribution of blood types in this community was any different than the rest of the United States. The test concerning blood type O was run using the sample data that led to the choice of that blood type in the first place. It is completely invalid to examine sample data, *then* decide which case will give you an extreme result and base your test on that case. ●

EXAMPLE 10-4   To decide whether a die was fair or not, the author rolled it 144 times, recording the frequency of each face. Here are the results.

Face	1	2	3	4	5	6
Observed number of times rolled	29	25	26	23	30	11

On the basis of this sample of rolls, would you say the die is fair? Use $\alpha = 0.05$.

Solution   As you can see, there seems to be a marked underrepresentation of sixes. But is it sufficient to be able to conclude that the die is unfair? A chi-square test provides a basis for an answer. The null hypothesis here is

$$H_0 : \text{The die is fair.}$$

If the null hypothesis is true, you would expect each face to occur an equal number of times; since there are six faces, this means each face should appear on $\frac{1}{6}$ of the rolls.

Table 10-2

Face	1	2	3	4	5	6	Total
Theoretical proportion	$\frac{1}{6}$	$\frac{1}{6}$	$\frac{1}{6}$	$\frac{1}{6}$	$\frac{1}{6}$	$\frac{1}{6}$	1
O	29	25	26	23	30	11	144
E	24	24	24	24	24	24	144
$O - E$	5	1	2	-1	6	-13	0
$(O - E)^2$	25	1	4	1	36	169	N/A
$\dfrac{(O - E)^2}{E}$	$\dfrac{25}{24}$	$\dfrac{1}{24}$	$\dfrac{4}{24}$	$\dfrac{1}{24}$	$\dfrac{36}{24}$	$\dfrac{169}{24}$	$\dfrac{236}{24} = \chi^2$

Thus, $\chi^2 = \dfrac{236}{24} \approx 9.83$.

Since there are 6 categories this time, $df = 6 - 1 = 5$. Consulting table 5, find,

$$\chi^2 = 9.8 \qquad P\text{-val} = 0.0811$$

$$\chi^2 = 9.9 \qquad P\text{-val} = 0.0781.$$

Thus, for $\chi^2 = 9.83$, $P\text{-val} = 0.0811 - \dfrac{3}{10}(0.003)$

$$\approx 0.0802 > 0.05 \ (= \alpha)$$

Fail to reject H_0.

The conclusion to reach is that the die may be fair.   ●

As in example 10–3, you could run a z-score test for face 6 and would probably find resounding evidence against the fairness of the die; but, as before, this evidence would be invalid, since there is no reason why that face should have been chosen in the first place. When the author is in a more superstitious frame of mind, rather than relying on sound statistical reasoning, he attributes his losses in Backgammon to his inability to roll a sufficient number of sixes!

More Than 10 Degrees of Freedom

In those rare situations when there are more than 10 degrees of freedom, and so more than 11 categories, table 5 cannot be used nor is it necessary. There is a very good normal approximation that gives accurate P-values for more than 10 degrees of freedom. It is

$$z = \sqrt{2 \cdot \chi^2} - \sqrt{2 \cdot df}, \tag{48}$$

from which the P-value is obtained from the normal table. For instance, if $\chi^2 = 20.0$, $df = 10$, table 5 gives P-val $= 0.0293$. Formula (48) gives

$$z = \sqrt{40.0} - \sqrt{20}$$

$$z = 1.85$$

$$A = 0.4678$$

$$P\text{-val} = 0.0322,$$

which is not bad agreement. The agreement becomes even better when $df > 10$.

To have more than 10 degrees of freedom, you would have to have more than 11 categories and this is, indeed, rare.

Exercise Set 10–1

1. If a χ^2-test using a simple hypothesis is performed on a sample involving 5 categories and this gives $\chi^2 = 12.63$, what is the resulting P-value?

2. If a χ^2-test using a simple hypothesis is performed on a sample involving 12 categories and this gives $\chi^2 = 18.32$, what is the resulting P-value?

 3. It is believed that the distribution of colors of Siamese cats in a large city is Seal point, 35%; Blue point, 30%; Chocolate point, 20%; Lilac point, 15%. To check this, a random sample of 150 Siamese cats is examined with these results.

Color	S.P.	B.P.	C.P.	L.P.
Number	65	40	26	19

Do these figures conform to the distribution of colors as it is believed to be? Use $\alpha = 0.05$.

4. The dean of admissions at a community college expects students to select the day of the week when they wish to register with unequal frequencies as follows:
Monday, 25%; Tuesday, 30%; Wednesday, 20%; Thursday 15%; Friday, 10%.

A random sample of registering students request the days as follows:

Day requested	M	T	W	Th	F
Number	52	70	62	30	36

Do these figures differ significantly from those that the dean would expect at the 1% significance level?

 5. A maker of loaded dice tells you that when rolled, it is twice as likely that a 4, 5, or 6 shows than a 1, 2, or 3. (The faces, 4, 5, and 6 are equally likely, as are the faces, 1, 2, and 3.) You decide to test this by rolling the die 450 times with these results.

Face	1	2	3	4	5	6
Number	42	58	45	92	115	98

What do you think of the dice-maker's claim at the 2% significance level?

 6. A large mathematics class at a university is divided into 8 tutorial sections, leaving it up to the students to decide which section to enroll in. Here is how the students decided.

Section	1	2	3	4	5	6	7	8
Number	18	23	21	17	22	15	19	25

Are these sections being chosen with approximately equal probability? Use $\alpha = 0.04$.

 7. A producer of mixed nuts claims that peanuts, cashews, hazelnuts, walnuts, and almonds are packed into cans marked "Mixed Nuts" in the ratio: 10:4:3:2:1, respectively. A random sample of a few of these cans are opened and the nuts counted with these results.

Nut	Peanuts	Cashews	Hazelnuts	Walnuts	Almonds
Number	132	41	36	21	10

Are these figures close enough to the ratio claimed by the producer at the 5% significance level?

 8. In a national study of teenagers who run away from home, it is claimed that of the females 40% are white, 25% are black, 20% are Hispanic, 10% are Asian, and 5% are Native American. A social worker at a runaway center decides to conduct her own study. Of a sample of runaways she has to deal with, she finds these results.

Ethnic Group:

White	Black	Hispanic	Asian	Native American
35	15	19	9	2

Assuming the sample is random, are these figures in line with the national study at a 5% significance level?

9. Supporters of the idea that women should have the choice of whether to have an abortion or not claim that 65% of Americans favor the legalization of abortion. Furthermore, they believe that this figure is the same regardless of the sex or marital status of the respondents. If random samples of 100 married men, 150 married women, 80 single men, and 50 single women showed, respectively, 54, 102, 59, and 32 who favored legalization of abortion, are these figures in line with the pro-choice supporters claim? Use $\alpha = 0.05$.

10. An ice-cream parlor in Southern California claims that customers buy the same amount of ice cream in any month of the year. Records over a random sample of previous years show the following figures in hundreds of gallons.

Month	J	F	M	A	M	J	J	A	S	O	N	D
Number	15	18	20	17	24	27	28	30	17	15	14	15

Are these figures in line with the parlor's claim at $\alpha = 0.10$?

Section 10–2 The χ^2-test for Composite Hypotheses

The foregoing discussion concerned the χ^2-test for a simple hypothesis, namely that the data fit a particular known distribution or set of proportions. When performing such a test, the expected proportions are known in advance of drawing the data, either because they are given to you directly, as in example 10–1, or because they follow indirectly from theoretical considerations pertaining to the null hypothesis, as in example 10–4.

There are other χ^2-tests for different kinds of hypotheses. One of the most important of these **composite hypotheses** concerns the association, if any, between two characteristics of a population. It is often referred to as the χ^2-test of independence. For instance, a population of high school students may be formed from students graduating from several (say 3 or 4) junior high schools. A counselor at the high school may want to know whether a student's performance in mathematics at the high school is *related to,* that is, *dependent on,* the particular junior high school that the student attended. Here, the population is being characterized in two different ways: mathematical performance and junior high school attended. The counselor could draw a random sample of these high school students and enter the results into a table like the one shown in table 10–3.

Table 10–3

			Grade in Mathematics					
			A	B	C	D	F	
Junior High School Attended		1	(a)	(b)	(c)	(d)	(e)	
		2	(f)	(g)	(h)	(i)	(j)	
		3	(k)	(l)	(m)	(n)	(o)	

This kind of table is called a **contingency table.** The squares with the lowercase letters of the alphabet in parentheses are called **cells.** The cells are where the results are recorded. For example, if the first student in the counselor's sample were receiving a B in mathematics while hailing from JHS #2, that student would be counted among the students in cell (g). Cell (n) would be for those students from JHS #3 who, at present, are receiving a grade of D in math. The lines reading across the contingency table are called **rows,** while those reading down are called **columns.** Notice that table 10–3 has three rows and five columns. Thus, it is labeled a 3 by 5 (or 3 $\times$ 5) contingency table.

The composite hypotheses that contingency tables are used to test are always stated in a similar fashion. In the tradition of presuming the defendant innocent until *proved* guilty, these hypotheses are always phrased to presume that one characteristic is *independent of,* or *unrelated to,* the other. Contingency table 10–3 would be used to test the hypothesis that high school performance in mathematics is independent of JHS attended.

According to the null hypothesis, the initial assumption, a high school student's performance in mathematics depends on causes other than the JHS that he/she attended. If the null hypothesis is wrong and the opposite turns out to be true, the counselor could predict how well the student would perform in math, with a better than even chance of being correct, by basing this prediction on knowing which JHS a student attended. Whether or not the above **independence hypothesis** is rejected depends on the value of χ^2 found in the hypothesis test.

It is for this reason that some statisticians refer to this χ^2-test for a composite hypothesis as the χ^2-test for *independence.*

EXAMPLE 10–5 Suppose you are interested in whether or not political preference among a university's students is related to religious preference. A random sample of 152 students is selected and the political and religious preference of each determined. The sample was then classified one way as Democrat (D), Republican (R), or Independent (I); and the other way as Christian (C), Jewish (J), Agnostic/Atheist (A), or Other (O). A 3 $\times$ 4 contingency table of observed sample frequencies can be formed.

This observed contingency table is shown in table 10–4.

Table 10–4   Observed Contingency Table

		Religious Preference				
		C	J	A	O	Totals
Political Preference	D	41	17	7	12	77
	R	22	6	1	3	32
	I	19	8	2	14	43
	Totals	82	31	10	29	152

Use these figures to determine whether political preference depends on religious preference or not.

Solution   The null hypothesis can be stated as follows:

H_0 : Political preference among this university's students is independent of religious preference.

The test will be illustrated at the 0.05 level of significance.

The cells of contingency table 10–4 are read as follows: Of the 152 students in the sample, 41 were Democrats and also Christian, 6 were Republicans and also Jewish, and so on. The final column and final row of the table show the row and column totals. These are called **marginal totals.** The row totals show the total for each political preference, regardless of religious preference; and the column totals show the numbers in the sample indicating a preference for each religion, regardless of political preference. Thus, for example, the second row total shows that there are 32 Republicans in the sample; and the third column total indicates that there are 10 Agnostics or Atheists in the sample. The sum of the row totals should always equal the sum of the column totals; this sum of totals, called the **grand total,** equals the sample size.

For the χ^2-test, an expected contingency table must be formed that parallels the table of actual, observed frequencies in the sample. The table of expected values has the same number of rows, columns, and therefore total cells as the observed contingency table. The values in the cells are determined based on the null hypothesis. If political preference is indeed independent of religious preference, would you not *expect* the same proportion of Christians, Jews, Agnostics/Atheists, and others to be Democrats? In the sample as a whole, 77 out of 152 respondents declared themselves Democrats; this is a proportion of $p = \frac{77}{152} = 0.5066$. Thus, you should expect 50.66% of the 82 Christians (or 41.5), 50.66% of the 31 Jews (or 15.7), 50.66% of the 10 Agnostics and Atheists (or 5.1), and 50.66% of the 29 indicating other religions (or 14.7), to be Democrats. Obviously, it is not realistic to talk about fractions of persons, but one decimal place is kept to make the calculation of χ^2 more accurate.

Similarly, 32 out of 152 were Republicans, which is a proportion of 21.1%, and 21.1% of 82 Christians is 17.3, and so on. Continuing in the same way across the other categories, a complete expected contingency table can be formed as shown in table 10–5.

Table 10–5   Expected Contingency Table

	C	J	A	O	Totals
D	41.5	15.7	5.1	14.7	77.0
R	17.3	6.5	2.1	6.1	32.0
I	23.2	8.8	2.8	8.2	43.0
Totals	82.0	31.0	10.0	29.0	152.0

Observe that the marginal and grand totals in this table are identical to the marginal and grand totals in the observed contingency table. This is as it should be. There should still be 77 Democrats, 82 Christians, and so on. However, there are differences in the interiors of the tables. The question is, as always: Are these differences sufficiently great to warrant rejecting the independence hypothesis?

The remainder of this test is carried out exactly as the χ^2-test for a simple hypothesis. First, a difference table is computed as shown in table 10–6. Cells in this table indicate differences between observed and expected frequencies $(O - E)$.

Table 10–6   Observed – Expected Contingency Table

	C	J	A	O	Totals
D	−0.5	1.3	1.9	−2.7	0.0
R	4.7	−0.5	−1.1	−3.1	0.0
I	−4.2	−0.8	−0.8	5.8	0.0
Totals	0.0	0.0	0.0	0.0	0.0

The table is constructed by subtracting corresponding cells from tables 10–4 and 10–5. Thus, for Christian Democrats, $41 - 41.5 = -0.5$, for Jewish Democrats, $17 - 15.7 = 1.3$, and so on. The marginal and grand totals in this table should always be zero, since cells in the expected contingency table that gain, with respect to the corresponding cells in the observed contingency table, must be balanced by other cells that lose, so that the marginal totals remain the same.

Using these differences $(O - E)$, χ^2 is calculated as $\sum \frac{(O - E)^2}{E}$:

$$\chi^2 = 0.006 + 0.108 + 0.708 + 0.496$$
$$+ 1.277 + 0.038 + 0.576 + 1.575$$
$$+ 0.760 + 0.073 + 0.229 + 4.102$$
$$= 9.948.$$

These values are found by calculations similar to the following one for Christian Democrats:

$$\frac{(-0.5)^2}{41.5} = \frac{0.25}{41.5} \approx 0.006.$$

To obtain a P-value, it is necessary to determine the number of degrees of freedom. The key to finding this quantity comes from the expected contingency table. Suppose only part of this table is filled in as shown in table 10–7.

Table 10–7

	C	J	A	O	Totals
D	41.5	15.7	5.1		77.0
R	17.3	6.5	2.1		32.0
I					43.0
Totals	82.0	31.0	10.0	29.0	152.0

Although this table is incomplete, the marginal and grand totals have been entered, because it is known that they *must* be the same as those in the observed contingency table. The remainder of the table can now be completed in a different way. The three figures in the Democrat row total 62.3, which *forces* the entry in the Democrat-Other cell to be $77.0 - 62.3 = 14.7$, which is precisely the same answer found by calculating 50.66% of 29. In fact, for the blank cells in table 10–7, only one specific value is allowable if the marginal totals are required to retain their indicated values. Given the cells that are filled, and the marginal totals as specified, the remaining (blank) cells cannot be selected freely. At the same time, you should satisfy yourself that if any one of the six completed cells is made to be blank, the table cannot then be completed without using the method discussed earlier. Thus, six entries in table 10–7 are "free" to vary, but once picked, "force" the other six entries. Therefore, here, $df = 6$.

More generally, once all cells in an expected contingency table except the last row and last column are completed, the remaining cells are forced by the requirement that the marginal totals be the same as in the observed contingency table. Thus, if a contingency table has R rows and C columns, it has $R \times C$ cells, and the number of degrees of freedom is the same as the number of cells excepting those in the last row and column. This will be given by $(R - 1) \times (C - 1)$. When dealing with contingency tables,

$$df = (R - 1) \times (C - 1).$$

In this example, $R = 3$ and $C = 4$. This means that $df = (3 - 1) \times (4 - 1) = 2 \times 3 = 6$, as was discovered above.

To complete this χ^2-test, consult the table for $\chi^2 = 9.948$ with $df = 6$, which gives P-val $= 0.125 > 0.05 (= \alpha)$. Therefore, the independence hypothesis cannot be rejected. The results of this test indicate that, at this university, it appears that political preference may be independent of religious preference. ●

EXAMPLE 10-6   To decide whether or not 2 cities in California have the same attitude toward a statewide ballot proposition, random samples were drawn from each city, and the citizens were asked their preference. The results are shown in table 10-8.

Table 10-8

	Yes	No	Don't Know	
City S	137	31	12	
City L	52	46	2	

Does the data suggest that there is a significantly different attitude to the proposition on the part of the voters in each city? If so, at what significance level?

Solution

H_0 : Attitudes toward the proposition in City S are the same as attitudes toward the proposition in City L.

Table 10-8 is the observed contingency table, to which the marginal totals must be added as shown in table 10-9.

Table 10-9

	Yes	No	Don't Know	Totals
City S	137	31	12	180
City L	52	46	2	100
Totals	189	77	14	280

This is a 2×3 contingency table, thus:

$$df = (R - 1) \times (C - 1) = 1 \times 2 = 2.$$

If you calculate 2 cells of the expected contingency table, the other 4 should be forced by the requirement that the marginal totals in both tables be identical. For City S–yes, find $189 \times \frac{180}{280} = 121.5$, and for City S–No, $77 \times \frac{180}{280} = 49.5$, giving the expected contingency table as shown in table 10–10 and the $O - E$ contingency table as shown in table 10–11,

Table 10–10   Expected Contingency Table

	Yes	No	Don't Know	Totals
City S	121.5	49.5	9.0	180.0
City L	67.5	27.5	5.0	100.0
Totals	189.0	77.0	14.0	280.0

Table 10–11 $O - E$ Table

	Yes	No	Don't Know	Totals
City S	15.5	−18.5	3.0	0.0
City L	−15.5	18.5	−3.0	0.0
Totals	0.0	0.0	0.0	0.0

which in turn gives,

$$\chi^2 = 1.977 + 6.914 + 1.000$$
$$+ 3.559 + 12.446 + 1.800$$
$$= 27.696.$$

For $df = 2$, this value χ^2 gives a P-value so small that it is not even included in the table. At a 5% significance level, or even at a 1% level or smaller, this value of χ^2 is highly significant. Therefore, H_0 is rejected; the test indicates, at the 0.05 level of significance, that attitudes toward the proposition in these two cities is different.   ●

**Summary of the χ^2-test for a Composite Hypothesis
(Test of Independence)**

1. The null hypothesis is set up by stating that one classification of the population is independent of the other.
2. The sample data is collected and put in the form of an $R \times C$ contingency table. The marginal totals for this table are calculated.
3. The marginal totals are used to construct an expected contingency table under the independence hypothesis. The two tables are then combined to form an $O - E$ table.
4. $\chi^2 = \Sigma \dfrac{(O - E)^2}{E}$ is calculated.
5. Table 5 is consulted to determine the P-value and the test is completed in the usual way. The number of degrees of freedom is determined by $df = (R - 1) \times (C - 1)$. If df is more than 10, use the normal approximation, with z determined in formula (48), page 332.

Why Not One Degree of Freedom?

You have probably noticed that table 5 does not have entries for $df = 1$. The reason for this is that if there is only one degree of freedom, methods other than the χ^2-test can, and should, be used.

First, for a simple hypothesis.

EXAMPLE 10–7 Test to see if a coin is fair if 56 heads and 44 tails are obtained in 100 tosses.

Solution

$$H_0 : \text{The coin is fair.}$$

An attempt at the use of the χ^2-test would work as follows:

Table 10–12

Category	H	T	Totals
Observed	56	44	100
Expected	50	50	100
$O - E$	6	−6	0
$\dfrac{(O - E)^2}{E}$	0.72	0.72	$1.44 = \chi^2$ $df = 2 - 1 = 1$

Since $df = 1$, table 5 does not give a P-value. However, the problem can be solved using a z-score test:

$$H_0 : P = 0.5 \quad \text{where } P \text{ is the proportion of heads.}$$

$$H_1 : P \neq 0.5$$

$$\alpha = ?$$

Sample results:

$$n = 100$$

$$p = \frac{56}{100} = 0.56$$

Using

$$z = \frac{p - P_H}{\sqrt{\dfrac{P_H Q_H}{n}}} \quad \text{get}$$

$$z = \frac{0.555 - 0.5}{\sqrt{\dfrac{(0.5)(0.5)}{100}}} \quad \text{(don't forget the continuity correction!)}$$

$$z = 1.1$$

$$A = 0.7287$$

$$P\text{-value} = 0.2713.$$

This P-value is extremely large and would lead to failing to reject H_0 at any reasonable significance level. The coin seems to be fair.   ●

If the continuity correction is not used, $z = 1.2$, which is precisely the square root of χ^2. In fact, this will always be true, and, hence, the χ^2-test is unnecessary and wrong, since the continuity correction is not applied. The same holds true for a composite hypothesis.

EXAMPLE 10–8   Suppose in example 10–6 there were no "Don't Know" responses and the results looked like table 10–13.

Table 10–13

	Yes	No	Totals
City S	137	43	180
City L	54	46	100
Totals	191	89	280

Since this is now a 2×2 contingency table, $df = 1 \times 1 = 1$. One cell should therefore completely determine the expected contingency table. For City S–Yes, find $191 \times \frac{180}{280} = 122.8$, and the rest of the computations are shown in tables 10–14 and 10–15.

Table 10–14   Expected Contingency Table

	Yes	No	Totals
City S	122.8	57.2	180.0
City L	68.2	31.8	100.0
Totals	191.0	89.0	280.0

Table 10–15 $O - E$ Contingency Table

	Yes	No	Totals
City S	14.2	-14.2	0.0
City L	-14.2	14.2	0.0
Totals	0.0	0.0	0.0

$$\chi^2 = 1.642 + 3.525 + 2.957 + 6.341$$

$$= 14.465$$

Again, a P-value cannot be determined, since $df = 1$. However, a z-score test for a difference between proportions can be used. Let P_1 be the proportion of voters in City S who favor the proposition, and let P_2 be that same proportion in City L. The problem can be solved by testing.

$$H_0 : P_1 = P_2$$

$$H_1 : P_1 \neq P_2$$

$$\alpha = ?$$

Sample results:

$$n_1 = 180 \qquad n_2 = 100$$

$$p_1 = \frac{137}{180} = 0.761 \quad p_2 = \frac{54}{100} = 0.54$$

Test using

$$z = \frac{p_1 - p_2}{\sqrt{p^*q^*\left(\frac{1}{n_1} + \frac{1}{n_2}\right)}} \qquad \text{where } p^* = \frac{191}{280} = 0.682$$

$$z = \frac{0.761 - 0.54}{\sqrt{(0.682)(0.318)\left(\frac{1}{180} + \frac{1}{100}\right)}}$$

$$z \approx 3.80$$

$$A = 0.4999$$

$$P\text{-value} = 0.0001, \text{ extremely small.}$$

Reject H_0. Attitudes toward this proposition differ in the two cities. ●

As with the simple hypothesis, the square root of the above value of χ^2, 3.8033, is almost identical to the above value of z. The discrepancy is due to round-off error.

The above two examples should convince you that you will never need to apply a χ^2-test with only one degree of freedom. Other, more accurate, statistical methods are available to you.

EXAMPLE 10–9   The table in the article at the beginning of the chapter was based on a sample of 226 cases. Ignore the first and last columns and test to see if the final verdict is independent of the first ballot of the jury, if that ballot is indecisive.

Solution   Ignoring the first and last columns leaves 69% (18% + 4% + 47%) of the cases, or 156 cases. Based on the percentages given, the observed contingency table is shown in table 10–16.

Table 10–16

		\multicolumn{4}{c}{Number of Guilty Votes on First Ballot}
		1–5	**6**	**7–11**	**Totals**
Final Verdict	Not guilty	37	5	5	47
	Hung jury	3	0	9	12
	Guilty	1	5	91	97
	Totals	41	10	105	156

The expected contingency table is shown in table 10–17.

Table 10–17

	1–5	6	7–11	Totals
Not guilty	12.4	3.0	31.6	47.0
Hung jury	3.1	0.8	8.1	12.0
Guilty	25.5	6.2	65.3	97.0
Totals	41.0	10.0	105.0	156.0

The $O - E$ table is shown in table 10–18.

Table 10–18

	1–5	6	7–11	Totals
Not guilty	24.6	2.0	−26.6	0.0
Hung jury	−0.1	−0.8	0.9	0.0
Guilty	−24.5	−1.2	25.7	0.0
Totals	0.0	0.0	0.0	0.0

$$\chi^2 = 48.80 + 1.33 + 22.39$$
$$+ 0.00 + 0.80 + 0.10$$
$$+ 23.54 + 0.23 + 10.11 = 109.30$$

This value of χ^2 is so large that it is significant at virtually any significance level you care to name. Obviously, the outcome of jury deliberations is strongly dependent on the outcome of the jury's first ballot.

Exercise Set 10–2

1. If a χ^2-test for a composite hypothesis is applied to a 3×3 contingency table and the resulting $\chi^2 = 12.6$, what is the corresponding P-value?

2. If a χ^2-test for a composite hypothesis is applied to a 4×5 contingency table and the resulting $\chi^2 = 21.2$, what is the corresponding P-value?

3. To test whether attitude toward nuclear power is dependent on how close one lives to a nuclear power plant, the following data was found for a random sample:

	For Nuclear Power	Against Nuclear Power	Don't Know
Less than 50 miles from a plant	33	24	3
50 to 100 miles from a plant	22	26	2
More than 100 miles from a plant	45	40	5

What can you conclude from this data at a 5% significance level?

4. To see if smoking habits are dependent on socioeconomic status, a random sample of people are classified by a sociologist as to SES and asked whether or not they smoked. Here are the results.

	Upper	Middle	Lower
Smokers	11	13	18
Nonsmokers	14	12	7

Test to see if smoking habits are dependent on SES at a 1% significance level.

5. Is happiness in life related to marital status? To answer this question, the following results were obtained:

	Married Men	Married Women	Single Men	Single Women
Very happy	18	9	10	3
Happy	15	12	21	15
Unhappy	8	15	16	12
Very unhappy	4	7	3	6

How would you answer the question, using $\alpha = 0.05$?

6. Are the number of good and defective items produced by the workers at a manufacturing plant related to which shift produced them? Here are the results of a sample survey.

	Day	Swing	Graveyard
Good	94	92	86
Defective	6	8	14

How would you answer the question, using $\alpha = 0.04$?

7. Fingerprints are classified into three broad groups: whorls, arches, and loops. To see if incidences of these are the same or different within different human races, the following data was obtained:

	European	African	Asian
Whorls	16	16	23
Arches	4	6	3
Loops	30	28	21

Test to see if fingerprint types depend on race with a $\alpha = 0.05$.

8. In 1980, a random sample of people born after 1940 were asked whether they had ever smoked marijuana. Here are the results.

	Teens	Twenties	Thirties
Yes	36	44	35
No	32	25	18

Is there a significant difference in the replies to the question between the different generations, using $\alpha = 0.01$?

9. Is delinquent behavior in high school related to birth order? A random sample of high school students was obtained with the following results:

	Oldest Child	Between	Youngest Child	Only Child
Shown delinquency	2	2	2	1
No delinquency	28	19	14	5

How would you answer the question at the 5% significance level?

10. In 1973, a draft lottery was held to determine the order in which 19-year-old men should be inducted into the armed forces. The lower the priority number assigned to each day of the year, the more likely it was that a man born in 1954 on that day would be drafted in the event of an emergency. Here are the figures, indicating how many low (1–182) and how many high (183–365) numbers were present for each month.

	J	F	M	A	M	J	J	A	S	O	N	D
High	12	14	14	16	13	12	17	17	14	20	17	16
Low	19	14	17	14	18	18	14	14	16	11	13	15

Was the draft lottery fair? Use $\alpha = 0.10$.

··

Section 10-3   Summary of Terms

cells 335
χ^2 distribution 326
χ^2-test 323
columns 335
composite hypotheses 334
contingency table 335
expected number 326

goodness-of-fit 328
grand total 336
independence hypothesis 335
marginal totals 336
observed number 326
rows 335
simple hypothesis 328

Section 10-4   Summary of Symbols

Symbol	Meaning
c	The numbers of categories in a χ^2-test for a simple hypothesis
C	The number of columns in a contingency table
χ	Greek letter chi used for the χ^2-test

E	Expected values in a χ^2-test
O	Observed values in a χ^2-test
R	The number of rows in a contingency table

Section 10–5   Summary of Formulas

Formula	Use
(47) $\chi^2 = \Sigma\dfrac{(O - E)^2}{E}$	Used to calculate χ^2 in both χ^2-tests
(48) $z = \sqrt{2 \cdot \chi^2} - \sqrt{2 \cdot df}$	Used with the normal table to determine a P-value for a χ^2-test if there are more than 10 degrees of freedom

Chapter 10 Review Exercises

1. A random sample of students majoring in English, all of whom had taken a class from each of 4 professors, were asked which of the instructors was their favorite. Here are the results.

Professor	A	B	C	D
Number	35	29	17	15

Test the hypothesis that these professors are equally popular, using a 2% significance level.

2. Gregor Mendel, while making his famous studies of genetics, crossed smooth, yellow peas with wrinkled, green peas. He theorized that the four types of offspring, smooth-yellow, smooth-green, wrinkled-yellow, wrinkled-green, should occur in the ratio: 9:3:3:1, respectively. In performing an experiment to test this, the following was found:

Type	Smooth-Yellow	Smooth-Green	Wrinkled-Yellow	Wrinkled-Green
Number	84	34	25	17

Are these figures in line with Mendel's theory with $\alpha = 0.05$?

3. In an effort to see if a computer program to generate random numbers is really generating *random* numbers, a count is made of each digit to see if each is occurring about equally. Here are the results.

Digit	0	1	2	3	4	5	6	7	8	9
Number	16	14	22	25	15	27	18	20	26	17

Does it appear that the computer is generating truly random numbers or should the program be adjusted? Use $\alpha = 0.05$.

4. The records of a random sample of criminals were examined to see if the crimes for which they were convicted were committed more often on a particular day of the week, or whether crimes were equally likely to be committed on any day of the week. Here are the results.

Day	M	T	W	Th	F	S	S
Number	21	31	32	24	34	36	22

Test to see if crimes are committed on all days of the week equally. Use $\alpha = 0.05$.

 5. The claim has been made that there is a higher degree of "math anxiety" among females than among males. A random sample of students was asked whether they had high, middling, or low anxiety at the thought of taking a math class. Here are the results.

Math Anxiety

	High	Middling	Low
Males	5	56	15
Females	23	32	7

Test to see if level of math anxiety is independent of sex at the 4% significance level.

 6. A large hospital found these results for a random sample of its patients.

Blood Pressure

	High	Normal	Low
Overweight	22	15	7
Normal	19	40	12
Underweight	7	18	10

At a 5% significance level, does the level of blood pressure appear to be related to weight?

7. A random sample of high school students were asked whether they intended to go to college and whether their parents had attended college. Here are the results.

	Father Attended	Mother Attended	Both Attended	Neither Attended
Child plans to attend	25	5	20	7
Child doesn't	16	5	2	32
Child doesn't know	22	6	12	28

Do these data indicate that there is a relationship between the child's intentions and the parents' background at the .05 level?

8. Are husband and wife's political preferences independent? To answer this question, the following data was gathered:

Husband's Preference

Wife's Preference	Democrat	Republican	Other
Democrat	96	14	38
Republican	22	15	51
Other	14	13	23

Now, how do you answer the question? Use $\alpha = 0.01$.

9. Two dice were rolled 720 times, producing the following totals:

Total	2	3	4	5	6	7	8	9	10	11	12
Frequency	27	27	72	58	116	133	82	89	44	54	18

Test the fairness of the dice at the 2% level.

10. To "grade on a curve," some instructors will award an F to those students who score less than 1.5 standard deviations below the mean, a D to those who score between 1.5 and 0.5 standard deviations below the mean, a C to those who score between 0.5 standard deviation below and 0.5 above, a B to those who score between 0.5 and 1.5 standard deviations above, and an A to those who score above 1.5 standard deviations above the mean. If the "curve" is a normal distribution, the mean and standard deviation of a large class is 72 and 13, respectively, and the instructor awards the grades below, is the instructor grading "on a curve" at the 0.05 level?

A	B	C	D	F
6	42	72	35	15

Hands-on Class/Student Project

Retrieve the data that you collected in chapter 1, work with the group with whom you worked in the past and perform the following exercises.

1. Set up a null hypothesis that states that the students at your college are a population that have a distribution of blood types identical to the distribution in the United States as a whole. (See section 10–1 for those percentages.)

2. Using the sample of blood types that you collected in chapter 1, perform a χ^2-test to test the hypothesis that you set up in exercise 1. Use a 0.05 significance level.

3. Compare your result of exercise 2 with other groups in the class.

4. Set up hypotheses to test the notion that average GPA in college is related to a student's score on the Scholastic Aptitude Test (SAT).

5. Set up a 3×3 contingency table, the three columns of which correspond to GPA as follows: lower than 2.5, between 2.5 and 3.5, higher than 3.5; and the three rows correspond to SAT score as follows: lower than 800, between 800 and 1200, higher than 1200. Fill in the cells of an observed contingency table of this type with the data that you collected in chapter 1.

6. Use the marginal totals of the table that you constructed in exercise 5 to form the expected contingency table under the null hypothesis set up in exercise 4.

7. Complete the χ^2-test using a 0.05 significance level.

8. Compare your result of exercise 7 to that of other groups in the class.

9. Discuss in your group and come up with a hypothesis that could be tested using a χ^2-test for a simple hypothesis (goodness-of-fit), gather data and conduct the test.

10. Discuss in your group and come up with a hypothesis that could be tested using a χ^2-test for a composite hypothesis using a contingency table, gather data and conduct the test.

11

Analysis of Variance

....

"Wisdom denotes the pursuing of the best ends
by the best means."

....

FRANCES HUTCHESON

....

Women's Life Satisfaction

Results and Discussion

The results of the analysis support our hypothesis that a woman's life satisfaction depends on the fit between her work and family *value orientation* and *role opportunities*. Specifically, two findings predominate:

1. Among women with high family values, and especially among those with low career values, married women are significantly more satisfied than single women. Yet, among those with low family values or high career values, the satisfaction level of married and single women does not differ significantly (Table 2). These

findings suggest that career interest reduces the importance of marriage to women's satisfaction. In other words, a career can, in a sense, substitute for marriage in enhancing satisfaction.

2. Among married women with high career values, paid employment makes a significant contribution to life satisfaction. Thus, for career oriented wives, periods out of the labor force exact a significant cost in overall satisfaction.

Table 2   One-way ANOVA of Life Satisfaction by Marital Status, by Family and Career Values

Marital Status	Mean Life Satisfaction Score	N	F-Statistic	Significance
		a. *High Family Values*		
Single	3.79	100	11.47	.0008
Married	4.18	495		
		b. *Low Family Values*		
Single	4.04	166	1.16	.2816
Married	4.14	297		
		c. *High Career Values*		
Single	4.05	219	3.01	.0831
Married	4.20	483		
		d. *Low Career Values*		
Single	3.43	53	19.79	<.0001
Married	4.14	353		

From *Sociology and Social Research: An International Journal*, University of Southern California, Los Angeles, CA.

....

Chapter Overview

Making Multiple Comparisons—II

The first part of the previous chapter dealt with the use of the χ^2-test to make comparisons between several (more than two) proportions. This chapter will deal with a powerful, statistical method of handling a comparison between several means.

To test the hypothesis $H_0 : \mu_1 = \mu_2$, you now have several weapons in your arsenal: the two-sample z-score test, the two-sample t-test, or the two-sample Wilcoxon test. Suppose, now, that you wish to test the hypothesis

$$H_0 : \mu_1 = \mu_2 = \mu_3 = \ldots$$

In other words, you may wish to know whether three or more population means are equal or not. One possibility that may occur to you is to test the means against each other in pairs, using the most appropriate of the three aforementioned tests. For instance, if there were exactly three means, μ_1, μ_2, μ_3, you could test μ_1 against μ_2, μ_1 against μ_3, and μ_2 against μ_3. This involves running the particular test chosen three times, and you may be no wiser when you're finished. It is possible, for example, that your sample data might show that $\mu_1 = \mu_2$ and $\mu_2 = \mu_3$, but that $\mu_1 \neq \mu_3$. How would you draw any meaningful conclusions from that? If more than three means are to be tested for equality, the number of possible pairs to be tested rises dramatically. With five means, there are 10 possible pairs, and with six, 15. Running a test 10–15 times is no picnic, and doing it would probably leave you none the wiser when you're finished anyway. What is needed is a single procedure that would enable you to decide on the validity of the hypothesis stating that all the means are equal, much as the χ^2-test accomplishes for proportions. Such a procedure goes by the strange name ANOVA. This is an acronym for analysis of variance, **the subject of this chapter.**

Section 11–1　Introduction to ANOVA

Variance, if you recall chapter 2, is a measure of dispersion and is simply the square of the standard deviation. At this point, you may be wondering how analyzing variance, a measure of dispersion, can help decide on the equality or inequality of means, which are measures of central tendency! To convince you that this is indeed possible, first take an overview of the ANOVA procedure.

Two typical situations where ANOVA may be used are as follows: (a) Several different populations are under study and it is desired to know whether their means are all equal or not. To this end, samples are drawn, one from each population, and the sample means calculated. (b) One population is under study and it is desired to know whether subjecting this population to one of several different treatments produces different results, measured by the population mean. To this end, one sample is drawn and treated as it has been up to that point (this is the control group), while other samples are drawn and each subjected to a different treatment (the experimental groups). All sample means are then calculated.

As with other statistical techniques already discussed, the application of ANOVA requires several assumptions to be made and satisfied by the populations and samples under study. In the case of ANOVA, there are three.

1. The distribution of each population involved in the analysis is a normal distribution.
2. All populations involved in the study should have equal variances (thus, equal standard deviations). That is, $\sigma_1{}^2 = \sigma_2{}^2 = \sigma_3{}^2 = \ldots$. If you are involved in a situation such as described in (b), the variance of the population should not be substantially altered by subjecting it to a different treatment.
3. All samples are randomly and independently drawn.

It was stated above that the null hypothesis under test states that all population means are equal. As always, the test begins with the assumption that H_0 is true. If assumptions 1 and 2 can also be made, then whether there are several different populations or one population subjected to several different treatments, all samples can be considered to come from one population. This is so because it is being assumed that all means, all standard deviations (the square root of the variance), and all shapes are identical. However, as you probably realize by now, it is quite possible that different random samples drawn from the same population produce different sample means. So, even though the null hypothesis may be true, the sample means will almost certainly turn out to be at least somewhat different. The question, as usual, is whether these means are significantly different (at some pre-chosen level) to warrant rejecting H_0.

Compare this to the two-sample Wilcoxon test.

The common variance promised by assumption 2 can be estimated in two different ways. Each of the several samples, being random, has a variance that is an estimate of the population variance, σ^2, say. If these several sample variances are averaged, the resulting quantity should be an even better estimate of σ^2. This estimate is called the **variance within groups.** The other way of estimating σ^2 is by considering the several sample means—they themselves form a distribution, having a mean, called the **grand mean,** and a standard deviation, which, in the notation of the central limit theorem (CLT), is labeled $SD_{\bar{x}}$. The square of this standard deviation, $SD_{\bar{x}}^2$, is the variance of the distribution. Now, according to part 2 of the CLT,

$$SD_{\bar{x}} = \frac{\sigma}{\sqrt{n}}$$

and by squaring,

$$SD_{\bar{x}}^2 = \frac{\sigma^2}{n}$$

and multiplying by n,

$$nSD_{\bar{x}}^2 = \sigma^2.$$

This, in fact, is only true if *all* possible samples of the same size n are drawn from the population. In running an ANOVA, you will get only a few of them. Thus, the variance of the sample means that you actually get an estimate of $SD_{\bar{x}}^2$, and by multiplying by n, the size of each sample, you will obtain an estimate of σ^2. This second estimate is called the **variance between groups.**

These two estimates of the same quantity should be approximately equal. However, if the population means under test are, in fact, substantially different, the sample means estimating them are likely to vary more than one would usually expect from sample fluctuations. This, in turn, would cause the variance between groups to be larger than the variance within groups. The degree to which it is larger will decide the issue of whether or not to reject the null hypothesis.

Can you now appreciate how analyzing variance can tell you something about equality of means?

Sir Ronald Fisher (1890–1962) was a British biologist who made extensive contributions to genetics and the design of experiments, also inventing ANOVA.

The late, great British statistician Sir Ronald Fisher discovered a method of comparing variances. Labeling the variance within groups by SD_{WG}^2 and the variance between groups by SD_{BG}^2, Fisher formed the ratio of these two quantities, later called the **F ratio** in honor of Fisher. Thus,

$$F = \frac{SD_{BG}^2}{SD_{WG}^2}.$$

Fisher discovered that this ratio followed an **F distribution** that looks like the one pictured in figure 11–1.

Figure 11–1

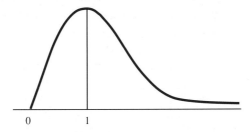

If H_0 is true, you would expect SD_{BG}^2 to be approximately equal to SD_{WG}^2, and this, in turn, would cause $F \approx 1$. It is possible that SD_{BG}^2 is smaller than SD_{WG}^2, and F will be smaller than 1. In this case, the sample means vary *less* than expected and you would wish to accept H_0. A more likely occurrence is that SD_{BG}^2 is larger than SD_{WG}^2, causing F to be larger than 1, and, therefore, its value will fall in the tail of the F distribution. The further in the tail that it falls, the more likely you will be to reject H_0.

The F distribution, like the t and χ^2 distributions, is actually a family of distributions, and the particular one you need depends on the concept of degrees of freedom. In this case, however, there are two sets of df. Since F is a ratio, it depends on the df of the numerator *and* the df of the denominator. This will be discussed further when discussing examples.

When conducting an ANOVA, the number of samples is likely to be small, and, often, the size of each sample is fairly small. For this reason, the $(n - 1)$ form of the formula for the variance is more precise than the n form. (Recall, from the discussion of the t-test, that you have $(n - 1)$ degrees of freedom, not n, when calculating the standard deviation or variance of a sample.) What this means is that, instead of using $s^2 = \dfrac{\Sigma d^2}{n}$, where $d = x - \bar{x}$, the formula used in ANOVA is

$$s^2 = \frac{\Sigma d^2}{n - 1}$$

or its algebraic equivalent,

$$s^2 = \frac{\Sigma x^2 - \dfrac{(\Sigma x)^2}{n}}{n - 1}.$$

EXAMPLE 11–1   A traveling salesman drives every day and he wonders whether he can get better mileage if he uses a different brand of gasoline. He decides to try each of four brands on each of five standard trips he makes, these trips being randomly chosen over a period of months. Table 11–1 gives the results in miles per gallon (mpg) rounded to the nearest whole number.

Table 11–1

	Brand			
	A	B	C	D
1	18	17	21	14
2	22	21	22	17
3	19	18	20	19
4	20	16	23	20
5	19	20	19	16

Should he conclude, at a 5% significance level, that his average mileage is the same or different depending on brand of gasoline used?

Solution   The data consists of 4 samples each of size $n = 5$. You should check for yourself that the following figures are correct.

$$\bar{x}_A = 19.6, \bar{x}_B = 18.4, \bar{x}_C = 21.0, \bar{x}_D = 17.2,$$

$$s_A^2 = 2.3, s_B^2 = 4.3, s_C^2 = 2.5, s_D^2 = 5.7$$

Remember to use the $(n - 1)$ form for the variance.

The sample means are different, but is that difference significant?

First, for the variance within groups, average the sample variances. Thus,

$$SD_{WG}^2 = \frac{2.3 + 4.3 + 2.5 + 5.7}{4} = 3.7.$$

Second, for the variance between groups, calculate the variance of the $\bar{x}$'s and then multiply by n, as explained above. To do this, note that you can find $SD_{\bar{x}}^2$ by using the $(n - 1)$ form of the formula for s^2, replacing x by $\bar{x}$ and n by k, the number of samples. (In this case, $k = 4$.) Thus,

$$SD_{\bar{x}}^2 = \frac{\Sigma \bar{x}^2 - \dfrac{(\Sigma \bar{x})^2}{k}}{k - 1}.$$

Here is the data again.

$\bar{x}$	$\bar{x}^2$	
19.6	384.16	
18.4	338.56	$SD_{\bar{x}}^2 = \dfrac{1{,}459.56 - \dfrac{(76.2)^2}{4}}{3}$
21.0	441.00	
17.2	295.84	
$\Sigma \bar{x} = 76.2$	$\Sigma \bar{x}^2 = 1{,}459.56$	$= 2.65$

Now, since $n = 5$, $SD_{BG}^2 = 5(2.65) = 13.25$. Finally,

$$F = \frac{13.25}{3.7} = 3.58.$$

This value of F is larger than 1; but how far into the upper tail of the F distribution is it? To decide this, you must know which F distribution is needed. This depends on the two sets of df, mentioned above. The numerator of the F ratio is SD_{BG}^2 and the number of df in its calculation is one less than the number of samples $= k - 1 = 4 - 1 = 3$. Thus,

$$df_{BG} = k - 1.$$

The denominator of the F ratio is SD_{WG}^2 and the number of df in its calculation is the number of samples multiplied by the df in each sample $= k(n - 1) = 4(5 - 1) = 16$. Thus,

$$df_{WG} = k(n - 1).$$

Table 6 in the appendix gives critical F-values for 5% and 1% significance levels for possible pairs of df. For 3 and 16 df, these tables give 3.2 for 5% significance and 5.3 for 1% significance. The interpretation of these values can best be seen from the graphs in figures 11–2 and 11–3.

Figure 11–2

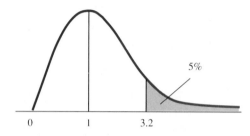

Figure 11–3

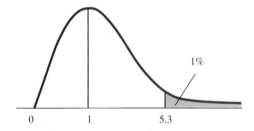

If the actual F ratio is larger than 3.2, there is less than a 5% chance that H_0 is true, and if it is larger than 5.3 less than a 1% chance that H_0 is true. The shaded regions of the graphs are often referred to as **rejection regions,** since it is when the F ratio falls into these regions that H_0 is rejected.

The F ratio for this example is 3.58, which does put it into the 5% rejection region but not in the 1%. Since the example called for a 5% significance level, reject H_0 and conclude that the brand of gasoline used by the salesman does make a difference in his mileage.

Summary and Source Tables for ANOVA

The above calculations seem brief enough; but they included the calculations of the mean and variance of each of the samples. If, as you were asked, you checked these calculations, you will realize that this, in itself, involves quite a bit of work. If the size of each sample (n) and the number of samples (k) is larger, the amount of work involved is even greater.

There exists a method that organizes and simplifies the work so that the same procedure can be applied to every problem requiring ANOVA. First, there are alternative formulas for the variances between and within groups.

The formula for the variance between groups could be written

$$SD^2_{BG} = n \left[\frac{\overline{\Sigma x^2} - \frac{(\Sigma \overline{x})^2}{k}}{k - 1} \right].$$

If $T = \Sigma x$ for each sample, then this can be rewritten without any $\overline{x}$'s.

$$SD^2_{BG} = \frac{\frac{\Sigma T^2}{n} - \frac{(\Sigma T)^2}{kn}}{k - 1}$$

The formula for the variance within groups could be written

$$SD^2_{WG} = \text{Mean of} \left[\frac{\Sigma x^2 - \frac{(\Sigma x)^2}{n}}{n - 1} \right] \text{ over all samples.}$$

This, in turn, can be rewritten as

$\Sigma(\Sigma x^2)$ means sum x^2 for each sample and then sum over all samples.

$$SD^2_{WG} = \frac{\Sigma(\Sigma x^2) - \frac{\Sigma T^2}{n}}{k(n - 1)}.$$

If you examine the alternative formulas for these variances, screened above, you will notice that the denominator of each expression is precisely the appropriate number of degrees of freedom. Because of this, the numerator of each has been given its own name: **sum of squares,** which is abbreviated to SS. Thus,

$$SS_{BG} = \frac{\Sigma T^2}{n} - \frac{(\Sigma T)^2}{kn}, \tag{49}$$

$$SS_{WG} = \Sigma(\Sigma x^2) - \frac{\Sigma T^2}{n}. \tag{50}$$

Formulas (49) and (50) can be calculated without finding the means or variances of the individual samples. Each can then be divided by its respective number of df to find the variances between and within groups, and these variances divided to form the F ratio.

Before reworking example 11-1 with these formulas, one further simplification can be made. In the discussion of coding procedures in chapter 2, it was pointed out that subtracting the same quantity from each piece of data lowers the mean by that amount but leaves the standard deviation (and therefore the variance) unchanged. Since the ANOVA procedure calls only for the calculation of variances, any quantity can be subtracted from each piece of data without altering the final answer.

EXAMPLE 11–2   Redo example 11–1 with all numbers reduced by 19.

Solution   Table 11–2 shows the initial calculations.

Table 11–2

	Brand								
	A		**B**		**C**		**D**		
	x	x^2	x	x^2	x	x^2	x	x^2	
	−1	1	−2	4	2	4	−5	25	
	3	9	2	4	3	9	−2	4	
	0	0	−1	1	1	1	0	0	
	1	1	−3	9	4	16	1	1	
	0	0	1	1	0	0	−3	9	
									Totals
$T = \Sigma x$	3		−3		10		−9		$\Sigma T = 1$
T^2	9		9		100		81		$\Sigma T^2 = 199$
Σx^2		11		19		30		39	$\Sigma(\Sigma x^2) = 99$
									$k = 4, n = 5$

$$\text{From (49), } SS_{BG} = \frac{199}{5} - \frac{1^2}{4(5)} = 39.75.$$

$$\text{From (50), } SS_{WG} = 99 - \frac{199}{5} = 59.2.$$

This is then usually summarized in a **source table** like table 11–3.

Table 11–3

Source	SS	df	SD²	F
BG	39.75	3	13.25	3.58
WG	59.2	16	3.7	

The SD^2, or variance, is found by dividing SS by df in each row. F is found by dividing the between-group variance by the within-group variance. Notice that the values of the variances and the F ratio are exactly as before, but with a more organized, simplified method.   ●

To code the data, any number may be chosen to subtract from each piece. To cement the procedure into your mind, you might rework example 11–1 using a number other than 19 to code the data. (See exercise 11 in the chapter exercises.) Your final source table should turn out identical to table 11–3.

Summary of ANOVA

Here is a step-by-step summary of the ANOVA procedure.

1. If appropriate and desired, code the data by subtracting the same number from each piece.
2. Compute x^2 for each piece of (coded) data x.
3. Calculate $T = \Sigma x$ for each sample and then sum T over all samples.
4. Calculate T^2 for each sample and sum T^2 over all samples.
5. Sum x^2 for each sample and sum these values over all samples, giving $\Sigma(\Sigma x^2)$.
6. Using formulas (49) and (50) and the formulas for degrees of freedom, complete the source table.

Source Table

Source	SS	df	SD^2	F
BG	$\dfrac{\Sigma T^2}{n} - \dfrac{(\Sigma T)^2}{kn}$	$k - 1$	$\dfrac{SS_{BG}}{df_{BG}}$	$\dfrac{SD^2_{BG}}{SD^2_{WG}}$
WG	$\Sigma(\Sigma x^2) - \dfrac{\Sigma T^2}{n}$	$k(n - 1)$	$\dfrac{SS_{WG}}{df_{WG}}$	

7. Look up the critical F-value in table 6 and compare it to the computed F ratio to decide whether or not to reject the null hypothesis. Finally, state the conclusion in words.

EXAMPLE 11–3   A psychologist conducts an experiment in memory retention as follows: She has 30 volunteers that she randomly divides into 3 groups of 10 each. They are all given a group of nonsense syllables to memorize. Then the volunteers are distracted, each group in a different way. After a certain length of time, they are all asked to recall the syllables, and their scores are recorded as a percentage correct. Here are the results.

	Group	
1	**2**	**3**
49	90	97
64	55	87
81	53	82
85	79	88
53	80	81
44	80	60
81	52	56
74	59	67
43	83	58
78	85	98

Do the different methods of distraction produce significantly different retention scores at the 1% significance level?

Solution   The hypothesis is $H_0 : \mu_1 = \mu_2 = \mu_3$, where the μ's are mean retention scores for different methods of distraction. Scores such as these are usually approximately normally distributed, with approximately equal variances. The samples were randomly selected and treated independently of each other. Thus, ANOVA is an appropriate procedure. The lowest score is 43 while the highest is 98. These two numbers have a mean of about 70. Thus the data would be made smaller by subtracting 70 from each score. Table 11–4 shows the initial calculations.

Table 11–4

		Group					
	1		2		3		
	x	x^2	x	x^2	x	x^2	
	-21	441	20	400	27	729	
	-6	36	-15	225	17	289	
	11	121	-17	289	12	144	
	15	225	9	81	18	324	
	-17	289	10	100	11	121	
	-26	676	10	100	-10	100	
	11	121	-18	324	-14	196	
	4	16	-11	121	-3	9	
	-27	729	13	169	-12	144	
	8	64	15	225	28	784	
							Totals
$T = \Sigma x$	-48		16		74		$\Sigma T = 42$
T^2	2,304		256		5,476		$\Sigma T^2 = 8{,}036$
Σx^2		2,718		2,034		2,840	$\Sigma(\Sigma x^2) = 7{,}592$

$$k = 3, n = 10$$

Source Table

Source	SS	df	SD²	F
BG	744.8	2	372.4	1.48
WG	6,788.4	27	251.4	

The 1% critical F-value, from table 6, is 5.5. Since the ratio is 1.48, which is smaller than 5.5, H_0 cannot be rejected and it must be concluded that, on the basis of these samples, there is no difference between the methods of distraction. In fact, the data is not even significant at the 5% level, because then the critical F-value is 3.4, and 1.48 is still less than that.   ●

Conclusion

The foregoing discussion of ANOVA is not the only application of the ANOVA technique. Modifications are necessary, for instance, if the samples are not all of equal size.

The article at the beginning of the chapter shows an application of a single factor ANOVA (also called one-way ANOVA) applied to samples of unequal sizes. Although you may not be able to see how they obtained the F statistics, are you now able to interpret the results? The last column, significance, can be obtained, probably using a computer, from the F-statistic and the numbers of degrees of freedom. Notice, for example, that highly significant differences between single and married women are obtained among those women who have high family or low career values, but not among those women who have low family or high career values.

Also, some real-life problems call for examining more than one factor affecting the population. The salesman, for example, may not only wish to examine the differences between gasoline brands. He may also wish to see if changing brands of oil affects his mileage and, moreover, may wish to examine this at the same time. A more complex ANOVA technique is then called for.

What is presented in this chapter is the simplest ANOVA procedure. The discussion of more complex techniques is left to more advanced statistics texts. Those interested are encouraged to consult these or to enroll in more advanced classes.

....

Section 11–2 Summary of Terms

analysis of variance 355
ANOVA 355
F distribution 357
F ratio 357
grand mean 356

rejection regions 360
source table 362
sum of squares 361
variance between groups 357
variance within groups 356

Section 11–3 Summary of Symbols

Symbol	Meaning
df_{BG}	The number of degrees of freedom between groups
df_{WG}	The number of degrees of freedom within groups
F	Symbol for the F ratio in ANOVA, named after Sir Ronald Fisher
SD^2_{BG}	The variance between groups
SD^2_{WG}	The variance within groups
$\Sigma(\Sigma x^2)$	The sum of x^2 for each sample summed over all the samples
SS_{BG}	The sum of squares between groups
SS_{WG}	The sum of squares within groups
T	Σx for each sample

Section 11–4   Summary of Formulas

Formula	**Use**
(49) $SS_{BG} = \dfrac{\Sigma T^2}{n} - \dfrac{(\Sigma T)^2}{kn}$	To find the sum of squares between groups in an ANOVA
(50) $SS_{WG} = \Sigma(\Sigma x^2) - \dfrac{\Sigma T^2}{n}$	To find the sum of squares within groups in an ANOVA

Chapter 11 Review Exercises

1. If the degrees of freedom between and within groups are 2 and 15, respectively, what are the critical F-values?

2. If the degrees of freedom between and within groups are 3 and 28, respectively, what are the critical F-values?

3. What are the critical F-values when 5 samples of size 7 each are used?

4. What are the critical F-values when 6 samples of size 10 each are used?

5. If 4 samples of size 12 each are used to test a hypothesis, and the variance between groups is estimated as 16.7, the variance within groups as 3.7, compute the F ratio and determine whether it is significant at the 5% level.

6. If 5 samples of size 6 each are used to test a hypothesis, and the variance within groups is estimated as 2.3, the variance between groups as 38.8, compute the F ratio and determine whether it is significant at the 1% level.

7. If 6 samples of size 11 each are used to test a hypothesis, and the sums of squares between and within groups are 72.3 and 223.5, respectively, complete a source table and make a decision about the hypothesis at the 5% level.

8. If 3 samples of size 9 each are used to test a hypothesis, and the sums of squares between and within groups are 144.6 and 74.3, respectively, complete a source table and make a decision about the hypothesis at the 5% level.

9. If 4 samples of size 10 each are used to test a hypothesis, and the following totals are calculated, complete a source table and make a decision about the hypothesis at the 5% level.
$$\Sigma T = 4, \qquad \Sigma T^2 = 320, \qquad \Sigma(\Sigma x^2) = 100$$

10. If 3 samples of size 4 each are used to test a hypothesis, and the following totals are calculated, complete a source table and make a decision about the hypothesis at the 1% level.
$$\Sigma T = 80, \qquad \Sigma T^2 = 2{,}378, \qquad \Sigma(\Sigma x^2) = 602$$

 11. Redo example 11–1 by first reducing each number by 18.

12. Redo example 11–3 by first reducing each number by 60.

13. Random samples of children were examined for mathematical ability in a controlled experiment. Group 1 had gone to a preschool, group 2 were encouraged to play with toys designed to promote mathematical ability, and group 3 (the control group) had neither of these advantages. Here are the results.

Group 1	Group 2	Group 3
8	9	3
7	5	8
7	6	4
9	10	5
6	8	4
4	7	2

Test to see if there are significant differences in the groups at the 5% level.

 14. Random samples of attorneys were selected from 4 different cities and their charges determined for standard legal procedures. Here are the results.

City A	City B	City C	City D
550	420	325	650
475	530	475	550
500	350	350	500
425	450	400	475
525	500	340	525

Is there a significant difference in attorneys' charges in these cities at the 1% level?

15. The following table gives the number of arrests for narcotics violations per 10,000 in the population for random samples of different types of communities.

Urban	Suburban	Rural
34	20	16
26	16	15
34	24	7
31	33	20
29	36	11
30	20	12

What conclusion can be reached about the average rates of arrest for narcotics violations in the three types of communities at $\alpha = 0.05$?

 16. An executive of a company producing 6-inch nails would like to see if 5 machines produce the same or different lengths. Random samples of 10 nails were taken from each machine with these results.

		Machine		
A	B	C	D	E
5.95	5.93	6.25	6.18	5.98
6.04	5.89	6.05	6.33	5.75
6.10	5.90	6.21	6.17	5.83
5.97	5.90	6.10	6.16	5.90
6.03	5.92	6.16	6.47	5.70
5.95	5.91	6.08	6.53	5.81
6.04	5.94	6.22	6.51	5.87
5.99	5.88	6.04	6.35	5.89
6.00	5.90	6.24	6.21	5.93
5.98	5.91	5.99	6.28	5.90

Is there a significance difference between the lengths of the nails produced by the 5 machines at $\alpha = 0.05$?

17. Random samples of volunteer subjects were subjected to 0, 24, 48, and 72 hours without sleep to test the effect of lack of sleep on reaction time. They were scored on a scale of 1 to 10, 10 being the fastest reaction time. Here are the results.

0	24	48	72
9	8	7	4
7	5	6	5
5	7	5	3
8	4	3	6
10	6	4	2
6	6	7	8

Do these results indicate a significant difference in reaction times depending on amount of time without sleep? Use $\alpha = 0.05$.

 18. Four statistics texts are to be compared for readability. To do this, four pages are chosen at random from each text, and the grade level required to read those pages determined. Here are the results.

Text A	Text B	Text C	Text D
17.5	14	9	8
15	13	14	16
17.5	14	10	7
16	12	12	9

Is there a significant difference in the mean readability of these texts at a 5% significance level?

19. Random samples of large, medium, and small businesses, all of which had failed, were studied and, among other things, the numbers of years they stayed in business were recorded.

Large	Medium	Small
4	6	5
10	4	1
5	1	2
6	3	3
8	5	0
3	7	6
9	2	5
7	4	4
4	5	4

Is there a significant difference in the mean lengths of time businesses stay in business depending on size? Use $\alpha = 0.05$.

20. Random samples of students were selected to learn a unit in English by different methods: traditional classroom instruction, video cassette tapes, audio cassette tapes, and self-instruction from the text. All students were then given the same test. Here are their scores.

Classroom	Video	Audio	Text Only
83	95	86	97
85	81	72	61
74	59	49	45
70	68	63	81
92	74	65	72
64	79	72	67
72	72	78	66
87	70	68	51
88	81	85	55
75	58	65	58

Test to see if there is a significant difference in learning by the different methods as shown by these figures. Use $\alpha = 0.05$.

Hands-on Class/Student Project

Working in the group with whom you worked in the past, collect the following materials: an 8-inch square piece of white paper, a pair of compasses to draw circles, and a set of darts. You are going to perform an experiment to see what effect, if any, your two eyes have on your aim. To accomplish this, try the following exercises.

1. Construct a target by placing the point of your compasses at the exact center of the paper and draw three circles, one with a 1″ radius, one with a 2″ radius, and the third with a 3″ radius. Pin this paper to a wall or bulletin board.

2. You are going to test the hypothesis that your aim is unaffected by what you do with your eyes.

3. Have one person in your group throw 25 darts at the target, from a distance of about 8 feet, keeping the right eye closed. Score these throws by recording 10 points if the dart is on or in the 1″ circle; 6 points if the dart is on or in the 2″ circle, but not in the 1″ circle; 2 points if the dart is on or in the 3″ circle, but not in either of the two smaller circles; 1 point if the dart lands on the paper, but not in any of the circles; and 0 points if the dart lands off the paper.

4. Repeat exercise 3, but this time close the left eye.

5. Repeat exercise 3, but this time keep both eyes open.

6. Perform an ANOVA to see if the means of the 3 sets of scores from exercises 3, 4, and 5 are significantly different (or not) at a 5% level (or 1% level).

7. Repeat with another member of your group.

8. Compare your results with other groups in the class.

9. If you did not get a significant difference in exercises 6 or 7, try repeating exercise 3 with both eyes closed!

10. Perform an ANOVA again with 4 sets of scores corresponding to: right eye closed, left eye closed, both eyes open, both eyes closed. Do you get a different result than in exercises 6 or 7?

12

Summary of Statistical Inference

"Our best is bad, nor bears Thy test; Still, it should be our very best."

ROBERT BROWNING

Surveys

Religion in America: A Rashomon Result

One more set of statistics that can be read several ways comes from a new survey of American attitudes and values. The official version is below at left; at right, an alternate—equally accurate—reading, based on the same data, by Carin Rubenstein, a social psychologist and associate editor of *Psychology Today.*

☐ "Forty-nine percent of Americans say they have made a personal commitment to Christ which they feel has changed their life."

☐ More than half of Americans (51 percent) say they have not made an important commitment to Christ.

☐ "Twenty-six percent of the American public over the age of 14, representing more than 45 million people, are highly religious."

☐ Fifty percent of the American public over the age of 14, representing more than 87 million people, have low religious commitment.

☐ "Blacks are far more likely to be highly religious than whites. . . . Women are more inclined than men to be highly religious. . . . Those with lower, rather than higher, incomes [and] . . . levels of education are more likely to be highly religious. . . . There is a steady increase in religious involvement of Americans as they grow older."

☐ In general, underprivileged groups and minorities— blacks, women, the poor, the uneducated, and the elderly—are the most religious.

☐ According to a press release, "The . . . report detects a religious current sweeping the United States today, finding America is a nation of people committed to religious beliefs."

☐ An alternate release might read: "The report detects strong religious faith only among underprivileged Americans, finding privileged America a nation of people largely indifferent to religion."

The findings are from *The Connecticut Mutual Life Report on American Values in the 80s: The Impact of Belief.* Information on the report is available at Connecticut Mutual Life, 140 Garden St., Hartford, Conn. 06115.

....

Chapter Overview
Of All These Tests, Which Do I Use?

The previous seven chapters have presented to you a wide variety of techniques of statistical inference. Having them in front of you can lead to confusion in simply having so many to choose from. This chapter will present an overview of these techniques and give you help in deciding which statistical tool is appropriate in various situations.

....

Section 12–1 Decision Procedures

The first decision that must be made can be summarized by the flowchart shown in figure 12–1.

Figure 12–1

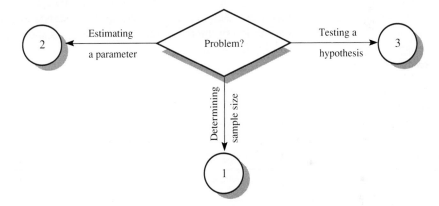

When confronting a problem, the first question you should ask yourself is, "What am I being asked to find?" The answer to this question falls into one of three general categories indicated.

1. Sometimes, before collecting sample data, you will either need to know or be asked how large a sample is required to obtain a certain amount of accuracy in estimation. The technique you use to solve this problem depends on the parameter you wish to estimate. See figure 12–2.

Figure 12–2

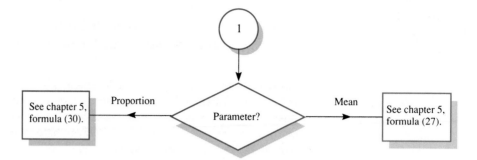

2. Once the size of the sample has been determined, the next issue to decide is whether you are required to estimate a parameter or test some claim made about a population. This is usually clear from the problem at hand—look for any reference to the parameter you are being asked to estimate. If there is none, then it cannot be a testing problem, since there is nothing for you to test. It must rather be a problem in estimation, and this is best handled by calculating a confidence interval estimate or CIE. Estimation problems fall into two categories shown in figure 12–3, which

Figure 12–3

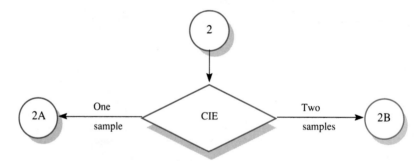

indicates the next issue that you have to confront. Are you being asked to estimate a single parameter, in which case only one sample is required, or are you being asked to estimate the difference between two parameters, when two samples are required?

2A. The previous chapters have discussed one-sample methods of estimating the mean (μ), the standard deviation (σ), and the proportion (P). A flowchart for deciding when to use what is shown in figure 12–4.

Figure 12–4

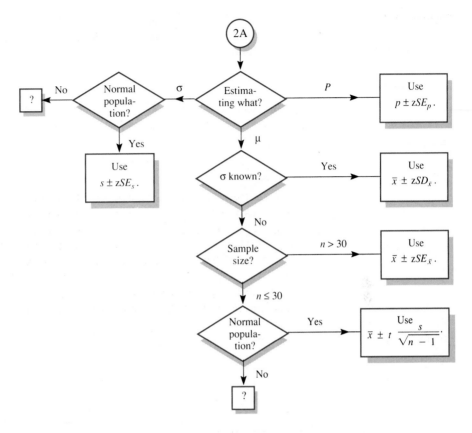

Proportion is the simplest. Sample size is not really an issue, since a small sample will not give you a CIE that is of any practical use. For large samples, the formula indicated is the one. A CIE for standard deviation can be constructed when the population follows a normal distribution. Otherwise, no satisfactory method has yet been devised. Hence, the question mark.

Mean is the most complex. Three issues come into play: the value of σ, the size of the sample, and, if the sample is small, whether the population is normal or not. A question mark has been placed in the box that represents the situation of having unknown σ, a small sample, and a population that is not normally distributed. You are on very shaky ground if you try to form a CIE in this situation. You are best advised to collect a larger sample. If this proves impossible, extreme caution is advised when quoting results.

The formulas for the standard errors can be found in the appropriate chapters.

2B. When to use which methods of two-sample estimation should be clear from the flowchart in figure 12–5. The missing formulas can be found in the relevant chapters.

Figure 12–5

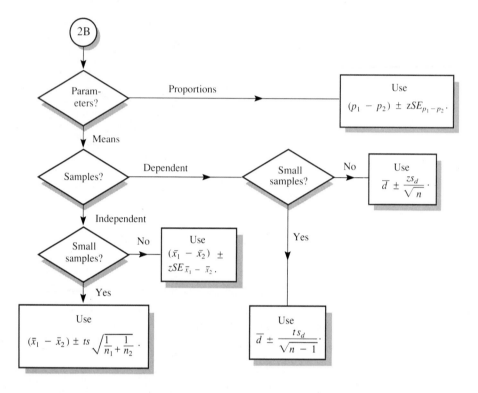

As with the one-sample techniques, proportions are a little easier. The formula given should only be used, and is only of practical value, if the two samples are large and independent.

3. If the problem confronting you requires you to see if something about a population is true or not, instead of requiring you to estimate a population parameter, then it is a problem of hypothesis testing. The previous chapters have discussed many hypothesis-testing techniques. To help you decide which of these many techniques to use, a flowchart to start you on your way is shown in figure 12–6.

Figure 12–6

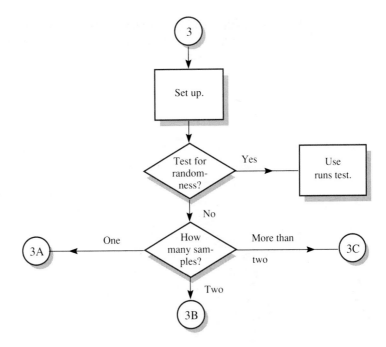

The first step in a hypothesis-testing problem, and the one performed before any sample data is collected, is the setup. This involves stating the null hypothesis, stating the alternative hypothesis (if necessary), deciding which test is to be used, and establishing the significance level. If the sample has already been collected and you are being asked to test its randomness, the runs test is available for accomplishing this. Otherwise, the remaining techniques of hypothesis testing fall into three categories, as shown in figure 12–6.

3A. One-sample techniques can be summarized by the chart in figure 12–7.

Figure 12–7

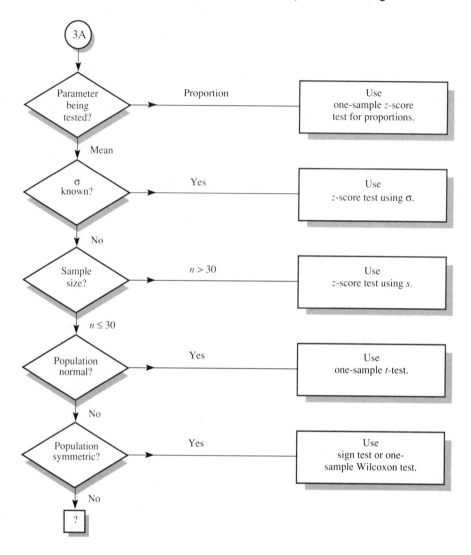

The trickiest one-sample testing problem is when σ is unknown, the sample is small, and the population is believed to be skewed. This situation is indicated by the question mark in figure 12–7.

There are several possible ways around this.

Consult previous statistical studies, if any exist in the area you are investigating, to see if a reliable estimate for σ can be obtained.

Obtain a larger sample, if possible.

Change the hypothesis to be about the median instead of the mean and use the sign or one-sample Wilcoxon test.

Change the data, by taking logarithms or square roots, for example, if that results in a more symmetrical population.

If none of these are possible, or help, consult a statistician. You will not be expected to know everything about statistics after one class!

3B. A hypothesis that you are asked to test may involve a comparison between two populations or two different treatments of the same population. The statement of the problem should make this clear. Usually, words of comparison are used: one population is "better" than another or one population parameter is "larger than," "smaller than," or "different from" another population parameter. In these cases, two-sample techniques are called for. These are exemplified by the flowchart shown in figure 12–8.

Figure 12–8

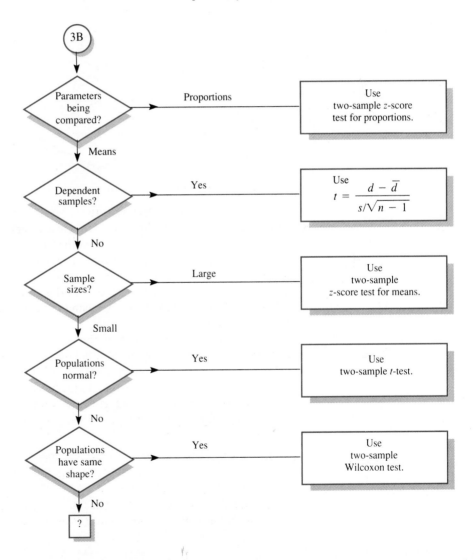

Again, you are stuck if the samples are small and the populations are deemed to have different shapes. This is rare, since populations that are being compared usually have some affinity. If this should occur, however, one possibility is to gather larger samples. If this proves impossible, techniques beyond the scope of this book are necessary.

One further assumption, required for both the two-sample t-test and the two-sample Wilcoxon test, is that the populations under study have equal standard deviations. This will be satisfied more often than not; but if you seriously doubt the equality of these standard deviations, other methods, beyond the scope of this book, must be used.

3C. The techniques remaining involve making comparisons when more than two samples are necessary. The flowchart for this situation is shown in figure 12–9.

Figure 12–9

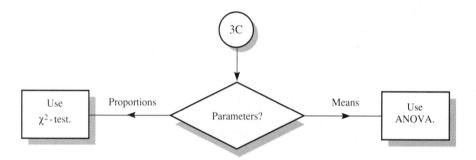

It should be remembered that two χ^2-tests were discussed. One, the χ^2-test for a simple hypothesis, is used when comparing several proportions to theoretical values. The other, the χ^2-test for a composite hypothesis, involves constructing contingency tables to test whether two factors affecting the population are related.

The ANOVA technique discussed in this book is the single-factor ANOVA. More complex ANOVA techniques exist but are, again, beyond the scope of this book.

You may feel overwhelmed by the number and variety of statistical techniques that can be employed. It is hoped that the previous flowcharts help you in arranging your ideas so that you can select the correct procedure for the problem at hand. These flowcharts could be fit together, like a jigsaw puzzle, to form one gigantic flowchart. By tracing the correct path through this huge chart, the correct procedure can be found.

EXAMPLE 12–1   Do doctors make more than dentists? To test this in Los Angeles, a random sample of 40 doctors' incomes and a random sample of 60 dentists' incomes were collected. Which statistical technique is appropriate here, given that incomes are usually skewed?

Solution   In the statement of the problem, the word "test" should immediately alert you to the fact that this is a hypothesis-testing problem. It could not be one of estimation—you are not being asked to estimate mean doctors' income or anything like that. Thus, in the flowchart in figure 12–1, you need branch 3. Moving to the flowchart in figure 12–6, after setting up the problem, you should arrive at branch 3B. This is so because two samples are required, one for doctors and one for dentists. Move to the flowchart in figure 12–8. The initial question in the problem can best be answered by comparing the means of the two populations. Moving down the flowchart to "means," notice that the next decision rests on the size of the samples. The problem calls for samples of 40 and 60, which can be considered large. Thus:

The fact that incomes are usually skewed does not enter into the discussion here. It would have, had the samples been small.

Use the two-sample z-score test for means.   ●

EXAMPLE 12–2   A random sample of 20 ninth-graders from a large school district scored a mean of 412 (out of 800) on a standardized test with a standard deviation of 89. What is the best technique available to you for estimating the mean score on this test of *all* ninth-graders in this school district?

Solution   Here, you are not being asked to test any claim about these test scores. Note that the word "test" in the statement of this problem has a totally different connotation than it had in example 12–1. You are being asked to *estimate*. Therefore, in the flowchart in figure 12–1, you need branch 2. The flowchart in figure 12–3 has itself two branches. Since there is, and you only require, one sample, branch to 2A and use figure 12–4. The problem calls for estimating a mean, so move down to μ. Do you know σ? No. How large a sample? Twenty is less than 30. Do you have a normal population? Well, test scores usually are. Thus:

$$\text{Use the CIE given by, } \bar{x} \pm t \, \frac{s}{\sqrt{n-1}} \, .$$
    ●

Some like the flowchart idea. Others may take an instant dislike to it. You are not forced to use it—you may have another way of organizing the ideas—one that works better for you. By all means, use it. Whichever method you use, you should now see how well it works by trying your hand at the exercises.

Presentation of Results

Rashomon was a 1950 Japanese movie depicting the same murder and rape from 4 different points of view. The viewer is left to decide which version, if any, is the truth.

However statistics are gathered or worked on, the way results are presented can be biased. Study the article at the beginning of the chapter. There the same statistics are presented in two different ways. An ethical researcher, one who is not biased toward one point of view or another, will try to present results in an objective manner, as far as is humanly possible.

Chapter 12 Review Exercises

For each of these problems, name the statistical procedure that is best suited to solving the exercise and list all formulas that are necessary. If an exercise cannot be solved using any of the techniques discussed in the text, so state and discuss what steps might be taken so that the exercise can be solved.

 1. You wish to investigate the difference between mean scores in algebra for philosophy majors and psychology majors on your campus. You only have time to find and interview at most 20 students in each major, but you may assume that algebra scores are normally distributed. How do you estimate the true difference in mean algebra scores between the two majors?

 2. A friend of yours claims that, because of the emphasis on health and nutrition classes, female nursing majors on your campus have a mean weight of less than 120 lb. You wish to test your friend's claim but do not want to spend more time than it takes to weigh 10–15 students, you have no idea what the population standard deviation is and you know that weights are usually skewed. How do you tackle the problem?

 3. Now your friend claims that female nursing students weigh less on the average than female physical education majors. Again, to test your friend's claim, you do not wish to spend more time than is necessary to weigh 10–12 students from each major. How do you solve this problem?

 4. You now wish to estimate the proportion of students on your campus who grew up in single parent homes. If you would like to be 95% confident that the error in your estimate is no more than 4 percentage points, how do you decide how many students to include in your random sample?

 5. The student government on your campus is considering charging a different student body fee for day students than for evening students. They would like to determine the difference between the mean incomes of these two groups and commission you to conduct the investigation. If they give you enough money to interview as many as 50 students from each group, how would you conduct the study?

 6. The basketball coach at your school is complaining because, he claims, the mean height of men in the community served by the college is only 5 ft 7 in. To test this claim, you plan to measure a random sample of men living in the surrounding community. Given that a population of human heights usually has a standard deviation of 3 inches, how would you design your experiment?

7. Is the fact that a student is a day or night student related to achievement in school? To answer this question, you plan to select a random sample of students and classify them as day student, night student, or combination day/night student. Then you will determine their overall GPA and classify them as under 2.0, 2.0 to 3.0, or above 3.0. How would you complete the analysis?

 8. While involved in the experiment described in exercise 6, you would like to kill two birds with one stone and use the sample results to estimate the true mean height in the community. How would you go about deciding how many men to include in your sample?

 9. At registration time, your college gives all students wishing to enroll in an English class a placement test. A random sample of students assigned to a remedial class on the basis of the test is given the test again after completion of the class. How would you estimate the true difference in test scores (before and after the class) if 25 students were selected for the study?

 10. You read a study of families in your area that states that their mean income is $18,000 with a standard deviation of $6,000. You are curious whether the mean income of families of students attending your college is more or less than $18,000. If you plan to use the study's standard deviation as the population value, how would you satisfy your curiosity?

 11. A women's PE instructor claims that female PE majors can do as many push-ups, on the average, as male PE majors. Yours is a large school, so you are able to test this claim using random samples of at least 40 students in each. How would the claim be tested?

 12. You would like to use the data of exercise 11 to estimate the true mean number of push-ups that female PE majors are able to do. How would you do it?

 13. A psychology teacher wishes to examine differences in stress levels among students. Random samples of full-time students who have no job, who have a part-time job, and who have a full-time job are tested for stress. How would you test whether there were different mean levels of stress among the three groups?

 14. Do first-year students who declare themselves as engineering majors do better on a mathematical aptitude test than those who declare themselves as computer science majors? To answer this question, random samples of about 20 from each major are selected and given the test. Given that scores on such a test are generally normally distributed, how would the question be answered?

 15. How would you estimate the true difference in percentages of students from two rival community colleges who transfer to four-year universities?

 16. The Raven Progressive Matrices Test is a psychological test of ability that is independent of literacy, since no words are used in the test. If the standard deviation of Raven scores is usually about 10 points, how would you estimate the mean Raven score for all students on your campus?

 17. Raven scores (see exercise 16) are usually normally distributed. If you did not know the standard deviation of Raven scores, how would you estimate it from the data collected in doing the problem in exercise 16?

18. A security officer on your campus feels that, based on observation, more female students own their own cars than males. If it is known that there are approximately equal numbers of male and female students enrolled, how would you test this?

 19. A college librarian wonders whether students would be more inclined to use the library if they were taken on a library orientation tour. To do this, a random sample of 50 students is observed for four weeks and the number of hours spent in the library by each student is recorded. The following semester the same 50 students were given a library orientation tour and observed for the corresponding four weeks. How would you estimate the true mean difference in hours spent in the library by students who have and have not had the tour?

 20. On a sociology class visit to a drug rehabilitation center, you are told that among heroin addicts the mean age when first experimenting with the drug is 20 years old, and, furthermore, the distribution of these ages is symmetric about 20, but too tight a distribution to be normal. If you are able to gain access to a random sample of 12 addicts, how would you test the claim that the mean age of first experimenting with heroin is indeed 20?

 21. A study of colleges and universities in your state reveals that of the incoming freshmen last year 22.6% had an A average in high school, 62.6% a B average, 14.7% a C average, and 0.1% a D average. How would you test whether the incoming freshmen at your college fit this pattern?

 22. You read in a botany book that the fully grown heights of certain plants are normally distributed with a mean of 2 ft 6 in. You are able to find 14 of these plants on campus and are assured by the head gardner that they are all fully grown. The botany book says nothing about the standard deviation of the heights of these plants. Assuming the sample you have found is random, how do you proceed to test the book's assertion?

 23. A student working in the admissions office says that the students entering the college this semester seem to be older than a year ago. If you are allowed to examine applications of students entering this semester and a year ago, and gather data on random samples of 100 students this year and 80 students last year, how would you find out if the student-worker is correct in her impression or not?

 24. A 20-year-old college has an administration that is mostly the original administration and faculty. The distribution of their ages, therefore, is left-skewed. It is desired to estimate the mean age from a random sample of 20 staff members with the population standard deviation unknown. How would you do it?

 25. The administration claims that the proportion of community college students who transfer to university is at least 25%. How would you test this?

26. You go to Las Vegas and record 50 consecutive numbers shown at a roulette game. How would you test if the game was producing a random selection of numbers?

 27. A physical education instructor claims that male PE majors have a mean waist measurement of 33 inches. If you can examine a random sample of 40 PE majors, how would you test this claim?

28. A new college opens and hires administrators and faculty of varying experience. It is determined that their ages are normally distributed. Despite the newness of the college, they claim that their mean age is older than the college in exercise 24. How would you test this claim if a random sample of 20 staff members from this college is compared to the sample in exercise 24?

 29. How would you estimate the mean age of the staff of the new college in exercise 28 using that sample of 20?

 30. How would you estimate the proportion of students at a community college who obtain an associate of arts (AA) degree?

Hands-on Class/Student Project

In this chapter, there is reference to 25 different techniques that were developed in previous chapters—techniques on determining sample size, on estimation, and on testing. Examine the data that you collected with your group in the project exercises for chapter 1. For each of the 25 techniques, make up a problem that would illustrate that technique using some of the data that you collected in chapter 1. You may cut down the 25 as follows: since the data is already collected, you may omit the techniques for determining sample size; also, you may omit any techniques that your teacher decided not to go over with you because of time considerations.

4

Analysis of Bivariate Data

....

"A poor relation is the most irrelevant thing in nature."

....

CHARLES LAMB

Correlation

....

" . . . but then had to find for them the right
relations, those that would bring them out."

....

HENRY JAMES

An Analysis of Delinquent Behavior Among U.S. Youth

The Social Context of Deterrence

Table 13–1	Group Violation Rates (GVR) and Act-Specific Indexes of Deterrence (r)†: 1972 National Survey of Youth		
Act		**GVR**	*r*
Marijuana		94.8	−.180
Drink		94.7	−.135
Entry		90.1	.007*
False I.D.		81.7	−.087
Drugs		81.0	−.095
Trespass		80.9	−.067
Vandalism		74.5	−.048
Truancy		71.2	−.036*
Assault		67.0	−.043*
Theft		56.1	−.064
Threat		55.1	−.020*
Car theft		54.5	−.070
Run away		47.2	.024*
Fraud		26.7	−.026*

Rho = .56

†r = product-moment correlation between perceived risk of arrest and self-reported delinquent acts, computed across subjects for each type of offense.
*Not significant at the .05 level.
From *Sociology and Social Research: An International Journal,* University of Southern California, Los Angeles, CA.

To understand the above table more fully, read this chapter.

....

Chapter Overview
Investigating Relations

In the discussion of the χ^2-test for a composite hypothesis, contingency tables were studied to determine whether two characteristics of a population or sample were related or independent. The Chi-square test can be used to assess the independence of any two population characteristics, such as religious preference and political preference, whether these characteristics are measured qualitatively or as numerical data that has been grouped, or aggregated into categorical ranges. It is also possible to classify a population or sample in two different ways quantitatively, and to study traits jointly, without grouping or categorizing the data. Quantitative data discussed thus far, referred to as univariate data, involves only one quantitative variable at a time. Data involving ungrouped numerical observations of two population traits is referred to as bivariate data.

The purpose of studying bivariate data is to see if one of the variables under study is related to the other and, if so, how close that relationship is.

A pediatrician may be interested in how closely heights of children at various ages relate to their respective weights.

A counselor at a college or university may be interested in how well scores on aptitude tests of incoming freshmen relate to scores in classes during their first year.

A farmer may be interested in how strongly associated are the amount of fertilizer placed on a field during seeding time and the yield at harvesttime.

An automobile dealer may be interested in how closely the age of a car is related to its resale value.

This chapter deals with a method of determining how well pairs of variables, forming bivariate data, are related.

....

Section 13–1 Graphical Methods

The first step recommended in a statistical problem is to get a "picture" of the data. One of the first procedures discussed for univariate data was graphing. Chapter 1 discusses histograms, frequency polygons, bar charts, and the like. None of these is appropriate for bivariate data, however. Rather, what is used is a **scatter diagram,** also known as **scattergram** or **scatter plot.**

Example 13–1 An educator develops tests to test language skills and arithmetic skills. She wishes to know how well language skill relates to arithmetic skill. To this end, she tests a random sample of nine elementary schoolchildren with these results.

Language skill,	x	83	38	47	56	23	90	75	87	89
Arithmetic skill,	y	95	70	34	66	45	100	58	71	68

Draw a scatter diagram for this data.

Solution Notice that, of the two variables referred to, one has been labeled x and the other y. These are the most common letters used for bivariate data. As with a histogram, two axes are used for a scatter diagram; but, unlike the histogram, the two axes are used, one each, for x and y. Then a point is plotted for each pair of values in the bivariate data. The scatter diagram for the given data is shown in figure 13–1.

Figure 13–1

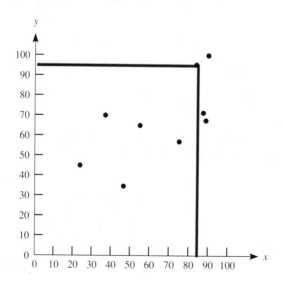

Notice that there are nine points—one for each pair of values in the bivariate data. The first pair, (83,95), has been indicated on the scattergram by lines parallel to the axes to show how it was plotted. You should check that each of the other eight points has been plotted correctly. ●

Once the scattergram has been plotted, it is an easy matter to see if there are any trends in the data. When each of x and y are roughly normally distributed, the data is said to have a **bivariate normal distribution** and the scattergram will fall into an oval shape. Thus, for the data of example 13-1, an oval shape has been drawn around the points of the scattergram as shown in figure 13-2.

Figure 13-2

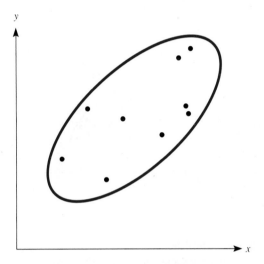

When this has been done, it can be seen that, in this example, the trend is from lower left to upper right. That is, as x increases, y tends to increase also. It seems that there is some **positive association** between language skills and arithmetic skills. What is meant by positive association, in this case, is that the better these children are at languages, the better they seem to be at arithmetic. Other scattergrams may exhibit **negative association,** which means, as x increases, y tends to decrease. For instance, as the age of a car increases, its value will tend to decrease. With the exception of vintage or "special" cars, this is generally the case. One further possibility is that x and y have no relationship at all, in which case the points of the scattergram will show no apparent upward or downward trend. Such a scattergram might result if you plotted heights of your classmates versus their scores on the first statistics test. There is no reason to suppose that the taller (or shorter) a student is, the higher he/she will score on the test.

A few scattergrams that might show up are shown in figure 13–3.

Figure 13–3

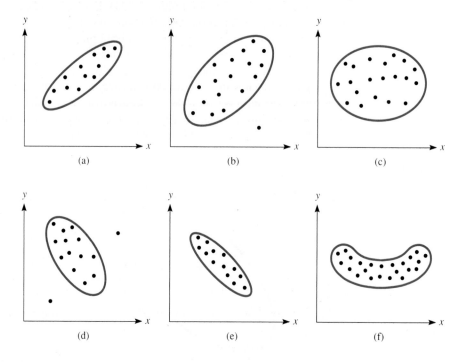

(a)   (b)   (c)

(d)   (e)   (f)

Notes:

(**a**) shows strong positive association.

(**b**) shows positive association, but not as strong as (a), and also notice the one point that does not fit the general pattern of the others. Such a point is known as an **outlier.**

(**c**) is an example of little or no association.

(**d**) shows some negative association with two outliers.

(**e**) shows strong negative association.

(**f**) shows strong association that cannot be classified as either positive or negative. There seems to be a curvilinear relationship between x and y, rather than a straight line (or linear) relationship.

Exercise Set 13–1

In these problems, draw a scatter diagram for the given data and decide what kind of association, if any, there is between the pairs of variables.

1.

x	6	4	2	5	1	3	7	4	1	5
y	13	8	6	11	4	7	15	9	3	9

2.

x	32	36	29	40	35	23	26	39
y	52	46	56	40	48	64	60	42

3.

x	65	72	68	71	75	67	76	73	71
y	90	95	45	51	50	62	45	63	83

4.

x	12	18	19	20	13	17	22	16	14
y	8	10	2	11	7	3	13	5	9

5. A random sample of elementary algebra students, showing the number of hours studied for an exam and the grade received

Hours studied, x	2	2	2	3	3	4	5	5	6	6	7	7	7	8
Grade received, y	57	63	70	72	69	75	73	84	82	89	91	81	100	96

6. Temperatures of water and surrounding air on a random sample of days at the beach

Air temperature, x	83	89	95	79	85	92	97	84
Water temperature, y	67	70	74	64	69	71	73	66

7. A psychologist runs rats through a maze, producing this data:

Number of times through maze, x	8	10	2	20	15	1	3	5	10	5
Number of errors, y	6	4	10	1	2	12	8	7	3	8

8. Several typists, applying for a job, were given the same material to type, with these results.

Speed (wpm), x	60	45	100	70	55	60	80	85	90	75
Errors, y	2	3	4	1	5	6	0	4	2	3

9. A ball is thrown up into the air from a height of 144 ft. The height after each second is computed.

Seconds, x	0	1	2	3	4	5
Height, y	144	192	208	192	144	0

10. Following is the distance from a sound source and the intensity of the sound.

Distance, x	1	4	9	16	25	36	49	64	81	100
Intensity, y	60	30	20	15	12	10	8.6	7.5	6.7	6

Section 13–2   The Measurement of the Relationship

It would be very unsatisfactory if statisticians and others could only refer to no association, some association, or strong association. Such expressions are extremely vague. Thus, statisticians sought and discovered a way of quantifying this association.

A first thought was to examine the spread of the points in the scattergram. After all, the more spread, the less association, and vice versa. When dealing with spread of univariate data, one measure of spread that was discussed was the variance given by

$$\frac{\Sigma d^2}{n} = \frac{\Sigma(x - \bar{x})^2}{n}.$$

This gives the spread of univariate data about its mean. Since, with bivariate data, two variables (x and y) are involved, a natural thing to try is to replace the quantity $(x - \bar{x})^2 = (x - \bar{x})(x - \bar{x})$ by the quantity $(x - \bar{x})(y - \bar{y})$. When this is done, a measure called the **covariance,** and abbreviated Cov, results. Thus,

$$\text{Cov}(x,y) = \frac{\Sigma(x - \bar{x})(y - \bar{y})}{n}.$$

Is the covariance a good measure of association? Consider this with the aid of two scattergrams, one showing positive and one negative association. Lines parallel to the axes at the means of x and y have been drawn in. See figures 13–4 and 13–5.

Figure 13–4

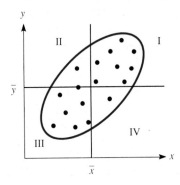

Figure 13–5

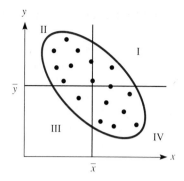

Recall, from algebra, that the product of two negatives is a positive.

The two lines at $\bar{x}$ and $\bar{y}$ divide each scattergram into four regions, called **quadrants,** that have been labeled with Roman numerals. The formula for covariance has the quantities $(x - \bar{x})$ and $(y - \bar{y})$ included in it. Now, the points of the scattergram in quadrant I are such that the values of x for those points are larger than $\bar{x}$ and the values of y for those points are larger than $\bar{y}$. Thus, both the quantities $(x - \bar{x})$ and $(y - \bar{y})$ must be positive and their product is positive. For the points in quadrant III, x is smaller than $\bar{x}$ and y is smaller than $\bar{y}$. Thus, both $(x - \bar{x})$ and $(y - \bar{y})$ are negative, making their product also positive. In quadrants II and IV, exactly one of the quantities $(x - \bar{x})$ and $(y - \bar{y})$ is positive while the other is negative. Thus, in both of these quadrants, the product $(x - \bar{x})(y - \bar{y})$ must be negative.

Now, figure 13–4, which demonstrates positive association, has many more points in quadrants I and III than it has in quadrants II and IV. When the products $(x - \bar{x})(y - \bar{y})$ are summed over all points, the contribution from the points in quadrants I and III will overwhelm the contribution from those in II and IV, causing the covariance to come out positive. Figure 13–5 exhibits the opposite phenomenon. There, the contribution to $\Sigma(x - \bar{x})(y - \bar{y})$ from the points in quadrants II and IV will overwhelm the contribution from the points in quadrants I and III, causing covariance to come out negative.

Thus, bivariate data that demonstrates positive association will have positive covariance, while bivariate data demonstrating negative association will have negative covariance. Moreover, the stronger the association, the larger (positive or negative) the covariance will be. This is so because with strong positive association there will be few points in quadrants II and IV resulting in few negative contributions to $\Sigma(x - \bar{x})(y - \bar{y})$. Similarly, for strong negative association, there will be few positive contributions to $\Sigma(x - \bar{x})(y - \bar{y})$, since there will be few points in quadrants I and III. Bivariate data exhibiting no (or very little) association will have a scattergram similar to figure 13–3(c) in section 13–1.

Then, there will be approximately equal numbers of points in each of the four quadrants, causing the covariance to be zero or close to it. This is as it should be—it would be consistent if the covariance is close to zero when there is little or no association in the data. Incidentally, graphs such as figure 13–3(f) in section 13–1 would also have covariance close to zero (there will still be approximately equal numbers of points in each of the four quadrants).

Therefore, covariance does measure the degree of association between the two variables of bivariate data. However, it only measures the degree of *linear* (or straight line) association.

Downstown   **By Tim Downs**

DOWNSTOWN © 1981
UNIVERSAL PRESS
SYNDICATE. Reprinted with
permission. All rights reserved.

Karl Pearson (1857–1936), considered one of the founders of modern statistics, was involved in radical politics in his youth, became a lawyer, and then became a professor of mathematics and statistics.

There is a disadvantage in using covariance to measure the strength of linear association between two variables. It is that its value depends on the units of measurement. For example, if you were studying the association (if any) between the heights of women and the heights of men whom they marry, you would get a different value for the covariance (using the same sample) depending on whether you measure the heights in inches or in centimeters. It is desirable to have the measure of association the same value irrespective of which units are used in collecting the data. The statistician Karl Pearson hit upon a method of creating a "dimensionless" measure of association. Since the standard deviation of any set of data is always in the same units as the data itself, Pearson thought of dividing the covariance by the two standard deviations since, then, what would result would not depend on any units. If x is in inches and y in pounds, say, the $(x - \overline{x})$ and s_x will also be inches and $(y - \overline{y})$ and s_y will be in pounds. Dividing a product of inches and pounds by another product of inches and pounds will produce a pure number. Pearson called this new quantity r for a sample and ρ (the lowercase Greek letter rho) for a population. Thus, for a sample,

$$r = \frac{\text{Cov}}{s_x s_y} = \frac{\Sigma(x - \bar{x})(y - \bar{y})}{n s_x s_y}.$$

Now, subtracting the mean from a piece of data and dividing the result by the standard deviation produces the z-score for that piece of data, if you recall the discussion of section 2-9. Thus,

$$z_x = \frac{x - \bar{x}}{s_x} \text{ and } z_y = \frac{y - \bar{y}}{s_y}.$$

The formula for r can now be written as

$$r = \frac{\Sigma z_x z_y}{n}. \tag{51}$$

Since z-scores are referred to, by some statisticians, as moments, and a preferred word for association is **correlation,** r's full title is **Pearson's product moment correlation coefficient.** Since this is somewhat of a mouthful, referring to r as the **correlation coefficient** is satisfactory.

The formula for the correlation coefficient as given by (51) always results in a number between -1 and $+1$. No matter how bizarre the data is, if a number outside this range results from an application of formula (51), an arithmetic mistake *must* have been committed. The correlation coefficient behaves just as the covariance did. Since standard deviations are always positive, dividing the covariance by them leaves the sign of the quantity unchanged. The interpretation of the value of r can be summarized by the diagram in figure 13-6.

Figure 13-6

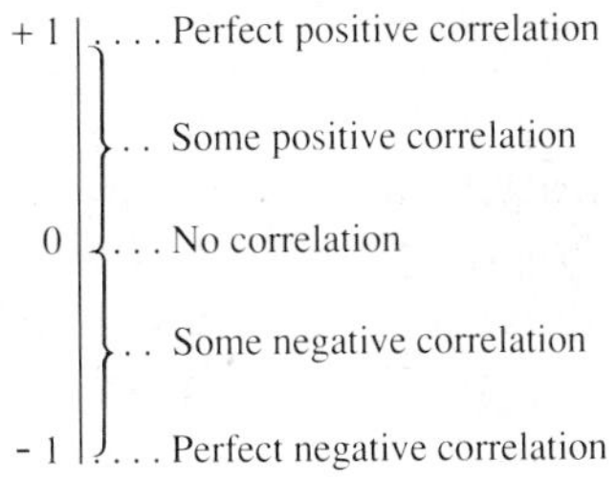

```
+1 |..... Perfect positive correlation

       }.. Some positive correlation

 0  |..... No correlation

       }.. Some negative correlation

-1  |..... Perfect negative correlation
```

The only time r is equal to ± 1 is if the points of the scattergram lie on a perfectly straight line. Even if only one point does not lie on the line, r will be between -1 and $+1$, exclusive of ± 1 themselves. The closer r is to zero, the weaker the association between the variables is, and the closer r is to $+1$ or -1, the stronger the association is.

Example 13–2   Calculate r for the data of example 13–1 using formula (51).

Solution   Recall that the scattergram showed some positive correlation but the points were not on a straight line. Therefore, you should expect r to be between 0 and 1, but not 0 or 1 themselves.

x	z_x	y	z_y	$z_x z_y$
83	0.75	95	1.39	1.04
38	−1.16	70	0.13	−0.15
47	−0.78	34	−1.68	1.31
56	−0.40	66	−0.07	0.03
23	−1.80	45	−1.13	2.03
90	1.05	100	1.64	1.72
75	0.41	58	−0.48	−0.20
87	0.92	71	0.18	0.17
89	1.00	68	0.03	0.03

$$\Sigma x = 588 \qquad\qquad \Sigma y = 607 \qquad\qquad \Sigma z_x z_y = 5.98$$
$$\bar{x} = 65.3 \qquad\qquad \bar{y} = 67.4$$
$$s_x = 23.6 \qquad\qquad s_y = 19.9$$

$$r = \frac{5.98}{9} \approx 0.66$$

The calculations were performed as follows: the mean and standard deviation of the x's and of the y's were calculated. These were used to convert each of the x's and each of the y's to z-scores. These z-scores were multiplied in pairs, these products summed, and this total divided by the sample size to give r. Note that r is between 0 and 1.   ●

If you check the calculations for yourself, you will notice that (a) quite a lot of work is involved and (b) a great deal of rounding takes place—there is rounding in the calculation of the means, standard deviations, z-scores, products of z-scores, and in the final division.

There exists a formula, algebraically equivalent to formula (51), that allows you to calculate r directly from the raw scores. It is

$$r = \frac{n\Sigma xy - (\Sigma x)(\Sigma y)}{\sqrt{n\Sigma x^2 - (\Sigma x)^2}\ \sqrt{n\Sigma y^2 - (\Sigma y)^2}}. \qquad (52)$$

Formula (52), despite its forbidding appearance, is not too difficult to apply.

EXAMPLE 13-3   Use formula (52) to calculate r for the data of example 13-1.

Solution

Table 13-2

x	x^2	y	y^2	xy
83	6,889	95	9,025	7,885
38	1,444	70	4,900	2,660
47	2,209	34	1,156	1,598
56	3,136	66	4,356	3,696
23	529	45	2,025	1,035
90	8,100	100	10,000	9,000
75	5,625	58	3,364	4,350
87	7,569	71	5,041	6,177
89	7,921	68	4,624	6,052
$\Sigma x = 588$	$\Sigma x^2 = 43,422$	$\Sigma y = 607$	$\Sigma y^2 = 44,491$	$\Sigma xy = 42,453$ $n = 9$

$$r = \frac{9(42,453) - (588)(607)}{\sqrt{9(43,422) - (588)^2}\ \sqrt{9(44,491) - (607)^2}}$$

$$r = 0.6630$$

Notice that to 2 decimal places this is the same answer as to example 13-2. This second method is more precise, however, since no rounding takes place until the very end.

Both formulas (51) and (52) are much easier to work with if you use a calculator. There are calculators in existence that allow you to enter the data, press one button and the correlation coefficient appears. Considering samples are often much larger than the one considered here, this is perhaps the easiest method of all!

Exercise Set 13-2

Calculate r, the correlation coefficient, for the following sets of bivariate data.

1.

x	1	2	3	5	6	7	8
y	1	2	3	5	6	7	7

2.

x	1	2	3	5	6	7	8
y	8	7	6	5	3	2	1

3.

x	6	4	2	5	1	3	7	4	1	5
y	13	8	6	11	4	7	15	9	3	9

4.

x	12	18	19	20	13	17	22	16	14
y	8	10	2	11	7	3	13	5	9

5. The following are heights of students and their grades on a statistics examination.

Height, x	60	62	63	64	66	68	68	69	70	72	73
Grade, y	83	65	92	71	44	75	70	69	96	55	60

6. A company monitors the amount it spends on advertising (in thousands of dollars) and its annual sales (in millions of dollars) over a period of years.

Ad $$, x	40	45	55	60	65	70	75	80
Sales, y	1.5	1.7	2.0	1.9	2.2	2.1	2.5	2.2

7. Following are the heights and weights of a sample of 10-year-old children, given in inches and pounds.

Height, x	59	56	53	51	56	51	58	51	54	52
Weight, y	106	86	82	60	73	54	86	60	74	65

There may be a small discrepancy between your answers to exercises 7 and 8, since the conversions were rounded.

8. Following is the same sample as in exercise 7, except in centimeters and kilograms. According to statements in the text, the value of r should be the same. Check this.

Height, x	149.9	142.2	134.6	129.5	142.2	129.5	147.3	129.5	137.2	132.1
Weight, y	48.1	39.0	37.2	27.2	33.1	24.5	39.0	27.2	33.6	29.5

9. A random sample of communities exposed to radioactive contamination was chosen and the number of deaths per 100,000 from cancer recorded.

Index of radioactivity, x	1.2	1.7	2.5	3.7	6.5	8.3	10.5
Death per 100,000 from cancer, y	110	130	150	160	180	210	200

10. In a psychological experiment on sleep, 15 subjects were allowed to sleep different lengths of time and the next day asked to solve a pencil and paper maze. The number of attempts at solving the maze was recorded.

Hours of sleep, x	7.4	4.9	7.5	8.5	7.6	7.2	8.0	6.5	9.5	6.1	6.0
Number of attempts, y	4	6	5	4	4	4	3	6	3	7	7
x	10.0	5.2	8.1	8.5							
y	4	8	3	4							

Section 13–3   A Nonparametric Alternative

Charles Spearman (1863–1945) was a psychologist deeply interested in the study of intelligence and how to measure it. He studied correlations of mental abilities and developed *theories* from these.

Long before Wilcoxon developed his tests based on ranks, a British psychologist, Charles Spearman, suggested that correlation between two variables could be measured using ranks. His reasoning went that if there was a strong positive correlation between two variables x and y, high-ranked values of x should correspond to high-ranked values of y and, similarly, low-ranked values to low-ranked. The differences, then, between the ranks of x and the ranks of y should be small. For negative correlation, high ranks for x should correspond to low ranks for y, and vice versa. The difference would then be large. Suppose these differences are labeled D. Then, Spearman's first thought was to measure $\Sigma|D|$, the sum of the absolute values of the differences. Much like the story behind standard deviation, this was later replaced in favor of ΣD^2. In fact, Spearman was able to prove that if x and y are replaced by their ranks, formula (51) becomes

$$r' = 1 - \frac{6\Sigma D^2}{n(n^2 - 1)}. \tag{53}$$

This measure of correlation is called **Spearman's rank correlation coefficient** or, more simply, the **rank correlation coefficient.** Notice that r' (read "r prime") is used in place of r, since there is no reason why r and r' should have the same value. Replacing raw data by their ranks changes things. Even so, the values of r and r' are often close. It matters little whether you rank from lowest to highest or vice versa—the same value for r' will result. The same method as for the Wilcoxon tests will be used here, that is, from lowest to highest.

Example 13-4   Calculate r', the rank correlation, for the data of example 13-1.

Solution   A neat way of performing the ranking is on the scattergram. See figure 13-7.

Figure 13-7

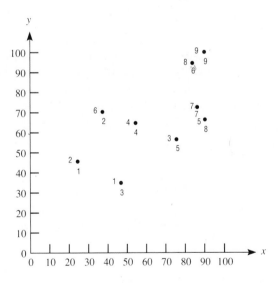

The x-ranks are written below the points, while the y-ranks are written to the left of the points. With the scattergram, ranking can be performed easily and quickly. Moving in order of the x-ranks, ΣD^2 can now be calculated.

$$\Sigma D^2 = 1 + 16 + 4 + 0 + 4 + 4 + 0 + 9 + 0$$

$$= 38$$

Thus,

$$r' = 1 - \frac{6(38)}{9(81 - 1)}$$

$$= 0.6833$$

$$\approx 0.68.$$

         ●

     Notice how close this value is to the value of r for the same data, 0.66, and much less work, too!

     Possible values of r' fall in the same interval as the possible values of r, that is, from -1 to 1. Moreover, the interpretations of those values are similar too.

Example 13–5   An athletic coach is interested in finding out how strong a correlation there is between amount of training and the time required for a runner to run a mile. To this end, he selects a random sample of athletes and has them keep a record of how many hours they train during the week prior to a race. He then has them run a mile, timing them to the nearest tenth of a minute. The results are shown in table 13–3.

Table 13–3

Amount of Training (x)	Time to Run a Mile (y)	Amount of Training (x)	Time to Run a Mile (y)
15	4.7	26	4.2
22	4.3	22	4.3
10	4.9	6	5.6
28	4.2	20	4.8
8	5.2	19	5.0
12	5.1	17	4.6
18	4.6	13	4.5
7	5.5	21	5.3
14	4.6	16	4.9
16	4.4	23	4.1

     Find (a) the correlation coefficient r and (b) the rank correlation coefficient r'.

Solution   The scattergram has been drawn in figure 13–8 and shows, apart from one outlier, a fairly strong negative correlation. The circled point indicates more than one point in the same place. In this example there are two points at (22,4.3). The ranks have been placed on the scattergram in preparation for part b. Ties are dealt with in the usual way. Since there are two points at (22,4.3), this is automatically a tie for both x and y. Although plotting a scattergram is not essential for the calculation of either r or r', it is a useful practice, since it provides a quick visual check for the calculations. In this example, if your calculations produced a positive correlation, or one close to zero, you would know there was something wrong.

Figure 13-8

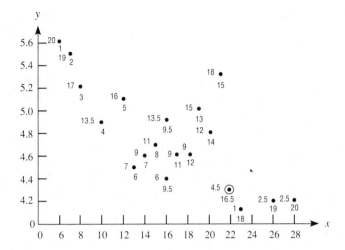

a. Verify for yourself that the following calculations are correct:
$n = 20$, $\Sigma x = 333$, $\Sigma x^2 = 6{,}267$, $\Sigma y = 94.8$, $\Sigma y^2 = 453.06$,
$\Sigma xy = 1{,}539.7$.
Then, using formula (52),

$$r = \frac{20(1{,}539.7) - (333)(94.8)}{\sqrt{20(6{,}267) - (333)^2} \sqrt{20(453.06) - (94.8)^2}}$$

$$= -0.7481$$

$$\approx -0.75.$$

b. $\Sigma D^2 = 361 + 289 + 196 + 90.25 + 121 + 1 + 4 + 9 + 12.25 + 16$
$+ 4 + 9 + 4 + 4 + 9 + 2(144) + 289 + 272.25 + 306.25$
$= 2{,}285.0$

and using formula (53),

$$r' = 1 - \frac{6(2{,}285.0)}{20(400 - 1)}$$

$$= -0.7180$$

$$\approx -0.72.$$

Again the agreement between r and r' is fairly good. ●

Correlation and Causation

There is a strong temptation among researchers who use statistics to infer that having found a strong correlation between two variables, a change in one *causes* a change in the other. Unless other, nonstatistical research points to that conclusion, this is not a valid deduction to make.

The formulas introduced in this chapter are purely mathematical and simply tell you whether one variable increases with another or as one increases, the other decreases. Being mathematical formulas, there is no possible way that they could tell whether a change in one variable *causes* a change in the other. If two variables increase together, causing you to find a positive correlation, the reason for this may be some third, hidden variable that you have not considered. To take somewhat of a ridiculous example, suppose in a small town that police salaries over several years were recorded and, at the same time, sales of liquor were monitored. Now, it is possible that over the same time period both these variables increase, and in such a way that there is a strong positive correlation between them. Does this mean that you can conclude that the police officers in this town are spending their extra salary on booze? Not at all! Other factors, not considered in the calculation of the correlation coefficient, have their effect. Police salaries rise, like other people's, because the cost of living is constantly rising. Sales of liquor may rise because the population of adults increases. The fact that the two may rise at approximately the same rate, causing the value of the correlation coefficient to be high, does *not* allow you to conclude that the increase in one *causes* the increase in the other.

Sometimes the issues are not as clear as in the example above. Several years ago, a correlation study was performed on pregnant women who smoked. It was found that there was a strong correlation between smoking and the incidence of low-birth-weight infants. Does smoking, then, cause the baby to be born underweight? A more detailed experiment was performed in which the smoking women were divided into those who smoked currently, those who had smoked and quit, and those who did not take up smoking until after the baby was born. It turned out that the highest incidence of low-birth-weight babies occurred among those women who did not take up smoking until after the baby was born. How can this be explained if you support the idea that smoking *causes* low birth weight? It is not smoking that causes low-birth-weight babies, but something else that women, who are destined to be smokers sometime in their lives, have in common.

A simple rule to remember is:

Correlation does not imply causation.

The study of correlation can be very useful and is, in fact, carried out extensively in the fields that use statistics. Researchers, however, should be very careful of the conclusions they draw when discovering a strong correlation.

Exercise Set 13-3

Calculate r', the rank correlation coefficient, for the following sets of bivariate data.

1.

x	10	8	6	9	5	7	11	8	5	9
y	15	10	8	13	6	9	17	11	5	11

2.

x	3.2	3.6	2.9	4.0	3.5	2.3	2.6	3.9
y	10.4	9.2	11.2	8.0	9.6	12.8	12.0	8.4

3.

x	4.5	5.2	4.8	5.1	5.5	4.7	5.6	5.3	5.2
y	11.0	11.5	6.5	7.1	7.0	8.2	6.5	8.3	10.3

4.

x	109	115	116	117	110	114	119	113	111
y	111	113	105	114	110	106	116	108	112

 5. The following data shows a measure of how much a sample of babies cry and their scores on an IQ test 3 years later.

Cry count, x	20	25	18	17	19	13	20	23	16	14	18	15	21
IQ score, y	152	150	140	130	122	120	112	115	110	106	102	100	92

 6. Following are heights and head circumferences of 1-year-old babies.

Height, x	30.4	28.7	29.1	28.2	29.7	30.6	28.3	20.1	30.0
Head circumference, y	18.7	18.4	18.1	18.2	18.6	18.5	18.6	19.0	18.3

x	29.5	27.8	28.1	28.9
y	17.9	19.1	19.2	17.8

 7. The following data are psychological test scores for pairs of nonidentical twins.

x	230	212	228	105	114	140	231	155	160	170	148
y	180	220	189	142	99	142	231	135	115	110	187

 8. The following are pairs of scores on the Raven test for pairs of identical twins.

x	52	39	46	29	45	47	37	42	42	40
y	48	44	44	38	52	47	43	39	49	41

 9. Here are the high/low temperatures for cities around the world in December 1981.
90/72, 32/28, 87/77, 57/48, 63/46, 36/23, 72/48, 81/66, 36/18, 25/03, 32/32, 45/41, 68/57, 28/18, 88/70, 27/23, 74/46, 0/−2, 61/55, 20/10, 90/70, 57/39, 46/32, 30/32

How well do the highs correlate with the lows?

 10. Two boxing judges ranked 10 welterweights as follows:

Boxer	1	2	3	4	5	6	7	8	9	10
Rank/Judge A	10	4	5	2	1	8	9	6	7	3
Rank/Judge B	9	1	5	3	2	7	10	4	8	6

How well do the judges correlate?

Section 13–4   Population Correlation

If bivariate data is available for a whole population, it is possible to calculate the population correlation coefficient ρ or the population rank correlation coefficient ρ' using the same formulas as were used in sections 13–2 and 13–3 for sample data. The question arises that if a random sample from a population of bivariate data is available, and if r is calculated for this sample data, what does this tell you, if anything, about ρ? Since the same formulas are used to calculate r or ρ, it is a natural consequence that r is the best *point estimate* for ρ. From the earlier discussion of estimation, you probably realize that a point estimate is not very informative. Despite the fact that r is the best point estimate for ρ, it is quite possible that r, calculated from a sample, is quite a bit different from the true value of ρ, the calculated value from the population of which the sample is a part. This is especially true if the sample size is small. The natural inclination, therefore, would be to attach a standard error SE_r to the estimate of ρ. This standard error, as in the past, would be an estimate of SD_r, the standard deviation of the frequency distribution of r. Unfortunately, no satisfactory formula has ever been developed for either SD_r or SE_r when details of the population bivariate distribution are unknown (which is usually the case).

The issues here are similar to those raised in the discussion of the estimation of σ. No satisfactory formula for SE_s or SD_s has been devised that is valid for *any* population.

It is not possible, therefore, to establish a reliable confidence interval estimate (CIE) for a population correlation coefficient ρ when only sample data is available.

On the other hand, you may wish to know whether a correlation discovered in a random sample implies that there is correlation in the parent population or not. The natural inclination would be to test the hypothesis

$$H_0 : \rho = 0.$$

In fact, statisticians originally tackled the problem in this way. However, they soon discovered that this formulation of the problem produced less than satisfactory results. Another, and better, approach is to examine the problem from the concept of independence.

Independence

In the discussion of the χ^2-test for a composite hypothesis, data was classified into two different (qualitative) ways. The test was then carried out to determine whether the two classifications were related in any way, or whether they were independent. It was discovered, for example, that political preference among certain university students was independent of their religious preference. Thus, knowing a particular student's political preference would not enlighten you at all as to this student's religious preference. On the other hand, it was also discovered that the attitude to a certain proposition on the ballot in California did depend on which city's voters were asked. Thus, knowing which city a particular voter lived in would give you a better than even chance of predicting his/her attitude toward the proposition.

The idea of independence can be extended to quantitative variables x and y in the following way:

The individual members of a bivariate population can be classified according to their x-values. That is, each one will fit into one or another x-class. The members within each of these x-classes will have a different set of y-values. If, however, the distribution of y-values is the *same* for each different x-class—that is, all these y distributions have the same mean, the same standard deviation, and the same shape—then y can be said to be *independent* of x. Obviously, then, knowing which x-value any particular individual had would tell you nothing about the corresponding y-value. The reverse would also be true: knowing a y-value would tell you nothing about the corresponding x-value. For example, a large class of statistics students could be classified according to height (x) and according to score on the first statistics exam (y). Common sense tells you that how well a person scores on a statistics exam has nothing to do with that person's height. Therefore, you might expect the distribution of scores on the exam to be approximately the same at each different height. Knowing a particular student's height would not give you any inside knowledge as to how well that student will do on the exam. These two variables, height and exam score, are independent.

It is a relatively easy matter to prove that if x and y are independent, then Pearson's product moment correlation coefficient (ρ) must be zero. One formula for ρ is

$$\rho = \frac{\Sigma(x - \mu_x)(y - \mu_y)}{N\sigma_x\sigma_y}.$$

If x and y are independent, then the distribution of y-values within each x-class is the same, and it will be the same as the distribution of y-values as a whole. For each x-class, therefore, y-values will have a mean equal to the mean of all y's, that is, μ_y. Thus, for each x-class $\Sigma(y - \mu_y) = 0$, since this is a property of the mean of any set of data values. Now, for each x-class $(x - \mu_x)$ is a constant value, which means $\Sigma(x - \mu_x)(y - \mu_y)$ will also be zero for each x-class, and this, in turn, implies that the sum $\Sigma(x - \mu_x)(y - \mu_y)$, taken over *all* x-classes, must be zero. Hence, ρ is equal to zero divided by a constant, which is zero.

The converse of the above statement is not true. If $\rho = 0$, it is *not necessarily true* that x and y are independent. The scatter diagram of some data that demonstrates this is shown in figure 13-9.

Figure 13-9

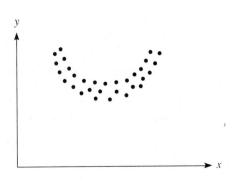

It is similar to the scattergram at figure 13–3(f). It was noted there that r or ρ, since they measure the strength of *linear* correlation, would be zero for a scatter diagram like this. But, here, is x independent of y? Definitely not! Knowing a particular value of x would give you a very good idea of the corresponding value of y. They are strongly dependent, albeit not linearly. To summarize, it can be said that

> If x and y are independent, then $\rho = 0$.
> If $\rho = 0$, then no conclusion about the dependence of x and y is possible.

For these reasons, statisticians have come to prefer to test the hypothesis

$$H_0 : x \text{ and } y \text{ are independent.}$$

This is a stronger hypothesis than the one mentioned above, equating ρ to zero, since the acceptance of this new hypothesis *implies* that $\rho = 0$. In contrast to this, accepting the hypothesis that $\rho = 0$ tells you nothing about the relationship of x and y.

An even better reason for preferring the independence hypothesis is that good tests *have* been devised for this (much stronger) hypothesis.

The *z*-score Test for Independence

Since r is the best point estimate for ρ, and since r, calculated from formula (51), is a kind of mean, the central limit theorem promises that r will be normally distributed with mean $= \rho$, provided n is not too small. Now, if it should turn out that x and y are independent, then $\rho = 0$. If you can show that r differs significantly from zero (at some prechosen significance level), then you can conclude that x and y are dependent. To discover this, you must find the P-value associated with the z-score given by

$$z = \frac{r - 0}{SD_r}.$$

But it was noted above that no satisfactory formula exists for SD_r for the general bivariate distribution. However, statisticians have discovered that, under the strong hypothesis of independence (which implies $\rho = 0$), SD_r is approximately equal to $\dfrac{1}{\sqrt{n-1}}$, regardless of the shape of the bivariate distribution. Substituting this into the above formula for z gives

$$z = \frac{r - 0}{1/\sqrt{n-1}}$$

or, upon simplification,

$$z = r\sqrt{n-1}. \tag{54}$$

The P-value can then be determined from the normal table and a decision about the independence hypothesis made.

Example 13-6   In example 13-1, an educator discovered that for a random sample of 9 schoolchildren, the correlation coefficient r between arithmetic and language skills was 0.66. What does this tell about the dependence of arithmetic and language skills in the population as a whole? Use a 0.05 level of significance.

Solution   If x is language skill and y is arithmetic skill, you are being asked to test the hypothesis

$$H_0 : x \text{ and } y \text{ are independent}$$

with

$$\alpha = 0.05.$$

The sample results give $n = 9$, $r = 0.66$. Therefore,

$$z = 0.66\sqrt{9-1}$$

$$z \approx 1.87$$

and, from table 1,

$$A = 0.4692$$

$$P\text{-val} = 0.0308 < 0.05 \ (= \alpha).$$

Reject H_0 and conclude that language and arithmetic skills are dependent in the population as a whole. ●

Notes:

    a. This z-score test is not very reliable with small samples such as in this example. It is more reliable if n is at least 20, or preferably 30.

    b. A one-tailed test was assumed here. The test was to see if x and y were independent or whether they related *positively*. If, before the sample was collected, it was deemed possible that x and y might be related positively *or* negatively, then a two-tailed test would be called for and, at the same significance level, it would not have been possible to reject H_0.

    c. The result was not significant at the 0.01 level.

A negative relationship between x and y, in this example, would mean that the better a child was in language, the worse he/she would be in arithmetic, and vice versa.

Example 13-7   In example 13-5, an athletic coach determined that the correlation coefficient between amount of training (x) and time to run a mile (y) was -0.75 for a random sample of 20 athletes. Are the two variables x and y dependent? Use a 0.01 level of significance.

Solution

   H_0 : Amount of training and time to run a mile are independent.

   $\alpha = 0.01$

A one-tailed test should be used here since there is no basis for assuming any possibility of a positive correlation. Nobody would argue that the more an athlete trains, the longer he/she would take to run a mile.
   Sample results:

$$n = 20, r = -0.75$$

$$z = -0.75\sqrt{20 - 1}$$

$$z \approx -3.27$$

$$A = 0.4995$$

$$P\text{-val} = 0.0005 < 0.01 \ (= \alpha).$$

Reject H_0 and conclude that time taken to run a mile does depend on amount of training.   ●

Example 13-8   In a study designed to test whether among married couples the husband's weight depended on the wife's weight, a random sample of 30 married couples produced the results in table 13-4.

Table 13-4

Wife's Wt (x)	Husband's Wt (y)	Wife's Wt (x)	Husband's Wt (y)
165	151	136	169
201	254	111	175
215	160	166	173
137	168	133	172
102	150	113	181
140	165	138	170
151	167	125	171
135	220	109	186
145	215	126	166
105	185	106	142
130	210	136	163
114	180	131	145
134	176	96	174
110	140	115	161
112	155	94	154

   At the 0.05 significance level, can it be concluded that in the parent population there is dependence between spouses' weights?

Solution   The scattergram for this data is given in figure 13–10, on page 419.

$$H_0 : \text{Husbands' and wives' weights are independent.}$$

$$\alpha = 0.05$$

Choose a two-tailed test here since it is not known, before starting, if there is a relationship, whether it is positive or negative.

To make the numbers smaller, reduce the wives' weights by 130 and the husbands' by 170. The results are shown in table 13–5. (This has no effect on the calculated value of r.)

x	y	x	y	x	y
35	−19	0	40	8	0
71	84	−16	10	−5	1
85	−10	4	6	−21	16
7	−2	−20	−30	−4	−4
−28	−20	−18	−15	−24	−28
10	−5	6	−1	6	−7
21	−3	−19	5	1	−25
5	50	36	3	−34	4
15	45	3	2	−15	−9
−25	15	−17	11	−36	−16

Verify for yourself the following computations: $n = 30$, $\Sigma x = 31$, $\Sigma y = 98$, $\Sigma x^2 = 22{,}563$, $\Sigma y^2 = 17{,}810$, $\Sigma xy = 6{,}815$.

From these figures,

$$r = \frac{30(6{,}815) - (31)(98)}{\sqrt{30(22{,}563) - (31)^2}\,\sqrt{30(17{,}810) - (98)^2}}$$

$$r \approx 0.34$$

$$z = 0.34\sqrt{29}$$

$$z \approx 1.83$$

$$A = 0.4664$$

$$P\text{-val} = 0.0336 > 0.025 \ (= \tfrac{1}{2}\alpha).$$

Fail to reject H_0. Based on this sample's results, there seems to be no dependence between wives' and husbands' weights.   ●

Exercise Set 13–4

In exercises 1–4, the sample correlation coefficient and sample size are given for a random sample of bivariate data. Test for the independence of the two variables in the population at the given significance level.

1. $r = 0.57$, $n = 25$, $\alpha = 0.05$. Two-tailed test.

2. $r = 0.78$, $n = 20$, $\alpha = 0.01$. One-tailed test.

3. $r = 0.25$, $n = 40$, $\alpha = 0.04$. One-tailed test.

4. $r = -0.30$, $n = 50$, $\alpha = 0.04$. Two-tailed test.

5. A random sample of 30 students' grades on the mid-term and final of a large statistics class were found to have a correlation coefficient of 0.31. Test to see if there is any relationship between the two grades at a 5% significance level.

6. In exercise 10 of exercise set 13–2, you were asked to find the correlation between quantity of sleep and number of attempts at solving a pencil and paper maze for a sample of subjects of a psychologist. You should have found $r = -0.82$ for $n = 15$. Assuming the sample is random, test to see if these two variables are dependent or not. How valid do you think your answer is?

7. The following data gives the heights of a random sample of fathers and the heights of their adult sons.

Father's height,	x	68	72	64	64	62	74	61	66	71	65
Son's height,	y	66	71	66	67	64	74	65	67	70	64
	x	73	68	62	72	63	70	72	64	67	68
	y	70	68	68	67	68	66	70	65	68	67

Test to see if son's height depends on father's height at the 5% significance level.

8. Is the frequency of chirping of a cricket related to the temperature? Test at the 1% level for the following random sample:

Chirps per second,	x	15	17	18	20	16	20	16	15	17	16
Temperature,	y	75	81	84	92	72	89	69	68	79	81
	x	17	14	21	18	22	17	13	15	18	19
	y	83	74	95	82	98	76	63	71	80	79

9. What is the smallest value of r, the correlation coefficient, that would give you a relation between x and y at the 5% significance level for a sample of 35 if there is a positive association?

10. What is the largest value of r, the correlation coefficient, that would show dependence between x and y at the 1% significance level for a sample of 45 if there is a negative association?

Section 13-5   The Nonparametric Alternative

The concept of ranking that is used in Spearman's rank correlation coefficient can be used to devise a test that is a nonparametric alternative to the z-score test for independence based on r. This is the **test for independence based on ΣD^2.**

When y is independent of x, then knowing x tells you nothing about y. When x's and y's are replaced by ranks, knowing an x-rank tells you nothing about the y-rank. Thus, the y-ranks can be considered to be randomly distributed with respect to the x-ranks. This idea can be used to obtain an exact P-value corresponding to a given value of ΣD^2. For a given sample size n, all combinations of the ranks of y, which are 1, 2, 3, . . . , n, can be considered against the ranks of x, also 1, 2, 3, . . . , n. Each possible combination, then, provides a value of ΣD^2. It is possible to obtain a P-value corresponding to a particular value of ΣD^2: calculate the probability of obtaining a value as small as, or smaller than, this value of ΣD^2 by chance. Tables have, in fact, been constructed giving P-values corresponding to different pairs of values of ΣD^2 and n, at least for small values of n.

When n is larger, there exists a z-score approximation that works quite well. It is

$$z = \frac{6\Sigma D^2 - n(n^2 - 1)}{n(n + 1)\sqrt{n - 1}} \qquad (55)$$

and the P-value can be obtained from the normal table. In formula (55), ΣD^2 should be corrected for continuity. When there are no ties, ΣD^2 is always an even number. The continuity correction, therefore, entails changing it to the appropriate odd number.

Example 13-9   Example 13-6 continued the discussion of example 13-1. There it was found for $n = 9$, $\Sigma D^2 = 38$. Test, using the test based on ΣD^2, to see if x and y are independent at the 0.05 significance level.

Solution

$\qquad H_0 : x$ and y are independent.

$\qquad \alpha = 0.05$

$\qquad$ Use one-tailed test, as in example 13-1.

With positive correlation, as it is here, the stronger the correlation is, the smaller the value of ΣD^2 will be. Therefore, it is necessary to know the probability of obtaining a value of ΣD^2 as small as 38 or *smaller*. This means that $\Sigma D^2 \le 38$ is being distinguished from $\Sigma D^2 \ge 40$, the next higher even number. Thus, the continuity corrected value of ΣD^2 is 39. So,

$$z = \frac{6(39) - 9(81 - 1)}{9 \cdot 10\sqrt{8}}$$

$$z \approx -1.91$$

$$A = 0.4719$$

$$P\text{-val} = 0.0281 < 0.05 \ (= \alpha).$$

Reject H_0 and conclude that language and arithmetic skills are dependent in the population.   ●

Notes:

a. The exact P-value for $\Sigma D^2 = 38$, $n = 9$, based on probability considerations, is 0.0252. The z-score approximation provided a P-value acceptably close to this. The approximation is much better when n is larger.
b. This rank-correlation test, using r', provided a more significant P-value than the z-score test for independence based on r. This nonparametric value is more reliable for such a small sample size.

Example 13–10   Example 13–7 continued the discussion of example 13–5. There $\Sigma D^2 = 2{,}285$ for a sample with $n = 20$. Using the test based on ΣD^2, determine whether or not x and y are independent at the 0.01 significance level.

Solution

$$H_0 : x \text{ and } y \text{ are independent.}$$

$$\alpha = 0.01$$

Use a one-tailed test, as in example 13–5.

When x and y are negatively correlated, as they are here, ΣD^2 becomes larger as the correlation becomes stronger. Therefore, it is necessary to know the probability of obtaining a value of ΣD^2 as large as 2,285 or *larger*. The reason ΣD^2 is an odd number here is because there were a number of ties. This means that $\Sigma D^2 \geq 2{,}285$ must be distinguished from $\Sigma D^2 \leq 2{,}284$, the next lower even number. Thus, the continuity corrected value of ΣD^2 is $2{,}284.5$. So,

$$z = \frac{6(2{,}284.5) - 20(399)}{20 \cdot 21 \cdot \sqrt{19}}$$

$$z \approx 3.13$$

$$A = 0.4991$$

$$P\text{-val} = 0.0009 < 0.01 \ (= \alpha).$$

Reject H_0 and conclude that the time taken to run a mile does depend on the amount of training.   ●

Example 13–11   Using the test based on ΣD^2, test to see if x and y for the data of example 13–8 are independent, using a 0.05 level of significance.

Solution   The scattergram with the associated ranks is shown in figure 13–10.

Figure 13–10

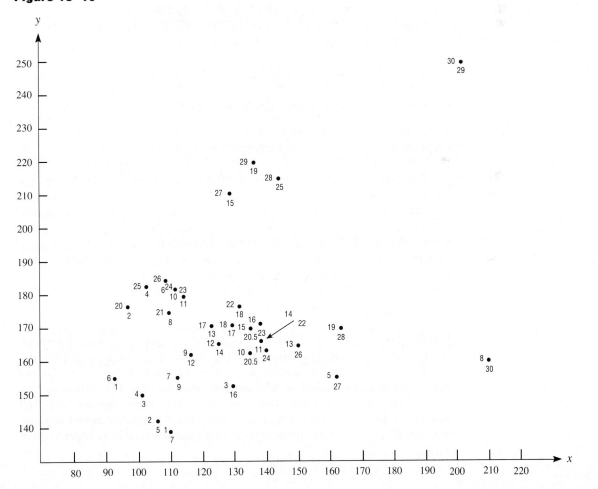

$H_0 : x$ and y are independent.

$\alpha = 0.05$

Use a two-tailed test as in example 13–8.

In order of the x-ranks, the sum of squared differences is

$$\Sigma D^2 = 25 + 324 + 1 + 441 + 9 + 400 + 36 + 169 + 4$$
$$+ 196 + 144 + 9 + 16 + 4 + 144 + 169 + 1$$
$$+ 16 + 100 + 110.25 + 30.25 + 64 + 49 + 169$$
$$+ 9 + 169 + 484 + 81 + 1 + 484$$

$\Sigma D^2 = 3{,}858.5$ and $n = 30$.

The scattergram suggests some positive correlation, which means correcting for continuity up to 3,859.25, halfway to the next higher even number, 3,860. Thus,

$$z = \frac{6(3{,}859.25) - 30(899)}{30 \cdot 31\sqrt{29}}$$

$$z \approx -0.76$$

$$A = 0.2764$$

$$P\text{-val} = 0.2236 > 0.025 \; (= \tfrac{1}{2}\alpha).$$

Fail to reject H_0. Based on this sample's results, there seems to be no dependence between wives' and husbands' weights. ●

The P-value based on ΣD^2 in example 13–11 is much less significant than the P-value based on r for the same data (calculated in example 13–8). The P-value calculated in example 13–11 is more reliable since the z-score test for independence based on r works best when the data follows closely to a bivariate normal distribution. Since weights are usually decidedly skewed, this assumption cannot be made with much confidence. Incidentally, $\Sigma D^2 = 3{,}858.5$ gives $r' = 0.14$, much smaller than the corresponding $r = 0.34$.

Which Test When?

When the population has a bivariate normal distribution and n is not too small, say $n \geq 20$, the z-score test for independence based on r is the more reliable test.

When the assumption of normality cannot be made, or when $n < 20$, the test for independence based on ΣD^2 becomes more reliable. This is true as long as there are not too many ties. With many ties, the P-value obtained from ΣD^2 becomes suspect. For this reason, if the assumption of normality cannot be made or n is small, accurate measurement is strongly recommended in an effort to avoid ties.

A final consideration is the amount of work involved. With small samples, once you have drawn the scattergram, which is a good idea in any case in these types of problems, it is a relatively small amount of work to perform the ranking and calculate ΣD^2 and the associated P-value.

Both tests have their place, and this is why both have been presented here.

The table at the beginning of the chapter was an analysis of a sample of 1,395 male and female youths drawn from the 48 contiguous states. They were each asked how many times they had committed any of 14 delinquent acts in the past 3 years. They were also asked how much of a risk they felt they were taking in committing each act. This risk was placed on a numerical scale and correlated with the number of times they had committed the act, to produce r. The GVR is a scale telling how likely it is that an act is committed with friends. The higher the number, the more often the act is committed in a group, while the lower the number, the more likely it is committed alone. Rho is Spearman's rank correlation coefficient for the correlation between GVR and r.

Example 13–12   Any r that has an asterisk in the table is said not to be significant at the 0.05 level. Verify that $r = -0.180$ is significant at the 0.05 level.

Solution $n = 1,395; r = -0.180; z = -0.180\sqrt{1,394} = -6.72$, a highly significant value.   ●

Exercise Set 13–5

In exercises 1–4, values of ΣD^2 and n are given for various random samples. Use formula (55) to find the associated P-value.

1. $\Sigma D^2 = 834, n = 15.$

2. $\Sigma D^2 = 2,080; n = 25.$

3. $\Sigma D^2 = 2,282; n = 30.$

4. $\Sigma D^2 = 4,680; n = 26.$

  5. The following data gives ages of automobile drivers and the amount of their insurance premiums for a random sample.

Age,	x	46	35	25	20	31	19	40	37	33	36
Premium,	y	420	510	650	780	550	850	400	525	500	450

Test, using ranks, to see if the amount of insurance premium depends on the age of the driver at a 2% significance level.

  6. If x is the number of divorces per 1,000 women and y is the number of juvenile delinquents per 1,000 persons under 18, test, using ranks, to see if these two rates are dependent at a 5% significance level, given the following data for a random sample of communities.

x	11	15	5	10	26	17	21	3	19	12	25	4
y	9	19	5	20	20	10	13	8	16	6	11	14

7. If x is the number of pounds of fertilizer used by a tomato grower and y is how many hundreds of pounds of tomatoes produced, test, using ranks, to see if these two amounts are dependent at a 1% significance level.

x	40	42	35	30	27	55	72	68	43	48	61	64	52
y	8	7	6	4.5	5	10	19	16	7.5	7	14	13	11

8. A random sample of subjects were given various amounts of alcohol and then tested for reaction time. Here is the data.

Alcohol level, x	0.035	0.04	0.01	0.025	0.06	0.08	0.095	0.10	0.12
Reaction time, y	1.4	1.8	0.8	1.6	2.0	2.4	3.2	3.1	3.5

x	0.05	0.065	0.07	0.085	0.09
y	2.3	2.2	2.5	2.8	2.7

Is reaction time related (in a negative way) to amount of alcohol consumed? Use $\alpha = 0.01$.

9. Given that there are no ties in the ranks, what is the smallest value of ΣD^2 that would produce significance at the 5% significance level for a random sample of 22?

10. What are the smallest and largest values of ΣD^2 for $n = 12$?

....

Section 13–6   Summary of Terms

Section 13–7   Summary of Symbols

Symbol	Meaning
$Cov(x,y)$	The covariance between variables x and y
D	Individual difference between ranks when computing a rank correlation coefficient
r	Pearson's product moment correlation coefficient for a sample
r'	Spearman's rank correlation coefficient for a sample
ρ	Pearson's correlation coefficient for a population
ρ'	Spearman's rank correlation coefficient for a population

SD_r	The standard deviation of a frequency distribution of r (Pearson's correlation coefficient for a sample)
s_x	The standard deviation of variable x
s_y	The standard deviation of variable y
z_x	The standard scores of variable x
z_y	The standard scores of variable y

Section 13–8   Summary of Formulas

Formula	**Use**
(51) $r = \dfrac{\Sigma z_x z_y}{n}$	To compute Pearson's product moment correlation coefficient
(52) $r = \dfrac{n\Sigma xy - (\Sigma x)(\Sigma y)}{\sqrt{n\Sigma x^2 - (\Sigma x)^2}\ \sqrt{n\Sigma y^2 - (\Sigma y)^2}}$	Alternative to formula (51) in which it is unnecessary to calculate means and standard deviations
(53) $r' = 1 - \dfrac{6\Sigma D^2}{n(n^2 - 1)}$	To compute Spearman's rank correlation coefficient
(54) $z = r\sqrt{n - 1}$	Used in the z-score test for independence
(55) $z = \dfrac{6\Sigma D^2 - n(n^2 - 1)}{n(n + 1)\sqrt{n - 1}}$	Used in the test for independence based on ΣD^2

Chapter 13 Review Exercises

1. Which shows a higher correlation, a set of data giving $r = 0.6$ or another set giving $r = -0.8$, or can you tell?

2. Which shows a higher correlation, a set of data giving $r = 0.7$ or another set giving $r' = 0.75$, or can you tell?

3. A study shows that the correlation between population density and noise level is 0.72. Can you conclude that a high concentration of people causes a high noise level?

4. A study shows that the correlation between temperature and sales of soda pop is 0.83. Can you conclude that hot weather causes an increase in the sales of soda pop?

 5. A sample of women who have lost weight produced the following statistics:

Total inches lost,	x	28	29	20	25	24	26	20	34	36
Total pounds lost,	y	18	19	9	16	27	12	11	30	31

Plot these points on a scattergram and estimate the value of the correlation coefficient.

 6. Calculate the correlation coefficient r for the data in exercise 5.

7. Calculate the rank correlation coefficient r' for the data in exercise 5.

8. A sociologist has developed a test of measuring how satisfied an employee is on the job. It was decided to see how well this "job satisfaction" index correlated with the number of days absent from the job. Here are the results for a random sample of employees of a company.

"Job satisfaction" index, x	70	68	68	90	81	78	80	88	85	71	62
Number of days absent, y	7	10	3	2	6	8	3	1	6	11	8
x	95	98									
y	2	0									

Draw a scattergram for this data and estimate the value of the correlation coefficient.

9. Calculate the correlation coefficient r for the data in exercise 8.

10. Calculate the rank correlation coefficient r' for the data in exercise 8.

11. The following data gives the weight of the mother and weight of her oldest child at age 10.

Mother's weight, x	147	156	111	247	179	123	180	112	125
Child's weight, y	64	70	52	75	80	63	82	47	61
x	143	140	112	157	132	144	143		
y	79	76	62	73	65	80	84		

Test, using r, whether these two variables are related. Use $\alpha = 0.05$.

12. Repeat exercise 11 using the test using ΣD^2.

13. A random sample of professional golfers were asked to keep a record of how many hours they practiced during the week prior to a big tournament. Then, each of their scores in the first round was noted. Here are the results.

Hours of practice, x	35	32	21	23	30	26	40	27	10	28
First round score, y	69	68	72	73	70	75	66	71	78	75
x	33	36	22	9	45	25	29	22	35	10
y	68	67	71	79	65	68	77	72	68	74

Test using ΣD^2, whether score in golf depends on amount of practice. Use $\alpha = 0.03$.

14. Repeat exercise 13 using the test using r.

Exercises 15–20 refer to the following data for a random sample of U.S. cities in the 1960s.

Median income, x	Rate of registration to vote, y	Voter turnout, z
1,600	39.1	13.8
2,100	69.7	59.5
1,500	61.2	46.9
1,900	33.8	25.6
2,600	68.1	54.0
1,350	48.2	28.0
1,900	69.8	54.0
1,650	66.6	52.5
2,200	90.5	70.4
1,450	43.6	22.4
1,800	77.7	69.8
1,700	73.8	65.9
1,850	70.7	46.5
2,150	96.4	85.2
1,400	81.1	44.6
1,600	79.8	66.8
1,300	61.8	53.9
2,000	88.4	87.2
1,800	55.6	45.9
2,050	77.0	64.2
1,550	81.9	69.3
1,250	32.1	24.2
1,500	73.6	62.5
1,750	64.7	47.8
1,050	46.5	31.2

 15. Test x versus y using r and $\alpha = 0.01$.

 16. Repeat exercise 15 using ΣD^2.

 17. Test x versus z using ΣD^2 and $\alpha = 0.04$.

 18. Repeat exercise 17 using r.

 19. Test y versus z using r and $\alpha = 0.05$.

 20. Repeat exercise 19 using ΣD^2.

21. For the table on delinquent behavior at the beginning of the chapter, show that $r = -0.036$ is not significant at the 0.05 level. (*Hint:* See example 13–12 and the paragraph before it.)

22. At the base of the table is given Rho = 0.56, where Rho is what this text has been calling r'. Rank GVR from smallest to largest. How are the r's ranked to produce $r' = 0.56$? (*Hint:* Try ranking with the signs taken into account and then try again, ignoring the signs.) Do you think the 0.56 figure is correct?

23. The category "Entry" is an exception. "Entry" means entering someone's property without permission. It has a high GVR and small, insignificant r. Find r' when this is deleted.

24. Test for independence between GVR and r, using the test using ΣD^2, when "Entry" is deleted.

Hands-on Class/Student Project

Working in the group with whom you worked in the past, retrieve the data that you collected in the Hands-on Class/Student Project for chapter 1.

1. Calling the students' SAT score (variable 6) the x-variable and the students' GPA (variable 5) the y-variable, draw a scattergram illustrating this bivariate data.

2. Calculate r, the sample correlation coefficient, between variables x and y.

3. Place ranks on the scattergram you drew in exercise 1.

4. Calculate r', the sample rank correlation coefficient, between variables x and y.

5. Test for independence of x and y using the z-score test developed in this chapter. Use a 5% significance level.

6. Repeat exercise 5 but this time with the test for independence using ΣD^2.

7. Now considering students' heights (variable 1) to be the x-variable, with GPA still being the y-variable, draw a scattergram for this new set of bivariate data.

8. Repeat exercise 2 for this x and y.

9. Placing ranks on the scattergram you drew in exercise 7, repeat exercise 4 for this x and y.

10. Compare all your results with those of other groups in the class.

14

Linear Regression

....

"Nulla dies sine linea." ("Not a day without a line.")

....

PLINY

Joyce E. Sirianna, Allan L. VanNess, and Daris R. Swindler, "Growth of the Mandible in Adolescent Pigtailed Macaques (Macca nemestrian)." Reprinted from *Human Biology,* Volume 54, No. 1 (February 1982) by permission of the Wayne State University Press.

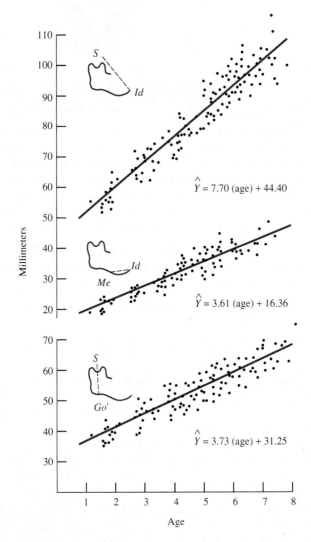

Chapter Overview
Analyzing Relations

In the previous chapter, it was discussed how to measure the correlation between the two variables in bivariate data, and how to determine, when finding a correlation in a sample, whether that implies there is correlation in the parent population. Suppose that a strong correlation is found. Then, knowing a particular value of *x* should give you a good idea of the corresponding value of *y*. At least, with strong correlation, the value of *y* that you can predict from a particular value of *x* should be more accurate than if you tried the same thing with bivariate data that exhibited little or no correlation. However, even with data exhibiting strong correlation, how do you choose which value of *y* to predict from a particular value of *x?* In other words, if two variables *x* and *y* are strongly related, *how* are they related?

The process that enables you to answer these and related questions is called regression analysis **or, more precisely,** linear regression analysis, **the subject of this chapter.**

Section 14–1   Linear Regression

Linear regression analysis was first studied in the last century by a man named Sir Francis Galton. His initial study was of fathers' heights and the heights of their sons. To no one's surprise, Galton found that tall fathers tend to have tall sons, while short fathers tend to have short sons.

What was somewhat of a surprise was that Galton discovered that those fathers who are well above the mean in height tend, on the average, to have adult sons who are shorter than they are, while fathers who are well below the mean in height tend to have sons who are taller than they are. Galton called this "regression toward mediocrity" and declared a law bearing that name. What that law states is that extreme values (high or low) of *x* are associated with values of *y* that tend to be less extreme than the corresponding values of *x*. Galton referred to this by saying that the *y*-values tend to regress toward the mean, and his use of that word stuck, until, today, the way *y* behaves, on the average, with respect to *x* is still called the **regression of *y* on *x*.**

HERMAN

© 1982 Universal Press Syndicate

**''I think he's even taller than
I was at his age.''**

If complete population data is available, a regression curve can be constructed in the following way:

The range of *x*-values can be divided into classes. Within each of these classes,
the mean of all corresponding *y*-values can be computed and this value plotted
as a point above the mean of the particular *x*-class. When this has been done for
each *x*-class, the resulting points can be joined to produce the regression curve.
Once this curve has been discovered, any value of *x* can be chosen and a value
of *y* predicted from it, such as has been done for the hypothetical regression curve
shown in figure 14–1.

Figure 14-1

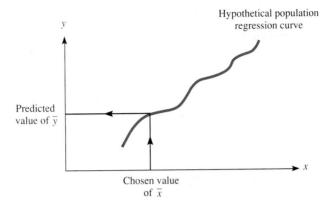

What is more, if the calculations have been performed correctly, then the predicted value of y should be the one that would produce the least amount of error, in the long run. This predicted value of y will be referred to, in this book, as $\hat{y}$, which can be read "y-hat."

What is more usual is that only sample data is available. The above method of constructing a regression curve, then, will be less reliable.

Because of the limited amount of data available in a sample, when the x-values are divided into classes, there are liable to be so few y-values in each class that a mean of these values has a strong possibility of being unreliable. In fact, with a small sample, it may turn out that some of the x-classes have no corresponding y-values at all, making the calculation of the mean impossible. It is possible, however, to use the sample data (even if the sample is fairly small) to *estimate* the regression curve, if certain assumptions are made.

The first assumption is that the regression curve is, in fact, a straight line. If each of the two populations (of x-values and y-values) follows a normal distribution *exactly,* then the regression curve *will be* a straight line, and if the population does not deviate too badly from the normal distribution, this assumption is still a reasonable one to make. It is because a straight line approximation of the relationship between y and x is being estimated that this process is referred to as *linear* regression analysis.

The second assumption is that the straight line (promised by the first assumption) passes through the point $(\bar{x}, \bar{y})$. This is based on the fact that the regression *curve* passes through the point (μ_x, μ_y). Since the regression curve is constructed by plotting a point for each x-class that corresponds to the mean y-value for that class, it makes common sense that the mean of *all* y-values (μ_y) should correspond to the mean of *all* x-values (μ_x). When dealing with sample data alone, the values of μ_x and μ_y are quite likely unknown. The best estimates of μ_x and μ_y are $\bar{x}$ and $\bar{y}$, respectively.

Since you will be involved in analysis of straight lines to determine the regression line, it is important that you know something about the algebra behind the graphing of straight lines, a topic that is usually covered in any beginning or intermediate algebra course. If your knowledge of this topic is sketchy, you are strongly encouraged to read and understand appendix A before reading any further in this chapter.

Finding the Regression Line

Suppose the scattergram for some sample data looks like the one in figure 14–2.

Figure 14–2

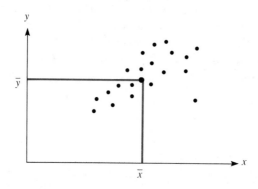

The point $(\bar{x}, \bar{y})$ has been plotted because it is known, from the second assumption above, that the line being sought passes through this point. But, there are an infinite number of lines that pass through any given point—somehow, the *one* line that is the regression line must be found. From the point-slope form of the equation of a straight line, it is known that the equation of any line through the point $(\bar{x}, \bar{y})$ is of the form

See appendix A if you require an explanation of this terminology.

$$y - \bar{y} = m(x - \bar{x}),$$

where m is the slope of the line. This is usually rewritten by transposing $\bar{y}$ to the right-hand side and replacing y by $\hat{y}$, since the result of applying this equation is a *predicted* value for y, not a real one. Thus, the equation becomes

$$\hat{y} = \bar{y} + m(x - \bar{x}).$$

This is not the final form, however, since m, the slope of the line, is not known yet.

In figure 14–3, an arbitrary line has been drawn on the scattergram of figure 14–2.

Figure 14–3

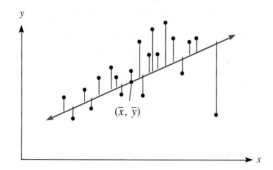

Vertical line segments have been drawn to illustrate the errors in prediction. The lengths of these segments represent the differences between actual values and predicted values of y, that is, $(y - \hat{y})$. These differences can be either positive or negative, so they are squared and the regression line is chosen to be the line that produces the least total squared differences. This process is called the **least-squares method** of fitting a straight line to the points of a scattergram. In this sense, the regression line will be the **best-fitting line** to the points of the scattergram. It can be shown, using algebra, that the value of m that minimizes $\Sigma(y - \hat{y})^2$ is

$$m = \frac{rs_y}{s_x},$$

where r is the correlation coefficient and s_x, s_y are the standard deviations of the two data sets. Substituting this value of m into the above equation for the line, the equation of the regression line, based on sample data, becomes

$$\hat{y} = \bar{y} + \frac{rs_y}{s_x}(x - \bar{x}). \tag{56}$$

In this equation, the quantities $\bar{x}$, $\bar{y}$, s_x, s_y, r are numbers that can be calculated from the sample data. The quantities x, $\hat{y}$ are not numbers but remain as variables in the equation, as is true for the equation of any line. After the equation has been determined, different values of x can be substituted into the equation, in turn, to find the corresponding predicted values for y. Finally, since a point $(\bar{x},\bar{y})$ is known to be on the line and the slope m is known, the regression line can be drawn on the scattergram.

Example 14–1 The bivariate data of example 13–1 showed the scores of elementary schoolchildren in tests of language skill (x) and arithmetic skill (y). (a) Find and (b) graph the equation of the regression line for the regression of y on x.

Solution In chapter 13, the following quantities were calculated: $\bar{x} = 65.3$, $s_x = 23.6$, $\bar{y} = 67.4$, $s_y = 19.9$, $r = 0.66$.

Using formula (56), the equation of the regression line is

Warning: Do not add the 67.4 to the 0.56. The 0.56 is multiplying the parentheses and multiplication is always done before addition.

$$\hat{y} = 67.4 + \frac{(0.66)(19.9)}{23.6}(x - 65.3)$$

or

$$\hat{y} = 67.4 + 0.56(x - 65.3).$$

This is the answer to part a. For part b, note that this equation is the equation of a line passing through the point (65.3,67.4) with slope = 0.56. To draw the line, first find the point, then go for a run of 10 and a rise of 5.6 (since slope = $\dfrac{\text{rise}}{\text{run}} = \dfrac{5.6}{10} = 0.56$). Figure 14–4 shows the line drawn on the scattergram.

Figure 14–4

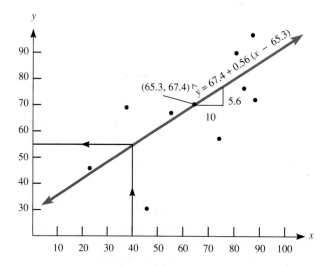

Notice that the line does not pass through even one of the nine points of the scattergram; but it is the line that "best fits" the nine points, in the sense that it is an "average" of the relationship between x and y that minimizes the sum of squared differences of actual y-values from their estimated y-values on the estimated regression line.   ●

Notice that in example 14–1, r = 0.66 and m = 0.56. Both are positive. It is always true that r, the correlation coefficient, and m, the slope of the corresponding regression line, have the same sign, that is, both positive, both negative, or both zero. This is true since $m = \dfrac{r s_y}{s_x}$ and the two standard deviations are always positive. It is also common sense that a scattergram exhibiting positive correlation should have a best-fitting line which has positive slope, while a scattergram exhibiting negative correlation should have a best-fitting line which has negative slope.

Example 14–2   Use the regression line found in example 14–1 to predict the arithmetic skill score of a child who scores (a) 85, (b) 40, (c) 65.3 on the language skill test.

Solution   Language skill score is x, while predicted arithmetic skill score is $\hat{y}$. Thus the problem is to find $\hat{y}$ for various values of x.

a. For $x = 85$, $\hat{y} = 67.4 + 0.56(85 - 65.3)$
                              $\hat{y} = 78.4$.

b. For $x = 40$, $\hat{y} = 67.4 + 0.56(40 - 65.3)$
                              $\hat{y} = 53.2$.

Figure 14–4 illustrates how the solution to this part could have been done using the graph alone. Starting at 40 on the x-axis, rise straight up to the line and then go horizontally across to the y-axis, reading off the answer of about 53. This method, although not as accurate, is very simple to do. Part a could have been done in the same way.

c. For $x = 65.3$, $\hat{y} = 67.4 + 0.56(65.3 - 65.3)$
                              $\hat{y} = 67.4$.

Thus, for $x = \bar{x}$, predict y as $\bar{y}$, as it should be.   ●

Exercise Set 14–1

1. If for a random sample of bivariate data it is found that $\bar{x} = 21.3$, $s_x = 4.7$, $\bar{y} = 29.2$, $s_y = 5.4$, $r = 0.78$, find the equation of the line of regression of y on x.

2. Use the line you found in exercise 1 to predict y when $x = 25$.

3. The following data is for a random sample of cars, giving age in years and how far driven in thousands of miles.

Age,	x	5	2	7	3	1	5	4	6	2	11	13
Miles driven,	y	72	31	102	58	19	84	42	62	38	124	156

Find the equation of the regression line that would enable you to predict y from x.

4. Draw a scattergram for the data in exercise 3 and draw the regression line on it.

5. Use the line found in exercise 3 to predict how many miles an eight-year-old car has been driven.

6. Each of the sample of cars described in exercise 3 was evaluated for resale value, with these results:

Age,	x	5	2	7	3	1	5	4	6	2	11	13
Resale value in $100s,	y	22	56	16	49	74	26	30	18	48	11	6

Find the equation of the regression line that would enable you to predict y from x.

7. Draw a scattergram for the data in exercise 6 and draw the regression line on it.

8. Use the line found in exercise 6 to predict the value of an eight-year-old car.

9. You are told that for sample bivariate data $r = 0.84$ and the equation of the regression line is

$$\hat{y} = 36.1 - 2.3(x - 10.1).$$

Comment on these results. *THIS IS WRONG – Should be positive*

10. You are told that for some bivariate data $r = -0.92$ and the predicted values for y are 56.2 and 72.3 for $x = 15$ and 25, respectively. Comment on these results.

FALSE *r should reflect trend*

R VALUE AND COEFFICIENT OF r ARE ALWAYS THE SAME SIGN

Section 14-2 Error in Prediction

In prediction, as in earlier estimations of population parameters, interval estimates provide a sense of how reliable our estimates are, something that is lacking when only point estimates are presented. In a similar fashion, it is more informative to attach a standard error to the prediction $\hat{y}$ rather than letting it stand alone.

Suppose you knew nothing about a population of men's heights except the mean μ_y. You then sit down and call a random sample of these men on the telephone. In each case, you predict the man's height, determine the man's true height, and compute the error. How should the predictions be made so that the least total amount of error is committed? The answer to this question is that the best course of action is to use μ_y as your prediction of the actual height y for each man whom you call. The standard error of each prediction is σ_y. Since heights generally follow a normal distribution, you would expect 68% of your predictions to be within one standard deviation of their true values, and 95% to be within two standard deviations.

An important objective of regression analysis is to use information about the relationship between y and a second variable x to improve one's ability to predict y. In the case of men's heights, your prediction might be more accurate if you knew something about parents' heights, and it could be determined that individual heights depend (for whatever reason) on parents' heights. With weak correlation, you cannot do much better with your predictions than having a standard error close to the standard deviation of the y's. As the correlation gets stronger, the value of the standard error of prediction gets closer to zero.

When using the regression line to make predictions, the extra knowledge imparted by the value of the correlation coefficient r should allow you to make more accurate predictions then when you knew only the mean of y. In fact, the stronger the correlation, the better the predictions should become. The equation of the line of regression of y on x was found to be

$$\hat{y} = \bar{y} + \frac{r s_y}{s_x}(x - \bar{x}). \tag{56}$$

When r is close to zero, $\hat{y}$ will be close to $\bar{y}$ for every value of x, and then the standard error of prediction should be close to s_y, the best estimate of the population standard deviation. When r is close to 1 or -1, the points of the scattergram will be close to lying on a straight line, and you should be able to make predictions with errors close to zero. It is for this reason that the standard error of prediction $SE_{\hat{y}}$ is *approximated* by

$$SE_{\hat{y}} = s_y\sqrt{1 - r^2}. \tag{57}$$

In this formula, when $r = 0$, $SE_{\hat{y}} = s_y$, and when r is close to zero, $SE_{\hat{y}}$ will not differ from s_y by much. When $r = \pm 1$, $SE_{\hat{y}} = 0$, showing no error in prediction, as it should be when all the points of the scattergram lie on a straight line. When r is close to ± 1, $SE_{\hat{y}}$ will be close to zero, showing very little error. Thus, formula (57) conforms with the ideas in the paragraph immediately before it.

Using these ideas, if an equation of a line of regression of y on x is discovered, and $\hat{y}$ computed for a particular value of x, then, instead of presenting $\hat{y}$ alone, couple it with $SE_{\hat{y}}$ as follows:

Corresponding to a particular value of x, the predicted value for y would be in the interval $\hat{y} \pm zSE_{\hat{y}}$, where, as usual, the amount of confidence you could claim for this result depends on the confidence level corresponding to the value of z.

There are formulas in existence that are more accurate than formula (57), but they are not considered in this book.

Actually, formula (57) works best for populations that follow closely to a normal distribution, and, even then, not so well in the tails of the distribution. For populations that deviate substantially from a normal distribution, the value obtained from formula (57) should be viewed with suspicion. Even when you are satisfied that what you are dealing with is a normal distribution, or close to it, you should be wary when applying formula (57) to predictions far from the center. Because it is often the case that strange things can happen with extreme (high or low) values, predictions can be much less accurate when dealing with these values than when dealing with values close to the mean.

It is for these reasons that formula (57) only provides a very rough approximation for the accuracy in predictions, and it should not be taken as something that is always super-accurate.

Example 14-3 Find the error in predictions for the data on language and arithmetic skills used in the previous examples in this chapter.

Solution Recall that $s_y = 19.9$ and $r = 0.66$. Using formula (57), get

$$SE_{\hat{y}} = 19.9\sqrt{1 - (0.66)^2}$$

$$= 14.95$$

$$\approx 15.0.$$

This result can be interpreted by saying that when predicting arithmetic skill score from language skill score, there is a 68% probability of being within 15 points of the actual value. You may be thinking that this is not very impressive; but, then, the correlation coefficient 0.66 is not that impressive. In fact, the quantity $\sqrt{1 - (0.66)^2}$ is equal to 0.75. Thus, the standard error of prediction is 75% of what it would have been had there been zero correlation. This, looking at it another way, is a mere 25% reduction.

In fact, $\sqrt{1 - r^2} = 50\%$ when $r \approx 0.87$. Thus, you can only get a 50% or more reduction in error of prediction when the correlation coefficient is 0.87 or higher.

Example 14–4 Usually in business, the larger quantity that you buy, the cheaper the unit price. Here is a sample of different amounts of a certain item, together with the unit price.

Number of items, x	1	2	4	7	10	15	20	30	40	50
Unit price, y	50	48	45	40	37	30	30	20	20	18

a. Find r.
b. Find the equation of the line of regression of y on x.
c. Predict y if $x = 12$.
d. Find the standard error of prediction.

Solution Verify for yourself that $n = 10$; $\Sigma x = 179$; $\Sigma x^2 = 5{,}795$; $\Sigma y = 338$; $\Sigma y^2 = 12{,}722$; $\Sigma xy = 4{,}326$.

a. $r = \dfrac{10(4{,}326) - (179)(338)}{\sqrt{10(5{,}795) - (179)^2}\sqrt{10(12{,}722) - (338)^2}}$

$r \approx -0.94$

b. $\bar{x} = 17.9,\ s_x = \sqrt{\dfrac{5{,}795}{10} - (17.9)^2} = 16.1$

$\bar{y} = 33.8,\ s_y = \sqrt{\dfrac{12{,}722}{10} - (33.8)^2} = 11.4$

Then, the equation of the regression line is

$$\hat{y} = 33.8 + \frac{(-0.94)(11.4)}{16.1}(x - 17.9)$$

or

$$\hat{y} = 33.8 - 0.67(x - 17.9).$$

c. When $x = 12$, $\hat{y} = 33.8 - 0.67(12 - 17.9)$
$\hat{y} = 37.8$.

d. $SE_{\hat{y}} = 11.4\sqrt{1 - (-0.94)^2}$
$= 3.89$

This can be used together with the solution to part c to say that when $x = 12$, predict that y is between $37.8 - 3.89$ and $37.8 + 3.89$, or in the interval 33.91 to 41.69, and you can be 68% confident of this result. •

This example illustrates the application of the formulas of this chapter when r is negative. Figure 14–5 shows the scattergram of the data with the regression line drawn in.

Figure 14–5

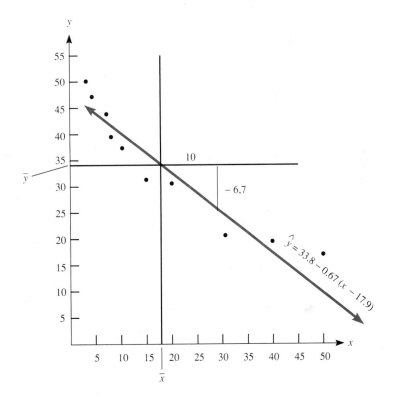

The Regression of x on y

The foregoing discussion and formulas have all applied to the regression of y on x. Sometimes it is desired to perform the regression the other way around, that is, the **regression of x on y**. A simple procedure for accomplishing this is to start from the beginning with the data interchanged, that is, naming the data that was previously labeled x, y; and naming the data that was previously labeled y, x. The same formulas would be used, but with a different regression line resulting.

Sometimes researchers wish to examine both regression lines on the same scatter diagram. In that case, the regression of x on y can be performed by using the previous formulas but replacing x by y and y by x everywhere they occur. Thus, the formula for the equation of the line of regression of x on y becomes

$$\hat{x} = \bar{x} + \frac{rs_x}{s_y}(y - \bar{y}) \qquad (58)$$

and the standard error of prediction becomes

$$SE_{\hat{x}} = s_x\sqrt{1 - r^2} . \tag{59}$$

This new regression line is constructed in a similar way to the original regression line except that instead of minimizing the vertical deviations, minimize the horizontal ones. Necessarily, a different line will result.

The graphs at the beginning of the chapter show three regressions of y on x. Here y represents, in each case, a different measurement of the mandible (or jaw) of a macaque, a species of monkey (each measurement is illustrated above the graph). Also, here x is the age of the macaque, as indicated in the equations below each graph.

The three graphs illustrate, quite effectively, how a macaque's jaw grows with age.

Exercise Set 14–2

1. If you have performed a regression of y on x and have found that $s_y = 4.6$, $r = 0.6$, what would the standard error of prediction $SE_{\hat{y}}$ be equal to?

2. If you have performed a regression of y on x and have found that $s_y = 7.2$, $r = -0.84$, what would the standard error of prediction $SE_{\hat{y}}$ be equal to?

3. A psychologist wishes to examine a sample of subjects' susceptibility to hypnotism. The same group is tested on two separate days to see if there is general agreement in the test scores. The following data was produced:

Test score on day 1, x	6	7	8	13	2	1	5	6	6	9	11	10
Test score on day 2, y	5	10	9	13	2	2	4	5	6	4	10	10
x	2	5	5	7	9	8	2	1	2	3	4	6
y	2	6	7	9	13	12	2	1	4	4	5	7
x	13	9	4	3	2	2						
y	13	11	5	4	4	3						

Calculate the correlation coefficient r.

4. For the data of exercise 3, find the equation of the line of regression of y on x.

5. Use the equation you found in exercise 4 to predict y when $x = 5$.

6. Find a 68% confidence interval for the predicted value of y found in exercise 5.

7. A sociologist gives a test of "intelligence" to a random sample of high school students and then another test in which the questions are stated in or are about black cultural terms. Both black and white students were included in the sample. Here are the results.

IQ test score, x	105	120	95	130	100	115	125	102	82
"Black" test score, y	88	76	109	68	110	79	80	92	120
x	110	108	95	132	122	92	86		
y	91	120	118	72	75	125	108		

Find the correlation coefficient r.

8. For the data of exercise 7, find the equation of the line of regression of y on x.

9. Use the equation you found in exercise 7 to predict the score on the black culture test when a student scores 108 on the IQ test.

10. Find a 95% confidence interval for the predicted value of y found in exercise 9.

....

Section 14–3 Summary of Terms

best-fitting line 435
error in prediction 438
least-squares method 435
linear regression analysis 431

regression analysis 431
regression of x on y 431
regression of y on x 431

Section 14–4 Summary of Symbols

Symbol	Meaning
m	The slope of a line
$SE_{\hat{x}}$	The standard error of prediction when performing a regression of x on y
$SE_{\hat{y}}$	The standard error of prediction when performing a regression of y on x
$\hat{x}$	The predicted value of x in a regression of x on y
$\hat{y}$	The predicted value of y in a regression of y on x.

Section 14–5 Summary of Formulas

Formula	Use
(56) $\hat{y} = \bar{y} + \dfrac{rs_y}{s_x}(x - \bar{x})$	The equation of the line of regression of y on x
(57) $SE_{\hat{y}} = s_y\sqrt{1 - r^2}$	This is used to find the standard error of prediction in a regression of y on x. It is not the most precise formula available and gives only a rough value for the SE.
(58) $\hat{x} = \bar{x} + \dfrac{rs_x}{s_y}(y - \bar{y})$	The equation of the line of regression of x on y
(59) $SE_{\hat{x}} = s_x\sqrt{1 - r^2}$	This is used to find the standard error of prediction in a regression of x on y. It is not the most precise formula available and gives only a rough value for the SE.

Chapter 14 Review Exercises

The following data gives the mean number of cigarettes smoked per person (x) and the number of deaths per 100,000 due to lung cancer (y) for a random sample of U.S. cities.

x	y
1,850	17
1,800	16
3,100	23
4,050	27
2,000	14
2,600	20
2,200	17
2,150	25
2,600	26
2,500	23
1,600	16
2,400	20
4,250	23
2,100	15
2,000	12
2,350	19
2,900	24
2,100	14
2,300	21
2,600	21
2,100	21
2,800	16
3,350	25
1,800	17
1,400	12

1. Find the equation of the line of regression of y on x.

2. Find the equation of the line of regression of x on y.

3. Draw a scattergram for the data and on it draw the line you found in exercise 1.

4. On the scattergram of exercise 3, draw the line you found in exercise 2.

5. Predict the number of deaths per 100,000 a city would have due to lung cancer if the mean number of cigarettes smoked per person is 2,000.

6. Predict the mean number of cigarettes smoked per person in a city where there were 24 deaths per 100,000 due to lung cancer.

7. Find the correlation coefficient r for the cigarette/lung cancer data.

8. Find the rank correlation coefficient r' for the cigarette/lung cancer data.

9. Find the standard error of the prediction of exercise 5.

10. Find the standard error of the prediction of exercise 6.

11. Do the above data indicate that smoking causes lung cancer?

12. Do the above data indicate that a susceptibility to lung cancer causes more people to smoke?

 13. Using the graphs and equations given at the beginning of the chapter on mandibular growth of macaques, predict how large the measurement from S to Id is when a macaque is 3 years old.

14. Repeat exercise 13 for the measurement Me to Id when a macaque is 5 years old.

Hands-on Class/Student Project

This is a continuation of the Hands-on Class/Student Project of the previous chapter.

1. Retrieve the scattergram drawn in exercise 1 of Hands-on Class/Student Project exercises for chapter 13, illustrating your sample of SAT scores and GPA's.
2. On this scattergram, draw lines parallel to the axes corresponding to the means of the two variables x and y.
3. Compute the equation of the line of regression of y on x.
4. Draw this line on your scattergram.
5. Using the equation that you computed in exercise 3, predict GPA for a SAT score of 650.
6. Attach a standard error to your prediction of exercise 5.
7. Compute the equation of the line of regression of x on y.
8. Draw the line that you computed in exercise 7 on your scattergram in a different color from the line that you drew in exercise 4.
9. Using the equation that you computed in exercise 7, predict SAT score for a GPA of 3.2.
10. Attach a standard error to the prediction of exercise 9.

Algebra Review of Straight Lines

The Rectangular Coordinate System

Points can be plotted in a plane using the following system:

Figure A–1

The quadrants are labeled with Roman numerals, see chapter 13.

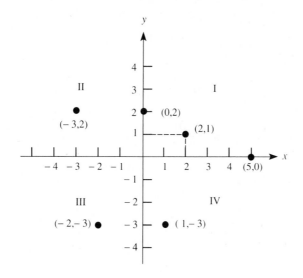

This system, called the rectangular coordinate system, consists of two number lines placed at right angles. The horizontal number line is called the *x-axis* and the vertical one is called the *y-axis*. Every point in the plane, referred to as the *xy*-plane, can then be labeled by an *ordered pair* of numbers. The pair is referred to as "ordered" since the order of the numbers matters—the point (2,1) is a *different* point in the *xy*-plane from (1,2). The first number in the pair, the *x-coordinate*, tells us how far to go in the *x* direction, while the second number, the *y-coordinate*, tells us how far to go in the *y* direction.

Equations

The equation $2x + 3 = x + 5$ has one solution, namely $x = 2$, since this is the only value of x that makes it true. The equation $2x + 3 = y + 5$, having two unknowns in it, has an infinite number of solutions. These solutions can be represented by ordered pairs, with the first number in the ordered pair corresponding to x and the second number corresponding to y. Some of the (infinite number) of solutions are

$$(1,0), (2,2), (3,4), (0,-2), (-1,-4), (-2,-6).$$

These ordered pairs are plotted as points in the xy-plane in figure A–2.

Figure A–2

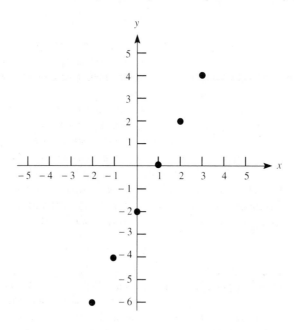

It is obvious that these points lie on a straight line. This straight line is called the *graph* of the equation $2x + 3 = y + 5$. What this means is that every ordered pair that is a solution of the equation also represents the coordinates of a point on the line, and, conversely, every point on the line has coordinates that form an ordered pair that is a solution of the equation.

Figure A–3

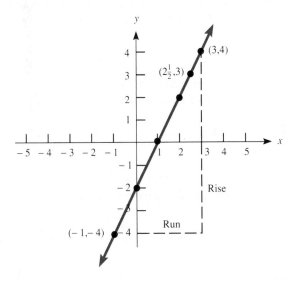

For instance, a point has been found on the line with coordinates $(2\frac{1}{2}, 3)$. Is this ordered pair a solution of the equation? Then,

$$2(2\frac{1}{2}) + 3 \text{ should equal } 3 + 5,$$

and it does, since both quantities are equal to 8.

Arrows have been drawn at each end of the line, to illustrate that the line, in fact, has no ends—it goes forever in both directions.

Solve the equation $2x + 3 = y + 5$ for y, to get

$$y = 2x - 2.$$

Notice that the graph crosses the y-axis at the point $(0, -2)$, and -2 is the last number in the above form of the equation. In the equation $y = mx + b$, it must always be true that $(0, b)$ is a solution of this equation. The point $(0, b)$ is the point where the line crosses, or intercepts, the y-axis. It is, therefore, called the y-*intercept*. The equation of a straight line can always be written in the form $y = mx + b$, where m and b are numbers. If there are any x^2, y^2, xy, or higher degree terms involved in the equation, then its graph is not a straight line. The value of b tells you that the y-intercept of the line is $(0, b)$. What, if anything, does the value of m tell you? To give the game away, it tells you the *slope* of the line. In mathematics,

$$\text{Slope} = \frac{\text{rise}}{\text{run}}.$$

In the graph above, two arbitrary points, (3,4) and $(-1,-4)$, on the line were chosen and the dotted lines drawn in. The "run" is measured in the x direction and, here, run = 4. The "rise" is measured in the y direction and, here, rise = 8. Therefore, slope = $\frac{8}{4}$ = 2. Notice that that is the value of m in the equation

$$y = 2x - 2.$$

The same value for m (that is, 2) would have been found if any other two points on the line had been chosen. Try it for yourself with a different pair of points.

Thus, the equation

$$y = mx + b$$

has a graph that is a straight line with slope = m and y-intercept $(0,b)$. It is therefore referred to as the *slope-intercept form* of the equation of a straight line.

Example A–1   Graph $x + 2 - y = 2x - 5$ by first putting the equation into slope-intercept form.

Solution

$$x + 2 - y = 2x - 5$$
$$x + 2 + 5 - 2x = y$$
$$y = -x + 7$$

Hence, $m = -1$, $b = 7$, which implies that slope = -1 and y-intercept is $(0,7)$. Now, -1 can be written as $-1/1$, so that for a run of 1, the rise is -1, which means a drop of 1. Starting at $(0,7)$, you can plot as many points as you wish by using the above rise and run.   ●

Figure A–4

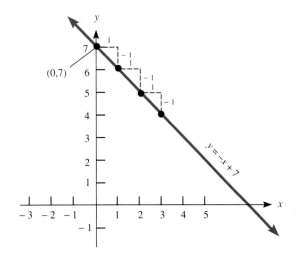

The Point-Slope Form

Suppose you know that the slope of a line is $\frac{3}{4}$ and that it passes through the point (1,2). What is its equation? Here is the graph.

Figure A–5

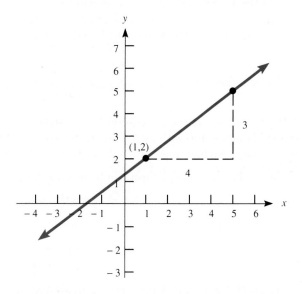

The line is easily constructed. From the given point (1,2), go for a run of 4 and then a rise of 3 (for the slope $\frac{3}{4}$) that produces what should be another point on the line. Join this point with the given point to give the line.

Since the slope is $\frac{3}{4}$, the line should have equation

$$y = \frac{3}{4}x + b,$$

but what is b? Now the point $(1,2)$ lies on the line. Hence its coordinates must be a solution of the equation. Hence,

$$2 = \frac{3}{4}(1) + b$$

giving

$$b = \frac{5}{4}$$

and so the equation is $y = \frac{3}{4}x + \frac{5}{4}$. This can be written,

$$y = \frac{3}{4}x - \frac{3}{4} + 2$$

$$2 = \frac{8}{4} \text{ and } \frac{-3}{4} + 2$$
$$= \frac{-3}{4} + \frac{8}{4} = \frac{5}{4}$$

or

$$y - 2 = \frac{3}{4}(x - 1).$$

The reason the equation has been written in this form is that now on the left is y minus the y-coordinate of the given point and, on the right, the given slope times x minus the x-coordinate of the given point.

This is always possible to do. If a line has slope $= m$ and passes through the point (p,q), then the equation of the line can be written

$$y - q = m(x - p)$$

and this is referred to as the *point-slope form* of the equation of a straight line.

Example A–2   Find the equation of the straight line having slope -0.63 and passing through the point $(2.4, 2.8)$ in both point-slope and slope-intercept forms. Also, graph the line.

Solution   The point-slope form can be written down almost without thinking.

$$y - 2.8 = -0.63(x - 2.4)$$

To obtain the slope-intercept form, multiply and rearrange.

$$y - 2.8 = -0.63x + 1.512$$
$$y = -0.63x + 4.312$$

To graph the line, plot the point $(2.4, 2.8)$ and use the fact that the slope $= -0.63$ to take a run of 1 corresponding to a drop of 0.63.   ●

Figure A–6

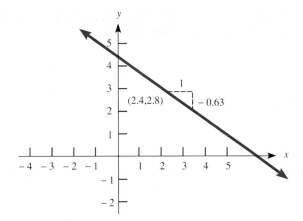

There is more to the graphing of straight lines than has been introduced here. This is a thumbnail sketch of the main points that you will need in your study of chapter 14. If you need more information about straight lines, please refer to almost any algebra text.

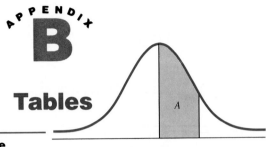

APPENDIX B

Tables

Table 1   The Normal Table

Note: To get *A* for a given value of *z*, insert a decimal point before the four digits. For example, *z* = 1.43 gives *A* = 0.4236.

z	0.00	0.01	0.02	0.03	0.04	0.05	0.06	0.07	0.08	0.09
0.0	0000	0040	0080	0120	0160	0199	0239	0279	0319	0359
0.1	0398	0438	0478	0517	0557	0596	0636	0675	0714	0753
0.2	0793	0832	0871	0910	0948	0987	1026	1064	1103	1141
0.3	1179	1217	1255	1293	1331	1368	1406	1443	1480	1517
0.4	1554	1591	1628	1664	1700	1736	1772	1808	1844	1879
0.5	1915	1950	1985	2019	2054	2088	2123	2157	2190	2224
0.6	2257	2291	2324	2357	2389	2422	2454	2486	2517	2549
0.7	2580	2611	2642	2673	2704	2734	2764	2794	2823	2852
0.8	2881	2910	2939	2967	2995	3023	3051	3078	3106	3133
0.9	3159	3186	3212	3238	3264	3289	3315	3340	3365	3389
1.0	3413	3438	3461	3485	3508	3531	3554	3577	3599	3621
1.1	3643	3665	3686	3708	3729	3749	3770	3790	3810	3830
1.2	3849	3869	3888	3907	3925	3944	3962	3980	3997	4015
1.3	4032	4049	4066	4082	4099	4115	4131	4147	4162	4177
1.4	4192	4207	4222	4236	4251	4265	4279	4292	4306	4319
1.5	4332	4345	4357	4370	4382	4394	4406	4418	4429	4441
1.6	4452	4463	4474	4484	4495	4505	4515	4525	4535	4545
1.7	4554	4564	4573	4582	4591	4599	4608	4616	4625	4633
1.8	4641	4649	4656	4664	4671	4678	4686	4692	4699	4706
1.9	4713	4719	4726	4732	4738	4744	4750	4756	4761	4767
2.0	4772	4778	4783	4788	4793	4798	4803	4808	4812	4817
2.1	4821	4826	4830	4834	4838	4842	4846	4850	4854	4857
2.2	4861	4864	4868	4871	4875	4878	4881	4884	4887	4890
2.3	4893	4896	4898	4901	4904	4906	4909	4911	4913	4916
2.4	4918	4920	4922	4925	4927	4929	4931	4932	4934	4936
2.5	4938	4940	4941	4943	4945	4946	4948	4949	4951	4952
2.6	4953	4955	4956	4957	4959	4960	4961	4962	4963	4964
2.7	4965	4966	4967	4968	4969	4970	4971	4972	4973	4974
2.8	4974	4975	4976	4977	4977	4978	4979	4979	4980	4981
2.9	4981	4982	4982	4983	4984	4984	4985	4985	4986	4986
3.0	4987	4987	4987	4988	4988	4989	4989	4989	4990	4990
3.1	4990	4991	4991	4991	4992	4992	4992	4992	4993	4993
3.2	4993	4993	4994	4994	4994	4994	4994	4995	4995	4995
3.3	4995	4995	4996	4996	4996	4996	4996	4996	4996	4997
3.4	4997	4997	4997	4997	4997	4997	4997	4997	4998	4998
3.5	4998	4998	4998	4998	4998	4998	4998	4998	4998	4998

The area, *A*, stays at 0.4998 until *z* = 3.62. From *z* = 3.63 to 3.90, *A* = 0.4999. For *z* > 3.90, *A* = 0.5000, to four decimal places.

Table 2 Student's t Distribution

Degrees of Freedom

t	1	2	3	4	5	6	7	8	9	10	11	12	13	14	15	16
0.1	0317	0353	0367	0374	0379	0382	0384	0386	0387	0388	0389	0390	0391	0391	0392	0392
0.2	0628	0700	0729	0744	0753	0760	0764	0768	0770	0773	0774	0776	0777	0778	0779	0780
0.3	0928	1038	1081	1104	1119	1129	1136	1141	1145	1148	1151	1153	1155	1157	1159	1160
0.4	1211	1361	1420	1452	1472	1485	1495	1502	1508	1512	1516	1519	1522	1524	1526	1528
0.5	1476	1667	1743	1783	1809	1826	1838	1847	1855	1861	1865	1869	1873	1876	1878	1881
0.6	1720	1953	2046	2096	2127	2148	2163	2174	2183	2191	2197	2202	2206	2210	2213	2215
0.7	1944	2218	2328	2387	2424	2449	2467	2481	2492	2501	2508	2514	2519	2523	2527	2530
0.8	2148	2462	2589	2657	2700	2729	2750	2766	2778	2788	2797	2804	2810	2815	2819	2823
0.9	2333	2684	2828	2905	2953	2986	3010	3028	3042	3054	3063	3071	3078	3083	3088	3093
1.0	2500	2887	3045	3130	3184	3220	3247	3267	3283	3296	3306	3315	3322	3329	3334	3339
1.1	2651	3070	3242	3335	3393	3433	3461	3483	3501	3514	3526	3535	3544	3551	3557	3562
1.2	2789	3235	3419	3518	3581	3623	3654	3678	3696	3711	3723	3734	3742	3750	3756	3762
1.3	2913	3384	3578	3683	3748	3793	3826	3851	3870	3886	3899	3910	3919	3927	3934	3940
1.4	3026	3518	3720	3829	3898	3945	3979	4005	4025	4041	4055	4066	4075	4084	4091	4097
1.5	3128	3638	3847	3960	4030	4079	4114	4140	4161	4177	4191	4203	4212	4221	4228	4235
1.6	3222	3746	3960	4075	4148	4196	4232	4259	4280	4297	4310	4322	4332	4340	4348	4354
1.7	3307	3844	4062	4178	4251	4300	4335	4363	4383	4400	4414	4426	4435	4444	4451	4458
1.8	3386	3932	4152	4269	4341	4390	4426	4452	4473	4490	4503	4515	4525	4533	4540	4546
1.9	3458	4026	4232	4349	4421	4469	4504	4530	4551	4567	4580	4591	4601	4609	4616	4622
2.0	3524	4082	4303	4419	4490	4538	4572	4597	4617	4633	4646	4657	4666	4674	4680	4686
2.1	3585	4147	4367	4482	4551	4598	4631	4655	4674	4690	4702	4712	4721	4728	4735	4740
2.2	3642	4206	4424	4537	4605	4649	4681	4705	4723	4738	4750	4759	4768	4774	4781	4786
2.3	3695	4259	4475	4585	4651	4694	4725	4748	4765	4779	4790	4799	4807	4813	4819	4824
2.4	3743	4308	4521	4628	4692	4734	4763	4784	4801	4813	4824	4832	4840	4846	4851	4855
2.5	3789	4352	4561	4666	4728	4767	4795	4815	4831	4843	4852	4860	4867	4873	4877	4882

Table 2 *continued*

Degrees of Freedom

t	17	18	19	20	21	22	23	24	25	26	27	28	29	30	∞
0.1	0392	0393	0393	0393	0394	0394	0394	0394	0394	0394	0395	0395	0395	0395	0398
0.2	0781	0781	0782	0782	0783	0783	0784	0784	0785	0785	0785	0785	0786	0786	0793
0.3	0928	1162	1163	1164	1164	1165	1166	1166	1167	1167	1168	1168	1168	1169	1179
0.4	1529	1531	1532	1533	1534	1535	1536	1537	1537	1538	1538	1539	1540	1540	1554
0.5	1883	1884	1886	1887	1889	1890	1891	1892	1893	1894	1894	1895	1896	1896	1915
0.6	2218	2220	2222	2224	2225	2227	2228	2229	2230	2231	2232	2233	2234	2235	2257
0.7	2533	2536	2538	2540	2542	2544	2545	2547	2548	2549	2550	2551	2552	2553	2580
0.8	2826	2829	2832	2834	2837	2839	2841	2842	2844	2845	2847	2848	2849	2850	2881
0.9	3097	3100	3103	3106	3108	3111	3113	3115	3116	3118	3120	3121	3122	3124	3159
1.0	3343	3347	3351	3354	3357	3359	3361	3364	3366	3367	3369	3371	3372	3373	3413
1.1	3567	3571	3575	3578	3581	3584	3586	3589	3591	3593	3595	3597	3598	3600	3643
1.2	3767	3772	3776	3779	3782	3785	3788	3791	3793	3795	3797	3799	3801	3802	3849
1.3	3945	3950	3954	3958	3962	3965	3968	3970	3973	3975	3977	3979	3981	3982	4032
1.4	4103	4107	4112	4116	4119	4123	4126	4128	4131	4133	4136	4138	4139	4141	4192
1.5	4240	4245	4250	4254	4258	4261	4264	4267	4269	4272	4274	4276	4278	4280	4332
1.6	4360	4365	4370	4374	4377	4381	4384	4387	4389	4392	4394	4396	4398	4400	4452
1.7	4463	4468	4473	4477	4481	4484	4487	4490	4492	4495	4497	4499	4501	4503	4554
1.8	4552	4557	4561	4565	4569	4572	4575	4578	4580	4583	4585	4587	4589	4590	4641
1.9	4627	4632	4636	4640	4644	4647	4650	4652	4655	4657	4659	4661	4663	4665	4713
2.0	4691	4696	4700	4704	4707	4710	4713	4715	4718	4720	4722	4724	4725	4727	4772
2.1	4745	4750	4753	4757	4760	4763	4766	4768	4770	4772	4774	4776	4777	4779	4821
2.2	4790	4794	4798	4801	4804	4807	4809	4812	4814	4816	4817	4819	4820	4822	4861
2.3	4828	4832	4835	4838	4841	4843	4846	4848	4850	4851	4853	4854	4856	4857	4893
2.4	4859	4863	4866	4869	4871	4874	4876	4877	4879	4881	4882	4884	4885	4886	4918
2.5	4885	4888	4891	4894	4896	4898	4900	4902	4903	4905	4906	4907	4908	4909	4938
2.6	4906	4910	4912	4914	4916	4918	4920	4921	4923	4924	4925	4926	4927	4928	4953
2.7	4924	4927	4929	4931	4933	4935	4936	4937	4939	4940	4941	4942	4943	4944	
2.8	4938	4941	4943	4945	4946	4948	4949	4950	4951	4952	4953	4954	4955	4956	
2.9	4950	4952	4954	4956	4957	4958	4960								

Note: The column headed by ∞ is for use when $df > 30$. It coincides with column 1 of table 1.

For missing entries, see the note on the next page, where values of A are given for larger values of t.

For $t > 2.9$ and $df > 16$, $A > 0.4950$.

457

Table 2 *continued*

	Degrees of Freedom															
t	1	2	3	4	5	6	7	8	9	10	11	12	13	14	15	16
2.6	3831	4392	4598	4700	4759	4797	4823	4842	4856	4868	4877	4884	4890	4895	4900	4903
2.7	3871	4429	4631	4730	4786	4822	4847	4865	4878	4888	4897	4903	4909	4914	4918	4921
2.8	3908	4463	4661	4756	4810	4844	4867	4884	4896	4906	4914	4920	4925	4929	4933	4936
2.9	3943	4494	4687	4779	4831	4863	4885	4901	4912	4921	4928	4933	4938	4942	4945	4948
3.0	3976	4523	4712	4800	4850	4880	4900	4915	4925	4933	4940	4945	4949	4952	4955	4958
3.1	4007	4549	4734	4819	4866	4894	4913	4927	4936	4944	4949	4954	4958			
3.2	4036	4573	4753	4835	4880	4907	4925	4937	4946	4953	4958					
3.3	4063	4596	4771	4850	4893	4918	4934	4946	4954							
3.4	4089	4617	4788	4864	4904	4928	4943	4953								
3.5	4114	4636	4803	4876	4914	4936	4950									
3.6	4138	4654	4816	4886	4922	4943										
3.7	4160	4670	4829	4896	4930	4950										
3.8	4181	4686	4840	4904	4937											
3.9	4201	4701	4850	4912	4943											
4.0	4220	4714	4860	4919	4948											
4.1	4239	4727	4869	4926	4953											
4.2	4256	4739	4877	4932												
4.3	4273	4750	4884	4937												
4.4	4289	4760	4891	4942												
4.5	4304	4770	4898	4946												
4.6	4319	4779	4903	4950												
4.7	4333	4788	4909													
4.8	4346	4796	4914													
4.9	4359	4804	4919													
5.0	4372	4811	4923													
5.8	4456	4858	4949													
6.4	4507	4882														
7.0	4548	4901														
10.0	4683	4951														
12.8	4752															
31.9	4900															
63.7	4950															

All missing entries in the table have $A > 0.4950$. The reason this value of A was chosen is that for $A > 0.495$, P-value < 0.005. Therefore, if the value you want for A is missing, you may assume significance at a significance level of 0.005 or higher.

For $t > 5.0$, selected values for t are given for $df = 1, 2,$ or 3, to show when significance is reached at various significance levels.

For example, $t = 7.0, df = 2$ gives $A = 0.4901$, P-value $= 0.0099 (< 0.01)$. Therefore, there is significance at the .01 level. Any value of t less than 7.0 would not be significant at the .01 level.

Table 3 Values of *t* for Selected Confidence Levels

df	0.68	0.80	0.85	0.90	0.925	0.95	0.975	0.98	0.99	0.995
1	1.82	3.08	4.17	6.31	8.45	12.7	25.5	31.8	63.7	127
2	1.31	1.89	2.28	2.92	3.44	4.30	6.21	6.97	9.92	14.1
3	1.19	1.64	1.92	2.35	2.68	3.18	4.18	4.54	5.84	7.45
4	1.13	1.53	1.78	2.13	2.39	2.78	3.50	3.75	4.60	5.60
5	1.10	1.48	1.70	2.02	2.24	2.57	3.16	3.36	4.03	4.77
6	1.08	1.44	1.65	1.94	2.15	2.45	2.97	3.14	3.71	4.35
7	1.07	1.41	1.62	1.89	2.09	2.36	2.84	3.00	3.50	4.03
8	1.06	1.40	1.59	1.86	2.05	2.31	2.75	2.90	3.36	3.83
9	1.05	1.38	1.57	1.83	2.01	2.26	2.69	2.82	3.25	3.69
10	1.05	1.37	1.55	1.81	1.99	2.23	2.63	2.76	3.17	3.58
11	1.04	1.36	1.55	1.80	1.97	2.20	2.59	2.72	3.11	3.50
12	1.04	1.36	1.54	1.78	1.95	2.18	2.56	2.68	3.05	3.43
13	1.03	1.35	1.53	1.77	1.94	2.16	2.53	2.65	3.01	3.37
14	1.03	1.35	1.52	1.76	1.92	2.14	2.51	2.62	2.98	3.33
15	1.03	1.34	1.52	1.75	1.91	2.13	2.49	2.60	2.95	3.29
16	1.03	1.34	1.51	1.75	1.90	2.12	2.47	2.58	2.92	3.25
17	1.02	1.33	1.51	1.74	1.90	2.11	2.46	2.57	2.90	3.22
18	1.02	1.33	1.50	1.73	1.89	2.10	2.45	2.55	2.88	3.20
19	1.02	1.33	1.50	1.73	1.88	2.09	2.43	2.54	2.86	3.17
20	1.02	1.33	1.50	1.72	1.88	2.09	2.42	2.53	2.85	3.15
21	1.02	1.32	1.50	1.72	1.87	2.08	2.41	2.52	2.83	3.14
22	1.02	1.32	1.49	1.72	1.87	2.07	2.41	2.51	2.82	3.12
23	1.02	1.32	1.49	1.71	1.86	2.07	2.40	2.50	2.81	3.10
24	1.02	1.32	1.49	1.71	1.86	2.06	2.39	2.49	2.80	3.09
25	1.01	1.32	1.49	1.71	1.86	2.06	2.38	2.49	2.79	3.08
26	1.01	1.31	1.48	1.71	1.85	2.06	2.38	2.48	2.78	3.07
27	1.01	1.31	1.48	1.70	1.85	2.05	2.37	2.47	2.77	3.06
28	1.01	1.31	1.48	1.70	1.85	2.05	2.37	2.47	2.76	3.05
29	1.01	1.31	1.48	1.70	1.85	2.05	2.36	2.46	2.76	3.04
30	1.01	1.31	1.48	1.70	1.84	2.04	2.36	2.46	2.75	3.03
∞	1.00	1.28	1.44	1.64	1.78	1.96	2.24	2.33	2.58	2.81

If you want to construct a 0.95 CIE with $df = 15$, use $t = 2.13$. Use the last line when $df > 30$. The values in it coincide with the normal table values.

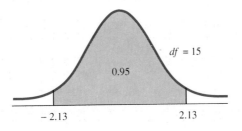

Table 4 Giving Critical Values of *R* for the Runs Test

All values are given at the 0.05 significance level. Read n_1 on the left and n_2 on top.

	5	6	7	8	9	10	11	12	13	14	15	16	17	18	19	20
2								2/6	2/6	2/6	2/6	2/6	2/6	2/6	2/6	2/6
3		2/9	2/9	2/10	2/10	2/10	2/10	2/10	2/10	2/10	3/10	3/10	3/10	3/10	3/10	3/10
4	2/9	2/9	2/10	3/10	3/10	3/10	3/10	3/10	3/10	3/10	3/10	4/10	4/10	4/10	4/10	4/10
5	2/10	3/10	3/11	3/11	3/12	3/12	4/12	4/12	4/12	4/12	4/12	4/12	4/12	5/12	5/12	5/12
6		3/11	3/12	3/12	4/13	4/13	4/13	4/13	5/14	5/14	5/14	5/14	5/14	5/14	5/14	5/14
7			3/13	4/13	4/14	5/14	5/14	5/14	5/15	5/15	5/15	6/16	6/16	6/16	6/16	6/16
8				4/14	5/14	5/15	5/15	6/16	6/16	6/16	6/16	6/17	7/17	7/17	7/17	7/17
9					5/15	5/16	6/16	6/16	6/17	7/17	7/18	7/18	7/18	8/18	8/18	8/18
10						6/16	6/17	7/17	7/18	7/18	7/18	8/19	8/19	8/19	8/20	9/20
11							7/17	7/18	7/19	8/19	8/19	8/20	9/20	9/20	9/21	9/21
12								7/19	8/19	8/20	8/20	9/21	9/21	9/21	10/22	10/22
13									8/20	9/20	9/21	9/21	10/22	10/22	10/23	10/23
14										9/21	9/22	10/22	10/23	10/23	11/23	11/24
15											10/22	10/23	11/23	11/24	11/24	12/25
16												11/23	11/24	11/25	12/25	12/25
17													11/25	12/25	12/26	13/26
18														12/26	13/26	13/27
19															13/27	13/27
20																14/28

Table 5 *P*-Values for χ^2

χ^2	Degrees of Freedom			χ^2	Degrees of Freedom					
	2	3	4		2	3	4	5	6	7
3.2	2019			7.2	0273	0658	1257	2062		
3.3	1920			7.3	0260	0629	1209	1993		
3.4	1827			7.4	0247	0602	1162	1926		
3.5	1738			7.5	0235	0576	1117	1860		
3.6	1653			7.6	0224	0550	1074	1797		
3.7	1572			7.7	0213	0526	1032	1736		
3.8	1496			7.8	0202	0503	0992	1676		
3.9	1423			7.9	0193	0481	0953	1618		
4.0	1353			8.0	0183	0460	0916	1562		
4.1	1287			8.1	0174	0440	0880	1508		
4.2	1225			8.2	0166	0421	0845	1456		
4.3	1165			8.3	0158	0402	0812	1405		
4.4	1108			8.4	0150	0384	0780	1355		
4.5	1054			8.5	0143	0367	0749	1307	2037	
4.6	1003	2035		8.6	0136	0351	0719	1261	1974	
4.7	0954	1951		8.7	0129	0336	0691	1216	1912	
4.8	0907	1870		8.8	0123	0321	0663	1173	1851	
4.9	0863	1793		8.9	0117	0307	0636	1131	1793	
5.0	0821	1718		9.0	0111	0293	0611	1091	1736	
5.1	0781	1646		9.1	0106	0280	0586	1051	1680	
5.2	0743	1577		9.2	0101	0267	0563	1013	1626	
5.3	0707	1511		9.3	0096	0256	0540	0977	1574	
5.4	0672	1447		9.4	0091	0244	0518	0941	1523	
5.5	0639	1386		9.5	0087	0233	0497	0907	1473	
5.6	0608	1328		9.6	0082	0223	0477	0874	1425	
5.7	0578	1272		9.7	0078	0213	0456	0842	1379	
5.8	0550	1218		9.8	0074	0203	0439	0811	1333	2002
5.9	0523	1166	2067	9.9	0071	0194	0421	0781	1289	1943
6.0	0498	1116	1991	10	0067	0186	0404	0752	1247	1886
6.1	0474	1068	1918							
6.2	0450	1023	1847							
6.3	0429	0979	1778							
6.4	0408	0937	1712							
6.5	0388	0897	1648							
6.6	0369	0858	1586							
6.7	0351	0821	1526							
6.8	0334	0786	1468							
6.9	0317	0752	1413							
7.0	0302	0719	1359							
7.1	0287	0688	1307							

All missing entries on this page give *P*-values > 0.20. See the next page for *P*-values corresponding to larger values of χ^2.

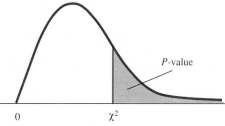

Table 5 *continued*

					Degrees of Freedom				
χ^2	2	3	4	5	6	7	8	9	10
11	0041	0117	0266	0514	0884	1386	2017		
12		0074	0174	0348	0620	1006	1512	2133	
13		0046	0113	0234	0430	0721	1118	1626	2237
14			0073	0156	0296	0512	0818	1223	1730
15			0047	0104	0203	0360	0591	0909	1321
16				0068	0138	0251	0424	0669	0996
17				0045	0093	0174	0301	0487	0744
18					0062	0120	0212	0352	0550
19					0042	0082	0149	0252	0403
20						0056	0103	0179	0293
21						0038	0071	0127	0211
22							0049	0089	0151
23								0062	0107
24								0043	0076
25									0053
26									0037

The missing entries in the top right-hand corner give P-value > 0.20. All other missing entries give P-value < 0.005.

Table 6   Critical Values of the *F* Distribution

$$\alpha = .05$$

						df_1						
df_2	1	2	3	4	5	6	7	8	9	10	11	12
1	161	200	216	225	230	234	237	239	241	242	243	244
2	18.5	19.0	19.2	19.2	19.3	19.3	19.4	19.4	19.4	19.4	19.4	19.4
3	10.1	9.6	9.3	9.1	9.0	8.9	8.9	8.8	8.8	8.8	8.8	8.7
4	7.7	6.9	6.6	6.4	6.3	6.2	6.1	6.0	6.0	6.0	5.9	5.9
5	6.6	5.8	5.4	5.2	5.1	5.0	4.9	4.8	4.8	4.7	4.7	4.7
6	6.0	5.1	4.8	4.5	4.4	4.3	4.2	4.2	4.1	4.1	4.0	4.0
7	5.6	4.7	4.4	4.1	4.0	3.9	3.8	3.7	3.7	3.6	3.6	3.6
8	5.3	4.5	4.1	3.8	3.7	3.6	3.5	3.4	3.4	3.4	3.3	3.3
9	5.1	4.3	3.9	3.6	3.5	3.4	3.3	3.2	3.2	3.1	3.1	3.1
10	5.0	4.1	3.7	3.5	3.3	3.2	3.1	3.1	3.0	3.0	2.9	2.9
11	4.8	4.0	3.6	3.4	3.2	3.1	3.0	3.0	2.9	2.8	2.8	2.8
12	4.8	3.9	3.5	3.3	3.1	3.0	2.9	2.8	2.8	2.8	2.7	2.7
13	4.7	3.8	3.4	3.2	3.0	2.9	2.8	2.8	2.7	2.7	2.6	2.6
14	4.6	3.7	3.3	3.1	3.0	2.8	2.8	2.7	2.6	2.6	2.6	2.5
15	4.5	3.7	3.3	3.1	2.9	2.8	2.7	2.6	2.6	2.6	2.5	2.5
16	4.5	3.6	3.2	3.0	2.8	2.7	2.7	2.6	2.5	2.5	2.4	2.4
17	4.4	3.6	3.2	3.0	2.8	2.7	2.6	2.6	2.5	2.4	2.4	2.4
18	4.4	3.6	3.2	2.9	2.8	2.7	2.6	2.5	2.5	2.4	2.4	2.3
19	4.4	3.5	3.1	2.9	2.7	2.6	2.5	2.5	2.4	2.4	2.3	2.3
20	4.4	3.5	3.1	2.9	2.7	2.6	2.5	2.4	2.4	2.4	2.3	2.3
21	4.3	3.5	3.1	2.8	2.7	2.6	2.5	2.4	2.4	2.3	2.3	2.2
22	4.3	3.4	3.0	2.8	2.7	2.6	2.5	2.4	2.3	2.3	2.3	2.2
23	4.3	3.4	3.0	2.8	2.6	2.5	2.4	2.4	2.3	2.3	2.2	2.2
24	4.3	3.4	3.0	2.8	2.6	2.5	2.4	2.4	2.3	2.3	2.2	2.2
25	4.2	3.4	3.0	2.8	2.6	2.5	2.4	2.3	2.3	2.2	2.2	2.2
26	4.2	3.4	3.0	2.7	2.6	2.5	2.4	2.3	2.3	2.2	2.2	2.2
27	4.2	3.4	3.0	2.7	2.6	2.5	2.4	2.3	2.2	2.2	2.2	2.1
28	4.2	3.3	3.0	2.7	2.6	2.4	2.4	2.3	2.2	2.2	2.2	2.1
29	4.2	3.3	2.9	2.7	2.6	2.4	2.4	2.3	2.2	2.2	2.1	2.1
30	4.2	3.3	2.9	2.7	2.5	2.4	2.3	2.3	2.2	2.2	2.1	2.1
40	4.1	3.3	2.9	2.7	2.5	2.4	2.3	2.3	2.2	2.1	2.0	2.0
50	4.0	3.2	2.8	2.6	2.4	2.3	2.2	2.1	2.1	2.0	2.0	2.0
60	4.0	3.2	2.8	2.5	2.4	2.2	2.2	2.1	2.0	2.0	2.0	1.9
70	4.0	3.1	2.7	2.5	2.4	2.2	2.1	2.1	2.0	2.0	1.9	1.9
80	4.0	3.1	2.7	2.5	2.3	2.2	2.1	2.0	2.0	2.0	1.9	1.9
100	3.9	3.1	2.7	2.5	2.3	2.2	2.1	2.0	2.0	1.9	1.9	1.8
120	3.9	3.1	2.7	2.4	2.3	2.2	2.1	2.0	2.0	1.9	1.9	1.8
∞	3.8	3.0	2.6	2.4	2.2	2.1	2.0	1.9	1.9	1.8	1.8	1.8

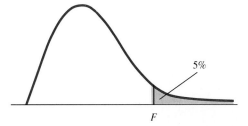

5%

F

Table 6   *continued*

df_2	15	20	24	30	40	$\alpha = .05$ df_1 50	60	75	100	120	∞
1	246	248	249	250	251	252	253	253	253	253	254
2	19.4	19.4	19.4	19.4	19.4	19.5	19.5	19.5	19.5	19.5	19.5
3	8.7	8.7	8.6	8.6	8.6	8.6	8.6	8.6	8.6	8.6	8.5
4	5.9	5.8	5.8	5.8	5.7	5.7	5.7	5.7	5.7	5.7	5.6
5	4.6	4.6	4.5	4.5	4.5	4.4	4.4	4.4	4.4	4.4	4.4
6	3.9	3.9	3.8	3.8	3.8	3.8	3.7	3.7	3.7	3.7	3.7
7	3.5	3.4	3.4	3.4	3.3	3.3	3.3	3.3	3.3	3.3	3.2
8	3.2	3.2	3.1	3.1	3.0	3.0	3.0	3.0	3.0	3.0	2.9
9	3.0	2.9	2.9	2.9	2.8	2.8	2.8	2.8	2.8	2.8	2.7
10	2.8	2.8	2.7	2.7	2.7	2.6	2.6	2.6	2.6	2.6	2.5
11	2.7	2.6	2.6	2.6	2.5	2.5	2.5	2.5	2.4	2.4	2.4
12	2.6	2.5	2.5	2.5	2.4	2.4	2.4	2.4	2.4	2.3	2.3
13	2.5	2.5	2.4	2.4	2.3	2.3	2.3	2.3	2.3	2.2	2.2
14	2.5	2.4	2.4	2.3	2.3	2.2	2.2	2.2	2.2	2.2	2.1
15	2.4	2.3	2.3	2.2	2.2	2.2	2.2	2.2	2.1	2.1	2.1
16	2.4	2.3	2.2	2.2	2.2	2.1	2.1	2.1	2.1	2.1	2.0
17	2.3	2.2	2.2	2.2	2.1	2.1	2.1	2.0	2.0	2.0	2.0
18	2.3	2.2	2.2	2.1	2.1	2.0	2.0	2.0	2.0	2.0	1.9
19	2.2	2.2	2.1	2.1	2.0	2.0	2.0	2.0	1.9	1.9	1.9
20	2.2	2.1	2.1	2.0	2.0	2.0	2.0	1.9	1.9	1.9	1.8
21	2.2	2.1	2.1	2.0	2.0	1.9	1.9	1.9	1.9	1.9	1.8
22	2.2	2.1	2.1	2.0	2.0	1.9	1.9	1.9	1.8	1.8	1.8
23	2.1	2.0	2.0	2.0	1.9	1.9	1.9	1.8	1.8	1.8	1.8
24	2.1	2.0	2.0	1.9	1.9	1.9	1.8	1.8	1.8	1.8	1.7
25	2.1	2.0	2.0	1.9	1.9	1.8	1.8	1.8	1.8	1.8	1.7
26	2.1	2.0	2.0	1.9	1.8	1.8	1.8	1.8	1.8	1.8	1.7
27	2.1	2.0	1.9	1.9	1.8	1.8	1.8	1.8	1.7	1.7	1.7
28	2.0	2.0	1.9	1.9	1.8	1.8	1.8	1.8	1.7	1.7	1.6
29	2.0	1.9	1.9	1.8	1.8	1.8	1.8	1.7	1.7	1.7	1.6
30	2.0	1.9	1.9	1.8	1.8	1.8	1.7	1.7	1.7	1.7	1.6
40	1.9	1.8	1.8	1.7	1.7	1.7	1.6	1.6	1.6	1.6	1.5
50	1.9	1.8	1.7	1.7	1.6	1.6	1.6	1.6	1.5	1.5	1.4
60	1.8	1.8	1.7	1.6	1.6	1.6	1.5	1.5	1.5	1.5	1.4
70	1.8	1.7	1.7	1.6	1.6	1.5	1.5	1.5	1.4	1.4	1.4
80	1.8	1.7	1.6	1.6	1.5	1.5	1.5	1.4	1.4	1.4	1.3
100	1.8	1.7	1.6	1.6	1.5	1.5	1.4	1.4	1.4	1.4	1.3
120	1.8	1.7	1.6	1.6	1.5	1.5	1.4	1.4	1.4	1.3	1.2
∞	1.7	1.6	1.5	1.5	1.4	1.4	1.3	1.3	1.2	1.2	1.0

Table 6 *continued*

						$\alpha = 0.01$						
						df_1						
df_2	1	2	3	4	5	6	7	8	9	10	11	12
1	4052	4999	5403	5625	5764	5859	5928	5981	6022	6056	6082	6106
2	98.5	99.0	99.2	99.2	99.3	99.3	99.3	99.4	99.4	99.4	99.4	99.4
3	34.1	30.8	29.5	28.7	28.2	27.9	27.7	27.5	27.3	27.2	27.1	27.0
4	21.2	18.0	16.7	16.0	15.5	15.2	15.0	14.8	14.7	14.5	14.4	14.4
5	16.3	13.3	12.1	11.4	11.0	10.7	10.4	10.3	10.2	10.0	10.0	9.9
6	13.7	10.9	9.8	9.2	8.8	8.5	8.3	8.1	8.0	7.9	7.8	7.7
7	12.2	9.6	8.4	7.8	7.5	7.2	7.0	6.8	6.7	6.6	6.5	6.5
8	11.3	8.6	7.6	7.0	6.6	6.4	6.2	6.0	5.9	5.8	5.7	5.7
9	10.6	8.0	7.0	6.4	6.1	5.8	5.6	5.5	5.4	5.3	5.2	5.1
10	10.0	7.6	6.6	6.0	5.6	5.4	5.2	5.1	5.0	4.8	4.8	4.7
11	9.6	7.2	6.2	5.7	5.3	5.1	4.9	4.7	4.6	4.5	4.5	4.4
12	9.3	6.9	6.0	5.4	5.1	4.8	4.6	4.5	4.4	4.3	4.2	4.2
13	9.1	6.7	5.7	5.2	4.9	4.6	4.4	4.3	4.2	4.1	4.0	4.0
14	8.9	6.5	5.6	5.0	4.7	4.5	4.3	4.1	4.0	3.9	3.9	3.8
15	8.7	6.4	5.4	4.9	4.6	4.3	4.1	4.0	3.9	3.8	3.7	3.7
16	8.5	6.2	5.3	4.8	4.4	4.2	4.0	3.9	3.8	3.7	3.6	3.6
17	8.4	6.1	5.2	4.7	4.3	4.1	3.9	3.8	3.7	3.6	3.5	3.4
18	8.3	6.0	5.1	4.6	4.2	4.0	3.8	3.7	3.6	3.5	3.4	3.4
19	8.2	5.9	5.0	4.5	4.2	3.9	3.8	3.6	3.5	3.4	3.4	3.3
20	8.1	5.8	4.9	4.4	4.1	3.9	3.7	3.6	3.4	3.4	3.3	3.2
21	8.0	5.8	4.9	4.4	4.0	3.8	3.6	3.5	3.4	3.3	3.2	3.2
22	7.9	5.7	4.8	4.3	4.0	3.8	3.6	3.4	3.4	3.3	3.2	3.1
23	7.9	5.7	4.8	4.3	3.9	3.7	3.5	3.4	3.3	3.2	3.1	3.1
24	7.8	5.6	4.7	4.2	3.9	3.7	3.5	3.4	3.2	3.2	3.1	3.0
25	7.8	5.6	4.7	4.2	3.9	3.6	3.5	3.3	3.2	3.1	3.0	3.0
26	7.7	5.5	4.6	4.1	3.8	3.6	3.4	3.3	3.2	3.1	3.0	3.0
27	7.7	5.5	4.6	4.1	3.8	3.6	3.4	3.3	3.1	3.1	3.0	2.9
28	7.6	5.4	4.6	4.1	3.8	3.5	3.4	3.2	3.1	3.0	3.0	2.9
29	7.6	5.4	4.5	4.0	3.7	3.5	3.3	3.2	3.1	3.0	2.9	2.9
30	7.6	5.4	4.5	4.0	3.7	3.5	3.3	3.2	3.1	3.0	2.9	2.8
40	7.3	5.2	4.3	3.8	3.5	3.3	3.1	3.0	2.9	2.8	2.7	2.7
50	7.2	5.1	4.2	3.7	3.4	3.2	3.0	2.9	2.8	2.7	2.6	2.6
60	7.1	5.0	4.1	3.6	3.3	3.1	3.0	2.8	2.7	2.6	2.6	2.5
70	7.0	4.9	4.1	3.6	3.3	3.1	2.9	2.8	2.7	2.6	2.5	2.4
80	7.0	4.9	4.0	3.6	3.2	3.0	2.9	2.7	2.6	2.6	2.5	2.4
100	6.9	4.8	4.0	3.5	3.2	3.0	2.8	2.7	2.6	2.5	2.4	2.4
120	6.8	4.8	4.0	3.5	3.2	3.0	2.8	2.7	2.6	2.5	2.4	2.3
∞	6.6	4.6	3.8	3.3	3.0	2.8	2.6	2.5	2.4	2.3	2.2	2.2

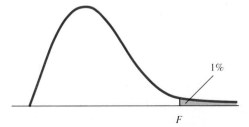

1%

F

Table 6 *continued*

						$\alpha = 0.01$						
						df_1						
df_2	15	20	24	30	40	50	60	75	100	120	∞
1	6157	6209	6235	6261	6287	6302	6313	6323	6334	6339	6366
2	99.4	99.4	99.5	99.5	99.5	99.5	99.5	99.5	99.5	99.5	99.5
3	26.9	26.7	26.6	26.5	26.4	26.3	26.2	26.3	26.2	26.2	26.1
4	14.2	14.0	13.9	13.8	13.8	13.7	13.6	13.6	13.6	13.6	13.5
5	9.7	9.6	9.5	9.4	9.3	9.2	9.2	9.2	9.1	9.1	9.0
6	7.6	7.4	7.3	7.2	7.1	7.1	7.1	7.0	7.0	7.0	6.9
7	6.3	6.2	6.1	6.0	5.9	5.8	5.8	5.8	5.8	5.7	5.6
8	5.5	5.4	5.3	5.2	5.1	5.1	5.0	5.0	5.0	5.0	4.9
9	5.0	4.8	4.7	4.6	4.6	4.5	4.4	4.4	4.4	4.4	4.3
10	4.6	4.4	4.3	4.2	4.2	4.1	4.1	4.0	4.0	4.0	3.9
11	4.2	4.1	4.0	3.9	3.9	3.8	3.8	3.7	3.7	3.7	3.6
12	4.0	3.9	3.8	3.7	3.6	3.6	3.5	3.5	3.5	3.4	3.4
13	3.8	3.7	3.6	3.5	3.4	3.4	3.3	3.3	3.3	3.2	3.2
14	3.7	3.5	3.4	3.4	3.3	3.2	3.2	3.1	3.1	3.1	3.0
15	3.5	3.4	3.3	3.2	3.1	3.1	3.0	3.0	3.0	3.0	2.9
16	3.4	3.3	3.2	3.1	3.0	3.0	2.9	2.9	2.9	2.8	2.8
17	3.3	3.2	3.1	3.0	2.9	2.9	2.8	2.8	2.8	2.8	2.7
18	3.2	3.1	3.0	2.9	2.8	2.8	2.8	2.7	2.7	2.7	2.6
19	3.2	3.0	2.9	2.8	2.8	2.7	2.7	2.6	2.6	2.6	2.5
20	3.1	2.9	2.9	2.8	2.7	2.6	2.6	2.6	2.5	2.5	2.4
21	3.0	2.9	2.8	2.7	2.6	2.6	2.6	2.5	2.5	2.5	2.4
22	3.0	2.8	2.8	2.7	2.6	2.5	2.5	2.5	2.4	2.4	2.3
23	2.9	2.8	2.7	2.6	2.5	2.5	2.4	2.4	2.4	2.4	2.3
24	2.9	2.7	2.7	2.6	2.5	2.4	2.4	2.4	2.3	2.3	2.2
25	2.8	2.7	2.6	2.5	2.4	2.4	2.4	2.3	2.3	2.3	2.2
26	2.8	2.7	2.6	2.5	2.4	2.4	2.3	2.3	2.2	2.2	2.1
27	2.8	2.6	2.6	2.5	2.4	2.3	2.3	2.2	2.2	2.2	2.1
28	2.8	2.6	2.5	2.4	2.4	2.3	2.3	2.2	2.2	2.2	2.1
29	2.7	2.6	2.5	2.4	2.3	2.3	2.2	2.2	2.2	2.1	2.0
30	2.7	2.6	2.5	2.4	2.3	2.2	2.2	2.2	2.1	2.1	2.0
40	2.5	2.4	2.3	2.2	2.1	2.0	2.0	2.0	1.9	1.9	1.8
50	2.4	2.3	2.2	2.1	2.0	1.9	1.9	1.9	1.8	1.8	1.7
60	2.4	2.2	2.1	2.0	1.9	1.9	1.8	1.8	1.7	1.7	1.6
70	2.3	2.2	2.1	2.0	1.9	1.8	1.8	1.7	1.7	1.7	1.5
80	2.3	2.1	2.0	1.9	1.8	1.8	1.8	1.7	1.6	1.6	1.5
100	2.2	2.1	2.0	1.9	1.8	1.7	1.7	1.6	1.6	1.5	1.4
120	2.2	2.0	2.0	1.9	1.8	1.7	1.6	1.6	1.5	1.5	1.4
∞	2.0	1.9	1.8	1.7	1.6	1.5	1.5	1.4	1.4	1.3	1.0

Table 7 Table of Random Numbers

31871	60770	59235	41702	89372	28600	30013	18266	65044	61045
87134	32839	17850	37359	27221	92409	94778	17902	09467	86757
06728	16314	81076	42172	46446	09226	96262	77674	70205	98137
95646	67486	05167	07819	79918	83949	45605	18915	79458	54009
44085	87246	47378	98338	40368	02240	72593	52823	79002	88190
83967	84810	51612	81501	10440	48553	67919	73678	83149	47096
49990	02051	64575	70323	07863	59220	01746	94213	82977	42384
65332	16488	04433	37990	93517	18395	72848	97025	38894	46611
42309	04063	55291	72165	96921	53350	34173	39908	11634	87145
84715	41808	12085	72525	91171	09779	07223	75577	20934	92047
63919	83977	72416	55450	47642	01013	17560	54189	73523	33681
97595	78300	93502	25847	19520	16896	69282	16917	04194	25797
17116	42649	89252	61052	78332	15102	47707	28369	60400	15908
34037	84573	49914	59688	18584	53498	94905	14914	23261	58133
08813	14453	70437	49093	69880	99944	40482	04254	62842	68089
67115	41050	65453	04510	35518	88843	15801	86163	49913	46849
14596	62802	33009	74095	34549	76634	64270	67491	83941	55154
70258	26948	60863	47666	58512	91404	97357	85710	03414	56591
83369	81179	32429	34781	00006	65951	40254	71102	60416	43296
83811	49358	75171	34768	70070	76550	14252	97378	79500	97123
14924	71607	74638	01939	77044	18277	68229	09310	63258	85064
60102	56587	29842	12031	00794	90638	21862	72154	19880	80895
33393	30109	42005	47977	26453	15333	45390	89862	70351	36953
92592	78232	19328	29645	69836	91169	95180	15046	45679	94500
27421	73356	53897	26916	52015	26854	42833	64257	49423	39440
26528	22550	36692	25262	61419	53986	73898	80237	71387	32532
07664	10752	95021	17030	76784	86861	12780	44379	31261	18424
37954	72029	29624	09119	13444	22645	78345	79876	37582	75549
66495	11333	81101	69328	84838	76395	35997	07259	66254	47451
72506	28524	39595	49356	92733	42951	47774	75462	64409	69116
09713	70270	28077	15634	36525	91204	48443	50561	92394	60636
51852	70782	93498	44669	79647	06321	04020	00111	24737	05521
31460	22222	18801	00675	57562	97923	45974	75158	94918	40144
14328	05024	04333	04135	53143	79207	85863	04962	89549	63308
84002	98073	52998	05749	45538	26164	68672	97486	32341	99419
89541	28345	22887	79269	55620	68269	88765	72464	11586	52211
50502	39890	81465	00449	09931	12667	30278	63963	84192	25266
30862	61996	73216	12554	01200	63234	41277	20477	71899	05347
36735	58841	35287	51112	47322	81354	51080	72771	53653	42108
11561	81204	68175	93037	47967	74085	05905	86471	47671	18456

List of Formulas

Chapter	Formula	Use
2	$(1)\ \bar{x} = \dfrac{\Sigma x}{n}$	To find sample mean from raw data
2	$(2)\ \mu = \dfrac{\Sigma x}{N}$	To find population mean from raw data
2	$(3)\ \bar{x} = \dfrac{\Sigma fm}{\Sigma f}$ or $\mu = \dfrac{\Sigma fm}{\Sigma f}$	To find the mean from a frequency distribution
2	$(4)\ \bar{x}' = \dfrac{\Sigma fx'}{\Sigma f}$ and $\bar{x} = C + w\bar{x}'$	To find the mean from a frequency distribution using coding
2	$(5)\ LTCL + r\left(\dfrac{w}{f+1}\right)$	To find the rth value in an interval, used in calculating a median from a frequency distribution
2	$(6)\ \bar{x} = \dfrac{\Sigma xw}{\Sigma w}$	To find the weighted mean
2	$(7)\ s^2 = \dfrac{\Sigma d^2}{n}$	To find the variance of a sample
2	$(8)\ s = \sqrt{\dfrac{\Sigma d^2}{n}}$	To find the standard deviation of a sample
2	$(9)\ s = \sqrt{\dfrac{\Sigma x^2}{n} - \left(\dfrac{\Sigma x}{n}\right)^2}$	Alternative to formula (8), often easier to apply
2	$(10)\ s = \sqrt{\dfrac{\Sigma fm^2}{\Sigma f} - \left(\dfrac{\Sigma fm}{\Sigma f}\right)^2}$	To find the standard deviation from a frequency distribution

2   (11) $s' = \sqrt{\dfrac{\Sigma f(x')^2}{\Sigma f} - \left(\dfrac{\Sigma fx'}{\Sigma f}\right)^2}$

and $s = ws'$

Formulas to find the standard deviation from a frequency distribution using the coding process

2, 3   (12) $z = \dfrac{x - \mu}{\sigma}$

To find the number of standard deviations a value x in a normal distribution is above or below the mean, also used to convert raw scores to standard scores in a population

2   (13) $z = \dfrac{x - \bar{x}}{s}$

To convert raw scores to standard scores in a sample

3   (14) $P(E) = \dfrac{\text{number of simple events in } E}{\text{number of simple events in } S}$

To find the probability (theoretical or empirical) of an event E

3   (15) $P(A') = 1 - P(A)$

To find the probability of the complement of an event; this is sometimes easier than finding the probability of an event directly

3   (16) $P(A \text{ or } B) = P(A) + P(B)$
            $- P(A \text{ and } B)$

3   (17) $P(A \text{ and } B) = P(A)P(B|A)$

Compound probability rules

3   (18) $\mu = \Sigma x \cdot P(x)$

To find the mean of a probability distribution

3   (19) $\sigma = \sqrt{\Sigma x^2 \cdot P(x) - [\Sigma x \cdot P(x)]^2}$

To find the standard deviation of a probability distribution

3   (20) $P(x) = w \cdot p^x \cdot q^{n-x}$, where w is found from Pascal's triangle or by using the formula

$w = \dfrac{n!}{x!(n - x)!}$

To find the probability of x successes in n trials of a binomial distribution

3   (21) $\mu = np$

To find the mean of a binomial distribution

3   (22) $\sigma = \sqrt{npq}$

To find the standard deviation of a binomial distribution

4   (23) $SD_{\bar{x}} = \dfrac{\sigma}{\sqrt{n}}$

To find the standard deviation of a frequency distribution of $\bar{x}$ for samples of size n drawn from a very large, or infinite, population

4   **(24)** $SD_{\bar{x}} = \dfrac{\sigma}{\sqrt{n}}\sqrt{\dfrac{N-n}{N-1}}$

Same as formula (23), but used when the population size N is not very large in comparison to n, the size of each of the samples

5   **(25)** $\bar{x} \pm zSD_{\bar{x}}$

A confidence interval estimate for μ

5   **(26)** $SE_{\bar{x}} = \dfrac{s}{\sqrt{n}}$

Formula for the standard error of the sample mean

5   **(27)** $n = \left(\dfrac{z\sigma}{E}\right)^2$

Formula for determining the sample size when estimating a sample mean; generally rounded up to the next whole number

5   **(28)** $SD_p = \sqrt{\dfrac{PQ}{n}}$

To find the standard deviation of the frequency distribution of p

5   **(29)** $p \pm zSE_p,$ where $SE_p = \sqrt{\dfrac{pq}{n}}$

A CIE for P, a population proportion

5   **(30)** $n = \dfrac{z^2}{4E^2}$

To determine a sample size when estimating a proportion and you have no idea of the rough size of P

5   **(31)** $SE_s = \dfrac{s}{\sqrt{2n}}$

To find the standard error when estimating a standard deviation of a population following a normal distribution

6   **(32)** $z = \dfrac{\bar{x} - \mu_H}{\sigma/\sqrt{n}}$

Formula for use with the z-score test using σ

6   **(33)** $z = \dfrac{\bar{x} - \mu_H}{s/\sqrt{n}}$

Formula for use with the z-score test using s

6   **(34)** $z = \dfrac{p - P_H}{\sqrt{\dfrac{P_H Q_H}{n}}},$ where p is corrected for continuity

Formula for use with the z-score test for proportions

7   **(35)** $t = \dfrac{\bar{x} - \mu_H}{s/\sqrt{n-1}}$

Formula for use with the one-sample t-test

7   **(36)** $\bar{x} \pm t\dfrac{s}{\sqrt{n-1}}$

A CIE for μ when sample size is less than 30

8   (37) $SE_{\bar{x}_1 - \bar{x}_2} = \sqrt{\dfrac{s_1^2}{n_1} + \dfrac{s_2^2}{n_2}}$ To find the standard error when estimating the difference between the means of two populations

8   (38) $SE_{p_1 - p_2} = \sqrt{\dfrac{p_1 q_1}{n_1} + \dfrac{p_2 q_2}{n_2}}$ To find the standard error when estimating the difference between the proportions in two populations

8   (39) $\bar{d} \pm t \dfrac{s_d}{\sqrt{n-1}}$ CIE for a dependent difference when the sample size is less than 30

8   (40) $z = \dfrac{(\bar{x}_1 - \bar{x}_2) - d_H}{\sqrt{\dfrac{s_1^2}{n_1} + \dfrac{s_2^2}{n_2}}}$ Formula used in the two-sample z-score test for means

8   (41) $z = \dfrac{p_1 - p_2}{\sqrt{p^* q^* \left(\dfrac{1}{n_1} + \dfrac{1}{n_2} \right)}}$,

where $p^* = \dfrac{n_1 p_1 + n_2 p_2}{n_1 + n_2}$ Formula used in the two-sample z-score test for proportions

8   (42) $t = \dfrac{(\bar{x}_1 - \bar{x}_2) - d_H}{s^* \sqrt{\dfrac{1}{n_1} + \dfrac{1}{n_2}}}$,

where $s^* = \sqrt{\dfrac{n_1 s_1^2 + n_2 s_2^2}{n_1 + n_2 - 2}}$

and $df = n_1 + n_2 - 2$ Formulas used in the two-sample t-test

8   (43) $(\bar{x}_1 - \bar{x}_2) \pm t s^* \sqrt{\dfrac{1}{n_1} + \dfrac{1}{n_2}}$ Formula giving a CIE for $\mu_1 - \mu_2$ when n_1, n_2 are small, the populations are normally distributed, and $\sigma_1 = \sigma_2$, they being unknown

9   (44) Mean of $V = \dfrac{1}{4} n(n+1)$

$SD_V = \sqrt{\dfrac{1}{24} n(n+1)(2n+1)}$ Formulas for use in the one-sample Wilcoxon test

9   (45) Mean of $W = n_2 \left(\dfrac{n_1 + n_2 + 1}{2} \right)$

$SD_W = \sqrt{n_1 n_2 \left(\dfrac{n_1 + n_2 + 1}{12} \right)}$ Formulas for use in the two-sample Wilcoxon test

9   (46) Mean of $R = \dfrac{2 n_1 n_2}{n_1 + n_2} + 1$

$SD_R = \sqrt{\dfrac{2 n_1 n_2 (2 n_1 n_2 - n_1 - n_2)}{(n_1 + n_2)^2 (n_1 + n_2 - 1)}}$ Formulas for use in the runs test when either or both n_1, n_2 are larger than 20

10   **(47)** $\chi^2 = \Sigma \dfrac{(O - E)^2}{E}$ Used to calculate χ^2 in both χ^2-tests

10   **(48)** $z = \sqrt{2 \cdot \chi^2} - \sqrt{2 \cdot df}$ Used with the normal table to determine P-value for a χ^2-test if there are more than 10 degrees of freedom

11   **(49)** $SS_{BG} = \dfrac{\Sigma T^2}{n} - \dfrac{(\Sigma T)^2}{kn}$ To find the sum of squares between groups in an ANOVA

11   **(50)** $SS_{WG} = \Sigma\left(\Sigma x^2\right) - \dfrac{(\Sigma T)^2}{n}$ To find the sum of squares within groups in an ANOVA

13   **(51)** $r = \dfrac{\Sigma z_x z_y}{n}$ To compute Pearson's product moment correlation coefficient

13   **(52)** $r = \dfrac{n\Sigma xy - (\Sigma x)(\Sigma y)}{\sqrt{n\Sigma x^2 - (\Sigma x)^2}\sqrt{n\Sigma y^2 - (\Sigma y)^2}}$ Alternative to formula (51) in which it is unnecessary to calculate means and standard deviations

13   **(53)** $r' = 1 - \dfrac{6\Sigma D^2}{n(n^2 - 1)}$ To compute Spearman's rank correlation coefficient

13   **(54)** $z = r\sqrt{n - 1}$ Used in the z-score test for independence

13   **(55)** $z = \dfrac{6\Sigma D^2 - n(n^2 - 1)}{n(n + 1)\sqrt{n - 1}}$ Used in the test for independence based on ΣD^2

14   **(56)** $\hat{y} = \bar{y} + \dfrac{rs_y}{s_x}(x - \bar{x})$ The equation of the line of regression of y on x

14   **(57)** $SE_{\hat{y}} = s_y\sqrt{1 - r^2}$ Used to find the standard error of prediction in a regression of y on x, not the most precise formula available and gives only a rough value for the SE

14   **(58)** $\hat{x} = \bar{x} + \dfrac{rs_x}{s_y}(y - \bar{y})$ The equation of the line of regression of x on y

14   **(59)** $SE_{\hat{x}} = s_x\sqrt{1 - r^2}$ Used to find the standard error of prediction in a regression of x on y, not the most precise formula available and gives only a rough value for the SE

Answers to Odd-Numbered Problems

Exercise Set 1–1

1. {H,T}   **3.** {0, 1, 2, . . ., 56(?)}   **5.** Population; finite   **7.** Sample   **9.** Qualitative

11. Quantitative   **13.** Discrete   **15.** Continuous

Exercise Set 1–2

1.

Class	f
5–9	4
10–14	4
15–19	6
20–24	6
25–29	4
30–34	4

3.

Class	f
36–40	2
41–45	3
46–50	4
51–55	4
56–60	5
61–65	4
66–70	7
71–75	1

(Answers may vary.)

5.

Class	f
11–13	2
14–16	4
17–19	8
20–22	7
23–25	4

(Answers may vary.)

7.

F
1
3
4
5
9
14
19
23
23
27

(a) w = 1
(b) 63–66
(c) 64

Exercise Set 1–3

1.

Cat.	f
S	8
C	4
R	10
P	7
A	2
F	3

3.

Cat.	f
J	9
K	8
L	7
U	6

5.

$f\%$	$F\%$
4	4
12	16
32	48
28	76
16	92
8	100

7.

$f\%$
46.7
33.3
15.0
3.3
1.7

9.

Class	f	$f\%$
1–20	13	32.5
21–40	17	42.5
41–60	1	2.5
61–80	3	7.5
81–100	3	7.5
101–120	2	5.0
121–140	1	2.5

Exercise Set 1–4

1.

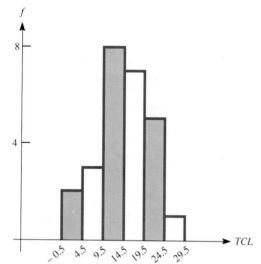

3.

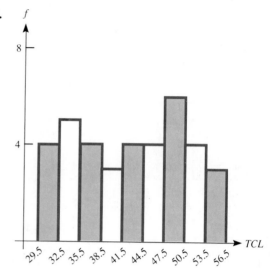

5.

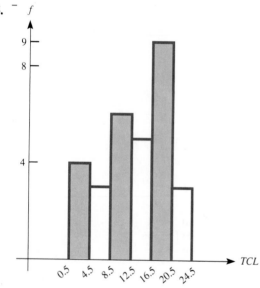

7.

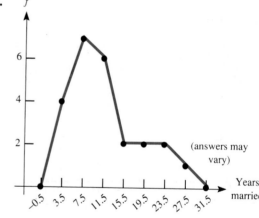

(answers may vary)

9. *f%* per $1,000

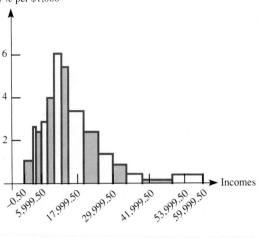

11 *f*

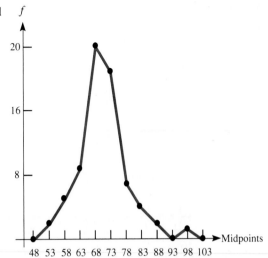

13.

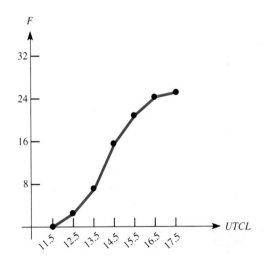

15.

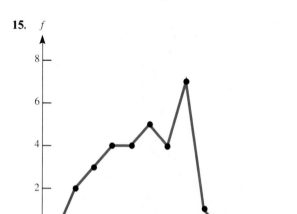

(a)

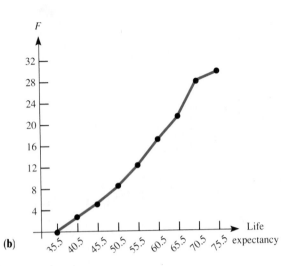

(b)

(c) Left-skewed

17.

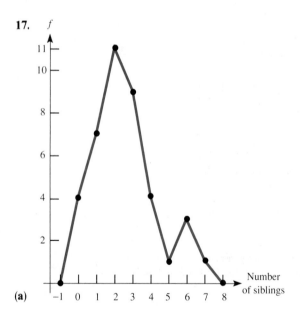

(a)

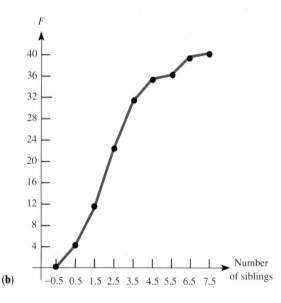

(b)

(c) Right-skewed

19. $f\%$

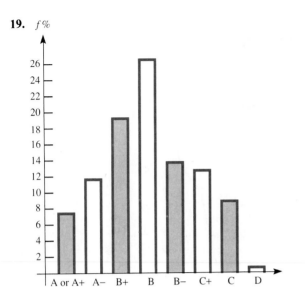

21.

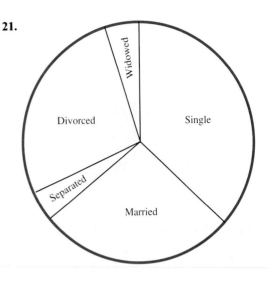

Chapter 1 Review Exercises

1. {1,2,3,4,5,6,7,8,9,10,11,12,13,14,15,16,17,18,19,20,21,22,23,24}   **3.** 1.5–20 lb (?)

5. $R = 187,901,984$; figure for China is almost certainly rounded.

7.

Int.	f
18–19	2
20–21	2
22–23	6
24–25	4
26–27	4
28–29	2

9.

Int.	f
1–25	14
26–50	22
51–75	9
76–100	3
101–125	0
126–150	0
151–175	1
.	.
.	.
.	.
351–375	1

11.

Int.	$f\%$	$F\%$
120–129	4.4	4.4
130–139	11.1	15.5
140–149	15.6	31.1
150–159	17.8	48.9
160–169	22.2	71.1
170–179	13.3	84.4
180–189	2.2	86.6
190–199	11.1	97.7
200–209	2.2	99.9

(Answers may vary.)
They were asked; 33% end in 0, 42% end in 5.

13.

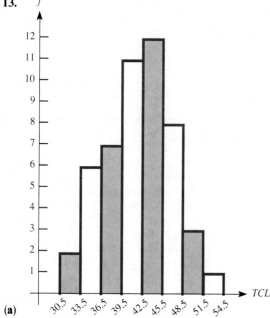

(a)

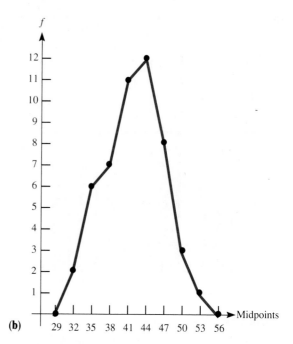

(b)

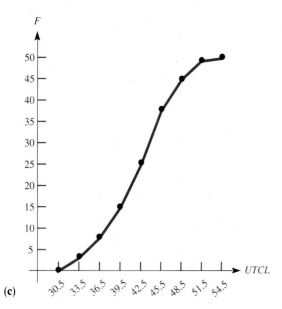

(c)

(d) Normal

15.

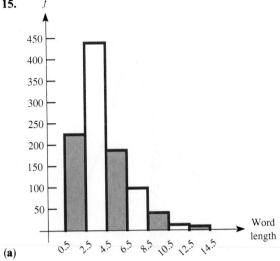

(a)

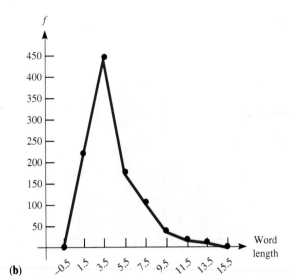

(b)

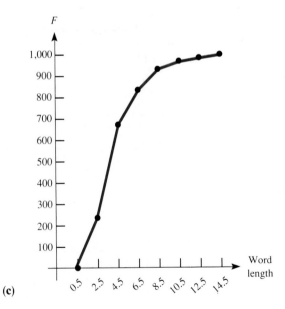

(c)

(d) Right-skewed

17.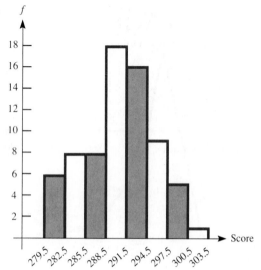

(a)

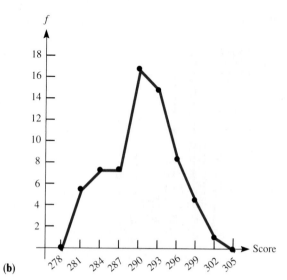

(b)

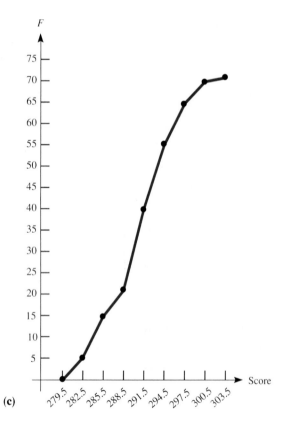

(c)

(d) Normal

19.

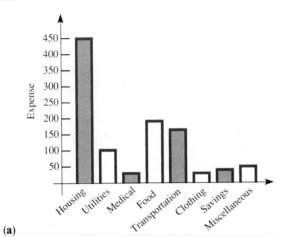

(a)

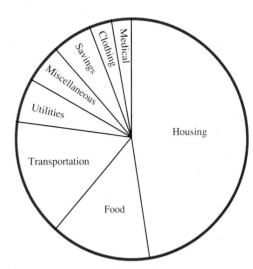

(b)

21. $f\%$ per °F

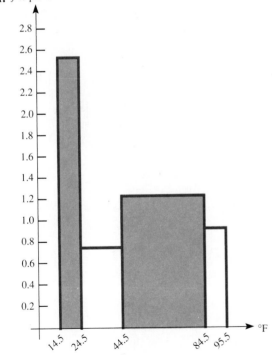

Exercise Set 2–1

1. 21   **3.** 6.5   **5.** 45.75 = 45.8   **7.** $\Sigma(\bar{x} - 7) = 0$, $\Sigma(\bar{x} - 8) = -8$; Yes, $\bar{x} = 7$ **9.** 8.4872

Exercise Set 2–2

1. After the data has been grouped, it is impossible to see which value occurred most frequently. A modal interval is designated instead.   **3.** 35–39   **5.** 36.5   **7.** 4,310,000   **9.** 23.38

Exercise Set 2–3

1. 18.2   **3.** 15.0   **5.** 23.6

Exercise Set 2–4

1. 3.1   **3.** 17.8   **5.** 106.7

Exercise Set 2–5

1. 3.4   **3.** 15.7   **5.** 13.2   **7.** 4.1   **9.** $\mu = 86.4$, $\sigma = 7.9$
11. $\bar{x} = 22$, $s = 4.1$; $\bar{x}$ has increased by 12, s is the same.

Exercise Set 2–6

1. 7.7   **3.** $\bar{x} = 65.5$, $s = 2.6$ **5.** $\mu = 5,166.2$, $\sigma = 3,214.6$ **7.** Calc.: $\bar{x} = 227.0$, $s = 17.9$;
f.d. ($w = 10$): $\bar{x} = 226.9$, $s = 17.9$ **9.** Calc.: $\bar{x} = 11.9$, $s = 4.8$; f.d. ($w = 3$): $\bar{x} = 11.8$, $s = 4.7$
11. $s^2 = 18.8$

Exercise Set 2–7

1. 34   **3.** at least 57   **5.** $\bar{x} = 76.8$, $s = 10.0$; 60% in $\bar{x} \pm s$, 100% in $\bar{x} \pm 2s$; 100% is more than 75%

7.

Class	f
0–5	4
6–11	11
12–17	10
18–23	6
24–29	1

$\bar{x} = 12.4$, $s = 6.1$;
62.5% in $\bar{x} \pm s$;
94% in $\bar{x} \pm 2s$

9. Verification

Exercise Set 2–8

1.

Value	18	81	66	53	5	93	27	64	36
Rank	8	2	3	5	9	1	7	4	6

3.

Value	155	157	165	163	155	150	154	164	165	168	167
Rank	6	8	12.5	10	6	1	4	11	12.5	16	15
Value	166	159	155	152	151						
Rank	14	9	6	3	2						

5. P_{72} **7.** P_{67} **9.** 58.5

Exercise Set 2–9

1. 0.67   **3.** -1.61 **5.** Bob: $z = 0.5$, Carl: $z = 1.5$; Carl   **7.** $\bar{x} = 232.6$, $s = 51.8$, $z = -0.73$
9. $\bar{x} = 34.2$

Chapter 2 Review Exercises

1. (a) 7.0 (b) 6.0 (c) No mode   **3.** (a) 15.7 (b) 15.5 (c) 13   **5.** (a) 4.2 (b) 4 (c) 5   **7.** (a) 100.7 (b) 99.6 (c) 90–99
9. (a) \$45.00 (b) \$44.06 (c) \$40–49.99   **11.** (a) 5,166.22 (b) 4,999.5   **13.** Answers will vary.
15. $\bar{x} = 25.1$, $s = 7.3$ **17.** $\bar{x} = 167.6$, $s = 63.5$ **19.** $\bar{x} = 4.5$, $s = 1.3$ **21.** $\bar{x} = 26.0$, $s = 7.9$
23. (a) 56% (b) 99%   **25.** Mercury: 1, Pluto: 2, Mars: 3, Venus: 4, Earth: 5, Neptune: 6, Uranus: 7, Saturn: 8,
Jupiter: 9   **27.** P_{78} **29.** $z = -0.81$

Exercise Set 3–1

1. 0.42   **3.** $\{HHH,HHT,HTH,THH,HTT,THT,TTH,TTT\}$ **5.** $\frac{3}{8}$ **7.** $\frac{9}{19}$ **9.** $\frac{1}{17}$ **11.** $\frac{1}{3}$ **13.** $\frac{1}{8}$
15. 6 to 5 for   **17.** 13 to 6 against

Exercise Set 3–2

1. $\frac{1}{4}$ **3.** $\frac{3}{5}$ **5.** $\frac{1}{12}$ **7.** $\frac{4}{9}$ **9.** $\frac{364}{365} \approx 0.997$; $\frac{(364)(363)}{(365)^2} \approx 0.992$ **11.** $\frac{1}{1,000}$ **13.** $\frac{1}{19}$ **15.** $\frac{3}{50}$
17. $\frac{113}{114} \approx 0.991$ **19.** (a) F (b) T

Exercise Set 3–3

1. X	P(X)
0	1/8
1	3/8
2	3/8
3	1/8

3. X	P(X)
3	1/10
4	1/10
5	2/10
6	2/10
7	2/10
8	1/10
9	1/10

5. X	P(X)
1	3/10
5	2/10
10	4/10
20	1/10

7. X	P(X)
0	0/15
1	1/15
2	2/15
3	3/15
4	4/15
5	5/15

It is a probability function.

9. $\mu = 1.5$, $\sigma = 0.87$

11. X	P(X)
0	1/16
1	4/16
2	6/16
3	4/16
4	1/16

$\mu = 2$, $\sigma = 1$

13. Six: $\mu = 3$, Seven: $\mu = 3.5$

Exercise Set 3–4

1. X	P(X)
0	1/16
1	4/16
2	6/16
3	4/16
4	1/16

3. $\mu = 2$, $\sigma = 1$ 5. 0.00024832   7. 0.47178   9. $\mu = 5$, $\sigma = 1.94$

Exercise Set 3–5

1. 0.4641   3. 0.6835   5. 0.0854   7. 0.0344   9. 0.7157   11. -1.645 and 1.645

Exercise Set 3–6

1. 0.8833   3. 7.78%   5. 1,127.2 hours   7. 16%   9. 0.234 to 0.264 inches   11. 0.925   13. P_8
15. 1.04   17. P_{92} 19. 63.4

Chapter 3 Review Exercises

1. {1,2,3,4,5,6,7,8,9,10}   3. $\frac{2}{27}$ 5. 0.518   7. $\frac{3}{8}$ 9. 631 to 32 against or approximately 19.7 to 1

11. X	P(X)
2	1/6
4	2/6
6	3/6

13. 0.195   15. 0.2462   17. 0.8413   19. 0.0217   21. $160 to $200
23. 0.0005   25. 37.7   27. P_{74} 29. 11.4

Exercise Set 4–1

1. Answers will vary.   **3.** 695 and 1,000   **5.** Stratified random sampling   **7.** Simple random sampling
9. Quota sampling

Exercise Set 4–2

1.

Sample	$\bar{x}$	Sample	$\bar{x}$	Interval	f
2,5,6,8,10	6.2	2,6,8,12,13	8.2	6.0–6.4	1
2,5,6,8,12	6.6	2,6,10,12,13	8.6	6.5–6.9	2
2,5,6,8,13	6.8	2,8,10,12,13	9.0	7.0–7.4	3
2,5,6,10,12	7.0	5,6,8,10,12	8.2	7.5–7.9	4
2,5,6,10,13	7.2	5,6,8,10,13	8.4	8.0–8.4	5
2,5,6,12,13	7.6	5,6,8,12,13	8.8	8.5–8.9	2
2,5,8,10,12	7.4	5,6,10,12,13	9.2	9.0–9.4	2
2,5,8,10,13	7.6	5,8,10,12,13	9.6	9.5–9.9	2
2,5,8,12,13	8.0	6,8,10,12,13	9.8		
2,5,10,12,13	8.4				
2,6,8,10,12	7.6				
2,6,8,10,13	7.8				

$\mu_{\bar{x}} = 8.0$ $\mu = 8.0$

$SD_{\bar{x}} = 0.94$ $\sigma = 3.7$ $\dfrac{\sigma}{\sqrt{n}}\sqrt{\dfrac{N-n}{N-1}} = 0.96$

3. 44.4%   **5.** 95.4%   **7.** 0.1625   **9.** 0.0125

Chapter 4 Review Exercises

1. Systematic sampling   **3.** Cluster sampling   **5.** 0.6826   **7.** 64.1 in.   **9.** 2 hr 34 min

Exercise Set 5–1

1. 72.0 to 74.4   **3.** 25.7 to 27.7   **5.** 7.0 to 7.4 lb   **7.** 41.0 to 45.4 min   **9.** $8.80 to $10.40

Exercise Set 5–2

1. 5.8 to 6.8   **3.** 3.2 to 4.0   **5.** 1.34 to 1.42   **7.** Not possible, since $n < 30$ and σ is not given (Incorrect
answer is 19.7 to 23.5)   **9.** 5.67 to 5.91 mm

Exercise Set 5–3

1. 59   **3.** 81   **5.** 78   **7.** 19   **9.** 44

Exercise Set 5–4

1. 0.067 to 0.133   **3.** 0.397 to 0.603   **5.** 0.140 to 0.260   **7.** 51.9 to 62.3%   **9.** 6.5 to 14.9%

Exercise Set 5–5

1. 157   **3.** 1,702   **5.** 625   **7.** 207   **9.** Exact calculation gives 93%, but they probably assumed 95%.

Exercise Set 5–6

1. 5.13 to 6.27   **3.** 2.39 to 3.21 in.; valid   **5.** $5,671 to $7,029; not valid   **7.** 74.5 to 91.5 lb

Chapter 5 Review Exercises

1. 12.0 to 20.4   **3.** 71.6 to 73.2   **5.** 1.036 to 1.044 in.   **7.** 131.3 to 137.9   **9.** 348   **11.** 0.41 to 0.55
13. 273   **15.** 1.82 to 3.18 mm; measurements are normally distributed.

Exercise Set 6–1

1. $H_0 : \mu = 65$ **3.** $H_0 : \mu = 30$ **5.** $H_0 : \mu = 65$ **7.** $H_0 : \mu = 1/2$ **9.** $H_0 : \mu = 12$
     $H_1 : \mu > 65$ $H_1 : \mu \neq 30$ $H_1 : \mu > 65$ $H_1 : \mu < 1/2$ $H_1 : \mu \neq 12$

Exercise Set 6–2

1. 0.0793   **3.** *P*-val. = 0.0051; no   **5.** *P*-val. = 0.0475 = probability of type I error; yes at 0.05 level
7. *P*-val. = 0.0681; no   **9.** *P*-val. = 0.0749; yes

Exercise Set 6–3

1. *P*-val. = 0.0485; no, don't reject.   **3.** *P*-val. = 0.0000; invalid   **5.** *P*-val. = 0.0084; the government office claim may be correct.   **7.** No, since the sample size is less than 30.   **9.** *P*-val. = 0.0028; yes, they are.
11. $\bar{x} = 7.29$, $s = 3.62$; *P*-val. = 0.3050; probably right

Exercise Set 6–4

1. 0.1170   **3.** *P*-val. = 0.0606; coin may be fair.   **5.** *P*-val. = 0.1292; no, he can't.   **7.** *P*-val. = 0.0885; she has less than 45% support; she shouldn't run.   **9.** *P*-val. = 0.1038; yes, it could.

Chapter 6 Review Exercises

1. $H_0 : P = 0.15$ **3.** $z = -5.50$; *P*-val. = 0.0000; the hypothesis that the mean is 1 should be rejected.
    $H_1 : P > 0.15$
5. *P*-val. = 0.0537; no, it isn't.   **7.** *P*-val. = 0.0032; the line appears to be weaker than the manufacturer claims.   **9.** *P*-val. = 0.2546; no, it is not.   **11.** 0.0016

Exercise Set 7–1

1. P-val. $= 0.0815$; the mean may be 18.   **3.** P-val. $= 0.3634$; the anthropologist's belief may be right.
5. P-val. $= 0.1447$; yes, the D.A. may be correct.   **7.** $\bar{x} = 80.4$, $s = 9.34$; P-val. $= 0.0060$; yes, it d~
9. $\bar{x} = 30.2$, $s = 4.2$; P-val. $= 0.4298$; the treatment has not been a success.

Exercise Set 7–2

1. (a) 2.11 (b) 2.11   **3.** 17.64 to 20.56   **5.** 51.6 to 55.6 inches

Chapter 7 Review Exercises

1. P-val. $= 0.2746$; boiling point could be 572° C.
5. 5.73 to 6.87 min   **7.** 18.8 to 20.4 mpg

Exercise Set 8–1

1. Verification   **3.** 0.1 to 4.1   **5.** Second college's mean is b~   ~d 0.8 years older than the first's.
7. -0.15 to -0.05 **9.** Between 17.2% in favor of the females a~   ~2% in favor of the males
11. 3.89 to 5.71   **13.** Between 0.43 lb in favor of the old and 3.09 lb in favor of the new

Exercise Set 8–2

1. $H_0: \mu_1 - \mu_2 = 0$ **3.** $H_0: \mu_1 - \mu_2 = 0$ **5.** $H_0: \mu_1 - \mu_2 = 0$ **7.** $H_0: P_1 - P_2 = 0$
$H_1: \mu_1 - \mu_2 > 0$ $H_1: \mu_1 - \mu_2 < 0$ $H_1: \mu_1 - \mu_2 \neq 0$ $H_1: P_1 - P_2 > 0$
9. $H_0: \mu_1 - \mu_2 = 6$
$H_1: \mu_1 - \mu_2 > 6$

Exercise Set 8–3

1. P-val. $= 0.0630$; the first mean is not significantly larger than the second.   **3.** P-val. $= 0.0516$; no, the
families in the first state don't spend significantly more.   **5.** P-val. $= 0.0004$; yes, they do.   **7.** $\bar{x}_1 = 304.00$,
$\bar{x}_2 = 321.83$, P-val. $= 0.2090$; no significant difference   **9.** P-val. $= 0.0049$; yes, there is a significant
difference.

Exercise Set 8–4

1. 0.0033   **3.** P-val. $= 0.3936$; no significant difference   **5.** P-val. $= 0.1736$; no significant difference in
support   **7.** P-val. $= 0.3015$; no, new treatment is not significantly better.   **9.** 267

Exercise Set 8–5

1. 0.0172   **3.** P-val. $= 0.0153$; yes, there is a significant difference.   **5.** P-val. $= 0.1862$; no significant
difference.   **7.** 0.2 to 5.0   **9.** 4.0 to 7.2 inches   **11.** Between 0.52 in favor of Group B and 1.10 in favor of
Group A   **13.** P-val. $= 0.2116$; Catholic family size is not at least 1 more than Protestant family size.

Chapter 8 Review Exercises

1. Country A has sugar consumption exceeding that of B by between 13.4 and 16.6 lb.   **3.** The difference is between 0.20 and 0.50 in favor of college-bound girls.   **5.** Mean monthly growth of fish using this food is between 0.85 and 1.31 cm.   **7.** P-val. $= 0.0437$; no, there is not a significant difference.   **9.** P-val. $= 0.1020$; there is not a significant difference in appeal.   **11.** P-val. $= 0.0228$; yes, college graduates do subscribe to more magazines.   **13.** P-val. $= 0.0138$; yes, drivers with alcohol were significantly slower.   **15.** 90% confidence   **17.** P-val. $= 0.0000$; the proportions are very different.

Exercise Set 9–1

1. P-val. $= 0.1151$; yes, the mean may well be 30.   **3.** P-val. $= 0.3936$; yes, the claim may be correct.   **5.** Since $p = 0.5$, $z = 0$, the boxes must conform to the manufacturer's guideline.   **7.** P-val. $= 0.1922$; the marijuana cigarette makes no difference.

Exercise Set 9–2

1. 0.0262   **3.** P-val. $= 0.0367$; yes, the hormone is effective.   **5.** P-val. $= 0.0060$; yes, they are keeping cars longer.   **7.** 61; P-val. $= 0.0455$

Exercise Set 9–3

1. 315, 28.98   **3.** P-val. $= 0.0250$; the display has the desired effect.   **5.** P-val. $= 0.0035$; supermarket shoppers spend more when hungry.   **7.** 0.0262   **9.** P-val. $= 0.0080$; yes, the informed students had a significantly higher mean GPA.

Exercise Set 9–4

1. Not random   **3.** Could be random   **5.** Not random   **7.** 26 and 42

Chapter 9 Review Exercises

1. Could be random   **3.** P-val. $= 0.2119$; children left to their own devices may begin walking before or after their first birthday.   **5.** P-val. $= 0.1292$; same conclusion as exercise 3   **7.** P-val. $= 0.0329$; special exercises and encouragement do appear to lead to a lower age when children start walking alone.

Exercise Set 10–1

1. 0.0136   **3.** P-val. $= 0.2027$; yes, they conform.   **5.** P-val. > 0.20; the dice-maker could be right.   **7.** P-val. > 0.20; yes, they are close enough.   **9.** P-val. > 0.20; yes, they are in line with pro-choice.

Exercise Set 10–2

1. 0.0137   **3.** P-val. > 0.20; attitude is independent of proximity to plant.   **5.** P-val. $= 0.0461$; yes, happiness is related to marital status.   **7.** P-val. > 0.20; fingerprint types don't depend on race. **9.** P-val. > 0.20; delinquent behavior is not related to birth order.

Chapter 10 Review Exercises

1. P-val. $= 0.0096$; professors are not equally popular.   **3.** P-val. > 0.20; the computer is generating random numbers.   **5.** P-val. < 0.0041; math anxiety is dependent on sex.   **7.** P-val. < 0.0042; yes, there is a relationship.   **9.** P-val. < 0.0037; the dice are not fair.

Chapter 11 Review Exercises

1. 3.7; 6.4   **3.** 2.7; 4.0   **5.** $F = 4.51$; significant at the 5% level

7.

SS	df	SD²	F
72.3	5	14.46	3.88
223.3	60	3.725	

Significant at the 5% level

9.

SS	df	SD²	F
31.6	3	10.53	5.57
68	36	1.89	

Significant at the 5% level

11. $F = 3.58$, as in text   **13.** $F = 4.67$; yes, there are significant differences between the groups at the 5% level. **15.** $F = 14.73$; there is a significant difference in narcotics violations between the different types of communities at the 5% level.   **17.** $F = 2.75$; these results indicate no significant difference in reaction time depending on the amount of time without sleep, at the 5% level.   **19.** $F = 4.45$; yes, there is a significant difference in mean lengths of time businesses stay in business depending on size, at the 5% level.

Chapter 12 Review Exercises

To save space, required formulas are given by number (see appendix C).

1. CIE for $\mu_1 - \mu_2$; formula (43) [with (42) for s^*]
3. Wilcoxon two-sample test; formula (45) and z-score test
5. CIE for $\mu_1 - \mu_2$; $(\bar{x}_1 - \bar{x}_2) \pm zSE_{\bar{x}_1 - \bar{x}_2}$ and formula (37)
7. χ^2-test for a composite hypothesis; formula (47) and $df = (R - 1) \times (C - 1)$
9. CIE for dependent differences; formula (39)
11. Two-sample z-score test for means; formula (40)
13. ANOVA; formulas (49) and (50) and source table, see chapter 11
15. CIE for $P_1 - P_2$: $(p_1 - p_2) \pm zSE_{p_1 - p_2}$ and formula (38)
17. CIE for σ; formula (31) and $s \pm z \cdot SE_s$
19. CIE for dependent differences; $d \pm z\dfrac{s_d}{n}$
21. χ^2-test for a simple hypothesis; formula (47), $df = c - 1$
23. Two-sample z-score test for means; formula (40)
25. One-sample z-score test for proportions; formula (34)
27. z-score test using s; formula (33)
29. CIE for μ; formula (36)

Exercise Set 13–1

1.

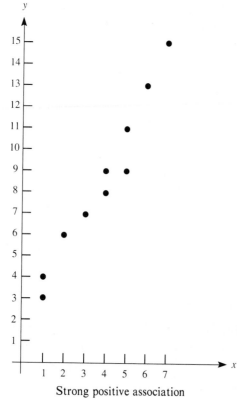

Strong positive association

3.

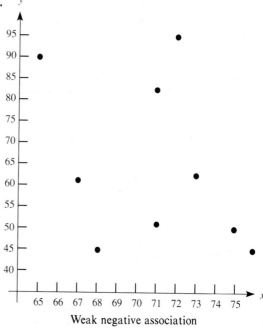

Weak negative association

5.

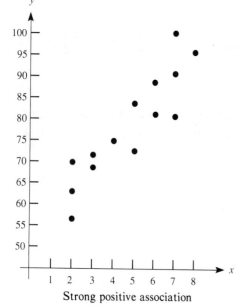

Strong positive association

7.

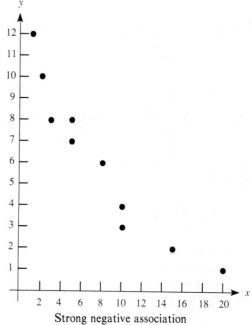

Strong negative association

9.

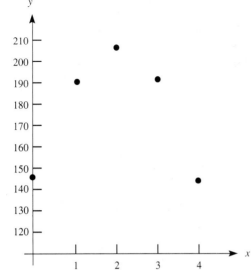

No linear association, possibly very weak negative

Exercise Set 13–2

1. $r = 0.99$ **3.** $r = 0.98$ **5.** $r = -0.28$ **7.** $r = 0.90$ **9.** $r = 0.94$

Exercise Set 13–3

1. $r' = 0.98$ **3.** $r' = -0.31$ **5.** $r' = 0.27$ **7.** $r' = 0.58$ **9.** $r' = 0.94$

Exercise Set 13–4

1. P-val. $= 0.0026$; dependent   **3.** P-val. $= 0.0594$; independent   **5.** P-val. $= 0.0475$; there is a relationship between the two grades   **7.** P-val. $= 0.0006$; son's height does depend on father's height.   **9.** 0.282

Exercise Set 13–5

1. 0.0336   **3.** 0.0040   **5.** P-val. $= 0.0036$; amount of insurance premium does depend on the age of the driver.   **7.** P-val. $= 0.0006$; number of pounds of tomatoes produced does depend on the amount of fertilizer used.   **9.** 2,408

Chapter 13 Review Exercises

1. $r = -0.8$ is a stronger correlation.   **3.** No, correlation does not imply causation.

5.

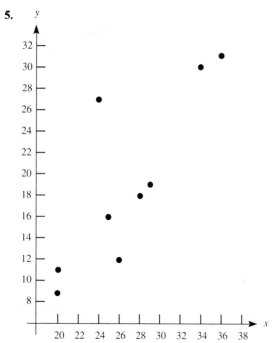

Actual $r = 0.81$

7. $r' = 0.81$ **9.** $r = -0.75$

11. P-val. $= 0.0122$; child's weight is related to mother's weight.

13. P-val. $= 0.0004$; score in golf does depend on amount of practice.

15. P-val. $= 0.0099$; y is dependent on x.

17. P-val. $= 0.0038$; z is dependent on x.

19. $r = 0.93$, P-val. $= 0.0000$; z is dependent on y.

21. P-val. $= 0.0901$; not significant at a 5% level.

23. $r' = 0.81$ when signs are ignored;
$r' = -0.82$ when signs are taken into account

Exercise Set 14–1

1. $\hat{y} = 29.2 + 0.90(x - 21.3)$ **3.** $\hat{y} = 71.6 + 10.7(x - 5.4)$ **5.** 99,400 miles

7.

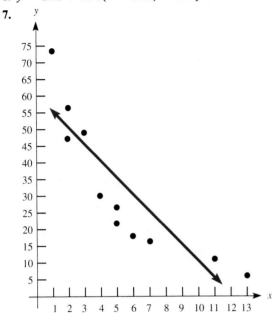

9. If r is positive, then slope is positive. So these results are inconsistent.

Exercise Set 14–2

1. 3.68   **3.** 0.89   **5.** 5.7   **7.** -0.87 **9.** 95.0

Chapter 14 Review Exercises

1. $\hat{y} = 19.4 + 0.0045(x - 2,436)$

5. 17.4   **7.** $r = 0.70$ **9.** 3.10

11. No, correlation does not imply causation.

13. 67.5

3.

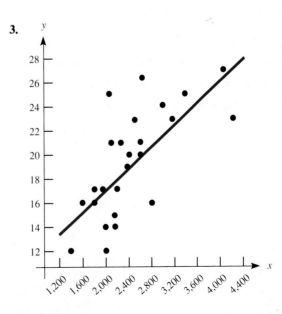

Index